Advances in
MARINE BIOLOGY

VOLUME 36

Advances in
MARINE BIOLOGY

The Biochemical Ecology of Marine Fishes

by

G.E. SHULMAN

Institute of Biology of the Southern Seas, Sevastopol, Republic of Ukraine

and

R. MALCOLM LOVE

Formerly Torry Research Station, Abbey Road, Aberdeen, Scotland
Present address: East Silverburn, Kingswells, Aberdeen AB15 8QL, Scotland

Series Editors

A.J. SOUTHWARD

Marine Biological Association, The Laboratory, Citadel Hill, Plymouth, England

P.A. TYLER

School of Ocean and Earth Science, University of Southampton, England

and

C.M. YOUNG

Harbor Branch Oceanographic Institution, Fort Pierce, Florida, USA

ACADEMIC PRESS
A Harcourt Science and Technology Company

San Diego   San Francisco   New York   Boston
London   Sydney   Tokyo

Academic Press Limited
24-28 Oval Road, London NW1 7DX, UK
http://www.hbuk.co.uk/ap/

Academic Press
A Harcourt Science and Technology Company
525 B Street, Suite 1900, San Diego, California 92101-4495, USA
http://www.apnet.com

ISBN 0-12-026136-7

A catalogue record for this book is available from the British Library

Typeset by Saxon Graphics Ltd, UK

Printed in Great Britain by MPG Books Ltd, Bodmin, Cornwall

99 00 01 02 03 04 MP 9 8 7 6 5 4 3 2 1

CONTENTS

1. Introduction

2. Adaptations of Fish

3. Strategies of Adaptation

4. Molecular and Metabolic Aspects of Life Cycles

5. The Metabolic Basis of Productivity and the Balance of Substance and Energy

6.   Indicators of Fish Condition

7.   Intraspecific and Interspecific Differentiation of Fish

8. Conclusions

Editor's Preface

This volume of *Advances in Marine Biology* is devoted to a comprehensive review of dynamic aspects of fish biology and their relationship to the environment and the world's fisheries. The authors adopt an integrative and holistic approach that looks for means of measuring the health and productivity of the fish populations. The condition of fish is assessed from relevant biochemical and physiological criteria, while production and energy flow in the marine ecosystem is followed through from cell to population level.

Such an approach was developed independently by the authors and their colleagues, who were prevented by international problems from becoming fully aware of one another's progress. With the introduction of openness in the former Soviet Union, it has been possible for the authors to meet and engage in full collaboration. This volume offers insight into hitherto inaccessible Russian data that concern principally the fatty fish and their lipid dynamics; to this is added complementary western studies on non-fatty fish, especially their proteins.

One of the main aims of the researches reviewed was to show how fish can be assessed for harvesting at the best time in their life cycles and in the correct condition for marketing, freezing and preserving. The book should interest both biochemists and fishery ecologists, and might be read with advantage by administrators who try to regulate the world's fisheries.

1. Introduction

Biochemistry, physiology and ecology are powerful branches of modern biology. They are essential to the understanding of factors that govern the functioning of vital processes at the molecular, subcellular, cellular and whole animal levels, and of the interactions between organisms, or groups of organisms, and the environment. These disciplines can be integrated as biochemical ecology and physiological ecology, emphasizing the influence of the environment. Biochemical and physiological ecology are themselves closely interrelated, so we can regard processes occurring at all levels of organization as an integrated whole.

The field is wide and requires the study of: (1) molecular, metabolic and functional principles of adaptation within and between species; (2) ecological factors influencing biochemical and functional evolution; (3) molecular and metabolic phenomena enabling differentiation between and within species – the establishment of criteria for taxonomic differentiation; (4) principles that regulate life histories and metabolic rhythms; (5) ecological specificity of substance and energy utilization by living systems; (6) bases of complex behaviour patterns in different localities; (7) population dynamics and productivity; (8) molecular and metabolic principles responsible for 'cementing' communities and biogeocoenoses ('ecological metabolism', Khailov, 1971); (9) the cycling of matter and energy within these systems; and (10) the effects of human activities on ecological systems and the choice of indicators for monitoring the 'condition' of organisms and ecological systems.

Many fundamental concepts in modern biology have been established through studies on aquatic organisms. Fish are of special interest to research workers, because some of their metabolic features characterized early vertebrates. Fish have also evolved numerous adaptations, which have permitted them not only to survive but also to thrive in recent times. The range of structural and functional adaptations and metabolic flexibility, combined with individual specializations, has resulted in an immense diversity of fish – more than 20 000 species – which greatly exceeds that of amphibians, reptiles, birds and mammals. As the final link in many food chains, fish can be reliable indicators of the condition of complex ecosystems. Studies on fish provide an understanding of the pathways of metabolic substances and of energy transformations in bodies of water, and they reveal specific features concerning the functioning of communities. They also enable us to assess the state of individual populations, as well as the ecosystem as a whole.

Fish have also been much studied because of their commercial importance as a major source of protein for mankind and domestic animals, and also of biologically active substances for medical purposes. Knowledge of the physiological and biochemical ecology of fish also increases our understanding of rhythms of vital activity and patterns of behaviour, besides helping us to assess the reproductive capacity, and so the productivity, of individual fish stocks. The importance of such studies in increasing the effectiveness of fisheries and aquaculture is obvious. These studies also contribute to our understanding of the effects of pollutants on fish. Uncontrolled industrial activity in recent decades has caused catastrophic pollution, not only of ponds, lakes and rivers, but also of vast expanses of the sea. Sadly, some pollutants have not only resulted in pathological conditions, but have affected the general ecology of whole areas. Biochemical ecology is therefore especially important in protecting fish stocks.

Individual physiological and biochemical studies on fish were undertaken in a small way at the end of the nineteenth century. These studies have developed into disciplines in their own right only in recent decades, as a result of progress in the fields of:

1. biochemical and functional adaptations of fish (Fry, 1957, 1971; Brockerhoff et al., 1963; Korzhuev, 1964; Bilinsky and Gardner, 1968; Leibson, 1972; Hochachka and Somero, 1973, 1984; Plisetskaya, 1975; Natochin, 1976; Johnston, 1977, 1982; Romanenko, 1978; Sargent, 1978; Romanenko et al., 1980, 1982, 1991; Kreps, 1981; Ackman, 1983; Polenov, 1983; Sidorov, 1983; Savina, 1992; Ugolev and Kuzmina, 1993);
2. population biochemistry (Altukhov, 1974, 1983; Kirpichnikov, 1978, 1987; Lukyanenko et al., 1991);
3. principles of ontogenesis (Milman and Yurovitsky, 1973; Neifakh and Timofeeva, 1977; Konovalov, 1984; Gosh, 1985; Ozernyuk, 1985; Zhukinsky, 1986; Novikov, 1993);
4. annual cycles (Fontaine, 1948; Hoar, 1953; Idler and Clemens, 1953; Gerbilsky, 1958; Barannikova, 1975; Beamish, 1978; Shatunovsky, 1980; Blaxter and Hunter, 1982; Wootton,1990);
5. active metabolism (Black, 1958; Brett, 1970, 1973, 1979; Matyukhin, 1973; Morozova, 1973; Yarzhombek, 1975; Johnston, 1982; Klyashtorin, 1982; Matyukhin et al., 1984; Belokopytin, 1993);
6. substance and energy balance, growth and productivity (Ivlev, 1939, 1966; Gerking, 1952, 1966, 1972; Karzinkin, 1952; Mann, 1965; Brett, 1983, 1986; Black and Love, 1986; Weatherley and Gill, 1987; Yarzhombek, 1996);

7. complex forms of behaviour (Malyukina, 1966; Pavlov, 1970, 1979; Protasov, 1978; Manteifel, 1980, 1987);
8. fish breeding (Shcherbina, 1973; Cowey and Sargent, 1979; Ostroumova, 1979; Sorvachev, 1982); and
9. ecological toxicology (Lukyanenko, 1976, 1987; Mikryakov, 1978; Malyarevskaya, 1979; Stroganov, 1979; Connell and Miller, 1984).

The present authors have themselves contributed to the solution of physiological and biochemical problems (Love, 1957, 1970, 1980, 1988; Shulman, 1960a, b, 1972a, b, 1974, 1978a; Shulman and Urdenko, 1989). For many years, at least until the publication of Shulman (1974), the two authors and their teams had been exploring a similar approach to fishery research, unknown to each other. The aims had been to bring the techniques of the biochemical laboratory to bear on problems of fish biology. Subsequently, they corresponded but were unable to meet because of the international situation. With the end of the 'cold war', it became possible for the authors to spend time in each other's laboratory. At present, Georgy Shulman continues work at the Institute of Biology of the Southern Seas, under great difficulties caused by shortage of funding, while the former laboratory of Malcolm Love, the Torry Research Station, was closed altogether in 1996 and is now an empty shell. The present collaboration is most appropriate because the studies of the two teams have complemented each other with almost no overlap. The Sevastopol group's main objects of study are species of fish, mostly fatty, which inhabit the Black Sea and neighbouring seas. Annual variations in the contents of body lipids have given powerful insights into many aspects of fish biology, particularly ecology and behaviour (Shulman, 1972a). The Aberdeen work focused mainly on the Atlantic cod and some other non-fatty species, concentrating mostly on body proteins. Together the two authors have been able to produce a fairly rounded picture, although many further approaches to the problems remain to be explored. Shulman's contribution makes a rich collection of literature from the former Soviet Union available to English-speaking scientists. It has not been the second author's intention to insert large parts of earlier reviews into the present text, but simply to extend the account of fatty fish biology to include non-fatty oceanic species and to supplement the review of Russian and Ukrainian work where appropriate.

The continuing aim of all the work has been to assess the condition or state of any fish in as much detail as possible – what Shulman's book in English translation calls 'its syndrome' (Shulman, 1974). Such information is invaluable in measuring the health or vigour of a population, the productivity of a fishing ground, the effects of pollutants, the prospects of fecundity in following years and the effectiveness of a particular regime of fish culture. Measuring the quality of eggs and the strength of the broodstock is an additional possibility. Modern hospital techniques permit subtle changes in the human body to be identified, leading to the diagnosis of early stages of disease. Were any doctor to base his diagnosis only on the weight loss of the patient, he would be limited to the point of ridicule. Even mediaeval practitioners drew on many other signs obtained by physical examination and detailed questioning. However, until fairly recent times, fishery biologists have been restricted to such a minimal criterion, in which the weight of the fish of a given length ('weight/length ratio') is measured. The weights of the liver or gonads in relation to the total body weight also yield some information, but again they are very limited. Biochemical studies have enlarged the horizons.

It is physically difficult to study the behaviour and health of fish in their natural state. The problem is only partially solved by capture and transfer to an aquarium because the capture process imposes great stresses on the fish and so modifies some of their physiology. Resting the captured fish for a few weeks before observing them brings its own problems, because the fish never regain their former state; for example, their stamina is now less than that of fish in the wild and, because of the different diet, their body composition alters also. It has to be accepted that most of one's information must come from the examination of recently captured fish, now dead, but the present account demonstrates that such material, at first sight unpromising, can yield a great deal of fascinating information. Some migratory fish look fitter than non-migratory ones: they seem more glossy and they flip furiously about on the deck after capture by trawling, while those from a stationary stock just lie down and die. This observation seems to relate to the rate at which nutrients can be mobilized to supply energy, but the notion of 'condition' is actually more closely associated with the absolute quantities of nutrients stored within the fish. For example, the size of the lipid store reveals whether or not a fish is ready to migrate and what its future fecundity is likely to be. The picture is complex, and the determination of a single constituent is insufficient to pinpoint the position of the fish along the pathway from fully fed to completely starved. In non-fatty fish, the first stage of starvation or spawning-depletion is a reduction in the lipids and carbohydrates of the liver. Only when these have fallen to a critical level does the fish begin to break down the proteins of its muscle tissue to use as a source of energy. A pattern of change in the lipid content of the liver therefore gives no information about the later stages of depletion, while studies on the muscle proteins miss the early stages: both constituents need to be measured.

There are other questions arising from study of the mobilization of energy resources. Is depletion still in progress? Have the depleted fish started to recover? If so, for how long have they been recovering? As we shall see, much of the substance of these questions can now be answered, so that fish of the same species at the same time of the year but from different habitats can be compared.

A subsidiary aim, but one of great importance in this sort of work, is to be able to modify the more sophisticated methods of observation so that they can be performed under rough field conditions, as on board ship, heaving about on the ocean. Some tissue samples can, of course, be frozen for examination ashore, but the post-mortem pH of the muscle can readily be measured at sea, the values giving a good picture of the carbohydrate reserves of the fish. A rough indication of the lipid content of the liver can be gained by observing its size and colour, and starvation or active feeding are indicated by the colour and size of the gall bladder.

Finally, as the title of this account suggests, it is shown that biochemical analyses of dead fish can reveal a great deal about the nature of the ground where the fish were caught – its average recent temperature, oxygenation, water flow, food supply and so on. The physiology of the fish is also inextricably linked to its biochemistry, so descriptions of biochemical phenomena can often more accurately be described as 'physiological–biochemical', but we have tried to avoid such an awkward phrase in the text. It should, however, be borne in mind. The Sevastopol group has carried out detailed lipid analyses on fish over a period of more than 30 years. The Aberdeen monthly survey of the carbohydrate resources of cod lasted for 10 years. Such extended periods are essential in order to place observations on a firm basis and to reveal additional phenomena that would not have been seen over a shorter period. The benefits of such work extend over the whole field of fisheries: future catching yields, fish culture, comparative ecology, quality of the flesh as food for mankind, and the effects of competitors or parasites, pollution or severe weather. Unfortunately, scientific research is becoming more and more politicized by administrators who do not appreciate the benefits of long-term research. Indeed, any research that will take more than a year or two to complete is now unlikely to receive funding. The resulting stream of second-class information represents an inefficient use of financial resources and will result in a backlog of more and more unanswered questions. The lesson may be learned eventually.

The Latin names of the fish mentioned in the text are listed in the Appendix, to reduce clumsiness in the text and improve its readability, especially where species are mentioned repeatedly. The only exceptions are fish without a generally used common name, or where two species have the same common name. The nomenclature is based on the *Catalog of Fishes*, produced by the California Academy of Sciences, 1997.

2. Adaptations of Fish

The metabolism of marine fish is influenced by the abundance, accessibility and composition of the food, by parasites, intra- and interspecies relationships and by the nature of the surrounding medium – temperature, pressure, illumination, gas content, salts and other substances, and perhaps radiation. The fish in turn influence some of these parameters, such as the quantity of food and the composition of the water. Let us consider how fish adapt to these influences.

2.1. TEMPERATURE

2.1.1. General Effects

Temperature influences the rates of chemical reactions. In theory, the range of temperatures tolerated by life forms is comparatively wide, but, in fact, each species shows characteristic, limited temperature preferences and tolerances.

Many effects of change in temperature are manifested through changes in the activity of metabolic enzymes. For example, Shchepkin (1978) found that the activity of succinate dehydrogenase in muscle and liver of several species of fish increased within the temperature range of 5°C to about 25°C, but that above this the activity of the enzyme declined – the heat was causing damage. Similar data were obtained by Emeretli (1994a) on succinate dehydrogenase activity of the mitochondria of the same tissues of round goby, scorpion fish and horse-mackerel from the Black Sea. The last two species, together with whiting and pickerel, show positive correlation between lactate dehydrogenase activity of the cytosol and the temperature. The activities of uricase, allantoicase and allantoinase in the liver of carp increase with a rise in temperature (Vellas, 1965). All enzymes are affected in the same way to some extent.

A rise in temperature accelerates the rate of development of eggs (Forrester and Alderdice, 1966) and reduces the period of gestation in live-bearing species (Kinne and Kinne, 1962). The growth rate of adults increases (Le Cren, 1958), as does the rate of emptying the gut after feeding (Hofer *et al.*,1982; Eccles, 1986) and the actual absorption of the food (Love, 1988, p. 51). The regeneration of nervous tissue following damage is also accelerated (Gas-Baby *et al.*, 1967). In addition, there are positive correlations between temperature, muscle contraction and swimming velocity (Wardle, 1980; Johnston, 1982).

None of these or similar observations is unexpected, nor are they of primary importance. More interesting are the ways in which fish adapt to changes in temperature.

2.1.2. Adaptation to Temperature

2.1.2.1.   *Enzymes and Metabolism*

The correlation between ambient temperature and enzymatic activity in muscle tissue is governed in some cases by the law of Van't Hoff and Arrhenius, with activity change with a change of temperature of 10°C (Q_{10}) approaching 2 (Prosser, 1979; Schmidt-Nielsen, 1979), but in other cases the law is not obeyed. Departure from the Van't Hoff coefficient in poikilothermic animals has been explained as 'temperature compensation', which can be achieved by a variety of mechanisms, each being appropriate to a specific temperature range (Bullock, 1955; Prosser, 1967; Newell, 1970; Brett, 1973; Hochachka and Somero, 1973; Klekowski *et al.*, 1973; Wallace, 1973; Slonim, 1979; Lozina-Lozinsky and Zaar, 1987; Ozernyuk *et al.*, 1993).

The fact that enzymes are heterogeneous and that each isoenzyme displays a characteristic optimum reaction temperature may be one mechanism that allows compensation for temperature change in many poikilotherms. A classic example is the oxygen consumption (i.e. metabolic rate) of Arctic (Scholander *et al.*, 1953) and Antarctic (Wohlschlag, 1960) species, which is considerably greater than that calculated from the Van't Hoff relationship. Love (1980, pp. 333–335), reviewing isoenzyme interplay at varying temperatures, concluded that the effect was probably limited to 'certain enzymes in specific organs of a few species'. Enzymic activity may change because of an increase in the actual amount of enzyme present, changes in its conformation or modification of the interrelation between the enzyme and various physico-chemical factors (Tsukuda, 1975; Hochachka and Somero, 1977, 1984).

The concept of temperature compensation of metabolism at the whole animal level in poikilotherms has been subject to strong criticism by Holeton (1974, 1980) and especially by I.V. Ivleva (1981), who believed that such 'compensation' was an artefact arising from inadequate acclimation. However, Ivleva's own experiments were carried out on aquatic invertebrates, not fish, and the data referred to standard metabolism rather than total or active. Note that total metabolism is that taking place in an animal in nature, active metabolism is that which supplies locomotory activity, standard metabolism is that observed in experiments, and basal metabolism is that in the resting state. Ivlev (1959) found that apparent adaptive reactions of animals are seen mostly in active, not standard or basal, metabolism. However, more recent work (Karamushko and Shatunovsky, 1993; Musatov, 1993) appears rather to favour the concept of I.V. Ivleva, that standard, not active, metabolism illustrates the adaptive reactions.

E.V. Ivleva (1989a) found that, during the winter, Black Sea horse-mackerel displayed increased thyroid activity. This is directly related to the intensity of energy metabolism. Other workers found enhanced growth of the follicular cells of the thyroid gland of brown trout and brook trout during periods of low temperature (Woodhead and Woodhead, 1965a,b; Drury and Eales, 1968), and increased thyroxine levels in the blood plasma (Eales *et al.*, 1982). On the other hand, Leatherland (1994) has demonstrated a close *positive* correlation between water temperature and the concentrations of both forms of thyroid hormone (thyroxine and tri-iodothyronine) in the plasma of brown bullhead.

Differences between species often dilute the impact of otherwise interesting correlations but, in the case of thyroid hormones, the dose level causes additional complications. Tri-iodothyronine enhances the incorporation of amino acids into tissues or enhances their catabolism, depending on the size of the dose of hormone given. A similar effect on weight gain/weight loss is seen according to the dose of the same hormone. A further difficulty with the conclusions of earlier workers is that the enhanced growth of the thyroid follicular cells does not necessarily mean greater secretion of the hormones into the blood stream. All that one can say is that greater quantities of hormones are being stored in the gland (Leatherland, personal communication). It has been considered that the thyroid hormone stimulates oxygen consumption in fish (Ruhland, 1969; Gabos *et al.*, 1973; Pandey and Mushi, 1976). However, having regard to all the data published so far, Leatherland (1994) has concluded that such an effect, if it exists, is small.

As the temperature falls, the number, volume and enzymatic capacity of mitochondria in the muscle tissues of fish increase (Jankowski and Korn, 1965; Wodtke, 1974; Johnston and Maitland, 1980; Dunn, 1988); the sizes of liver cells and their nuclei also decrease (Campbell and Davis, 1978), while the number of muscle cytochromes increases (Sidell, 1977; Demin *et al.*, 1989). Johnston and Horne (1994) have shown that the proportion of each muscle fibre occupied by mitochondria in herring larvae is greater at lower rearing temperatures. The same finding has been reported in adult carp (Rome *et al.*, 1985). Distinct temperature compensations are observed for glycolytic enzymes (phosphofructokinase, aldolase, lactate dehydrogenase), those in the hexose monophosphate shunt (6-phosphogluconate dehydrogenase), the Krebs cycle and electron transfer (succinate dehydrogenase, malate dehydrogenase, cytochrome oxidase, succinate cytochrome-*c*-reductase, NADH cytochrome-*c*-reductase), protein synthesis (aminoacyltransferase) and Na^+K^+-ATPase (Eckberg, 1962; Krüger, 1962; Freed, 1965; Smith *et al.*, 1968; Somero *et al.*, 1968; Caldwell, 1969; Haschemeyer, 1969; Hazel and Prosser, 1970; Smith and Ellory, 1971; Hazel, 1972; Hochachka and Somero, 1973; Wodtke, 1974; Shaklee *et al.*, 1977; Campbell and Davis, 1978; Tirri *et al.*, 1978; Stegeman, 1979; Sidell, 1980; Jones and Sidell, 1982; Christiansen, 1984; Bilyk, 1989; Romanenko *et al.*, 1991; Klyachko and Ozernyuk, 1991; Yakovenko and Yavonenko, 1991; Karpov and Andreeva, 1992). Other enzymes and authors are listed by Love (1980, Table 20).

Not only do the activities of the enzymes increase, but their catalytic efficiencies (enzyme–substrate affinities, as defined by the Michaelis constant, K_m) increase also (Cowey, 1967; Assaf and Graves, 1969; Baldwin, 1971; Hochachka and Somero, 1973; Johnston and Walesby, 1977; Valkirs, 1978; Walsh, 1981; Graves and Somero, 1982; Kleckner and Sidell, 1985; Klyachko and Ozernyuk, 1991). In addition, their concentrations often increase (Sidell *et al.*, 1973; Hazel and Prosser, 1974; Shaklee *et al.*, 1977).

Johnston *et al.* (1994) compared the oxidative activities of mitochondria from Antarctic fish with those of a species (*Oreochromis alcalicus grahami*) which inhabits a hot spring at 42°C. Results showed that the evolutionary adjustment of mitochondrial respiration to temperature was surprisingly small, the rates of oxygen consumed by a given volume of mitochondria being quite similar in all the species examined. Most of the 'compensation' for temperature was therefore achieved via variations in the relative mitochondrial volume.

The results of studies using a wide range of oxidizable substrates showed that, in fact, there was a modicum of temperature compensation, but that it was nothing like the 'metabolic cold adaptation' originally envisaged. Some thermal compensation was also achieved by increasing the density of cristae within the mitochondria, but this was of less importance than actual mitochondrial density.

Many enzymes, however, do not show temperature compensation in their activities (Eckberg, 1962; Precht, 1964; Vellas, 1965; Hebb *et al.*, 1969; Hazel and Prosser, 1970; Shkorbatov *et al.*, 1972; Hochachka and Somero, 1973; Penny and Goldspink, 1981). Among these are catalase, peroxidase, acid phosphatase, D-amino acid oxidase, choline acetyltransferase, acetylcholinesterase and glucose-6-phosphate dehydrogenase. What is more, the same enzyme can show opposing temperature dependencies when examined in different organs and tissues of fish (Romanenko *et al.*, 1991).

Conclusions can be invalidated by the use of unrealistic experimentation. For example, the enzymatic activities of other fish species studied differ according to their temperature of habituation, as we have seen already in the work of Johnston *et al.* (1994). In Black Sea whiting, which prefer cold water, the lactate dehydrogenase (LDH) activity of red and white muscle tissues is lower (in all seasons except winter) than that in Black Sea horse-mackerel, pickerel and scorpion fish which prefer warm water. However, the data were obtained during enzyme incubation at 25°C, when a comparison carried out at the temperatures normally experienced in the sea by the different species would have been more appropriate. Use of a standard incubation temperature of 25°C in many cases does not reflect the ambient conditions of the species studied. When it did, the LDH activity of whiting displayed rates either equal to or greater than those recorded in warm-water fish (Figure 1). We believe that this gives a definite example of temperature compensation, bearing in mind that individual phenomena in different species may differ because of factors other than the one which interests us. Similar results were obtained in a study of LDH in cod from the northern seas (Karpov and Andreeva, 1992). In the Arctic species, navaga and polar cod, the estimates of K_m have been lower than in the cod, haddock and saithe from warmer water.

It is probably unwise to draw firm conclusions from comparisons between different species, but a possible further example of temperature compensation is the finding that the activity of cytochrome oxidase in vermilion rockfish,

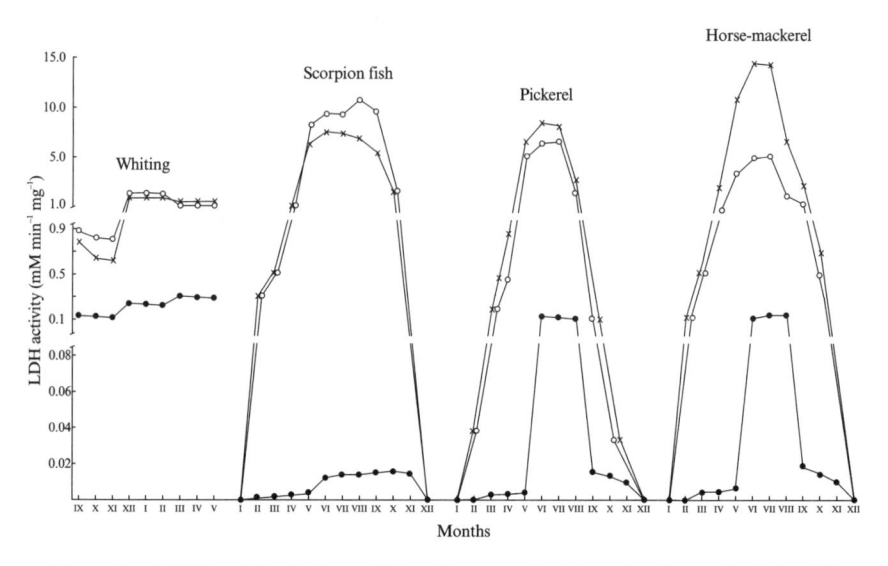

Figure 1   LDH activity in liver (●), white muscle (×) and red muscle (○) of four species of Black Sea fish, measured by incubating at the environmental temperature characteristic of each species.

which favours low temperatures, is higher than that of brown rockfish, which inhabits warmer water (Wilson et al., 1974). A relationship between the activities of tissue enzymes and the temperature tolerance of the fish has been demonstrated by Hochachka and Somero (1973), Klyachko and Ozernyuk (1991) and Klyachko et al. (1992), and for the digestive enzymes by Kuzmina (1985) and Ugolev and Kuzmina (1993) using incubation temperatures close to those of the natural habitat.

Measurements of oxygen uptake and the activities of enzymes connected with aerobic metabolism are insufficient to demonstrate that temperature compensation in poikilothermic organisms does not take place. Firstly, as the water temperature decreases, the blood–oxygen affinity increases in fish, while P_{50} (the pressure, in millimetres of mercury, of oxygen required to half-saturate haemoglobin) declines (Klyashtorin and Smirnov, 1983; Soldatov and Maslova, 1989). This may be related to changes in the ATP concentration within the erythrocytes or to transformation of the haemoglobin system, or both. Secondly, it should not be forgotten that the anaerobic components of energy metabolism often run counter to aerobic components and show less dependence on ambient temperature.

In carp, at low ambient temperatures, the entire structure of energy metabolism is transformed (Arsan, 1986; Romanenko et al., 1991). The tricarboxylic acid cycle is inhibited, oxidation is detached from phosphorylation and glycolysis proceeds more intensively. In this case, the intermediate and terminal products of energy metabolism (oxaloacetate, pyruvate, lactate, etc.) increase. Both carp and silver carp lay down greater stores of triacyl-glycerols during severe winters than under more moderate conditions (Brizinova, 1958; Faktorovich, 1958; Sidorov, 1983). These changes may represent a preparation for living under anoxic conditions should the surface of the water freeze over (Love, 1980, p. 271).

Reduced oxidative phosphorylation coupled with increased glycolysis in ATP synthesis during the adaptation of fish to cold has been reported by other authors also (Eckberg, 1962; Freed, 1965; Gubin et al., 1972; Ramaswama and Sushella, 1974; Wells, 1978; Bilyk, 1989).

The hexose monophosphate (Helly, 1976) and pentose phosphate (Hochachka and Hayes, 1962; Yamaguchi et al., 1976; Walsh, 1985; Malinovskaya, 1988; Kudryavtseva, 1990) shunts have also been found to increase in importance. The activity of transketolase, the enzyme which inhibits the peptide-phosphate pathway, is greater in fish from cold water, e.g. trout and smelt, than in those from warm water (Kudryavtseva, 1990). In the Black Sea horse-mackerel, a sharp decline in adenine nucleotide content (AMP, ADP and ATP) in white and red muscle tissues and in liver occurs at low temperature (Trusevich, 1978). In this case, the ATP is mostly resynthesized by glycolysis. The increase in the glucose content of the blood of fish at low ambient temperatures may be of the same nature (Prosser, 1967;

Hochachka and Somero, 1973; Shelukhin *et al.*, 1989). In adapting to cold, the mullet *Liza* sp. enhances the content of neutral lipids in the muscle by inhibiting aerobic catabolism (Soldatov, 1993). It is therefore no accident that Mongolian grayling from cold water contain more lipids than their counterparts from warmer water (Lapin and Basaanzhov, 1989).

The levels of anabolites and catabolites in fish are closely connected with the ambient temperature. Protein synthesis (Ray and Medda, 1975; Haschemeyer *et al.*, 1979; Haschemeyer, 1980; Berezhnaya *et al.*, 1981; Saez *et al.*, 1982) and growth rate associated with protein synthesis (Ryzkov, 1976) proceed more intensively at higher temperatures. In contrast, the enhanced synthesis of neutral lipids noted above is in fact often found in fish adapted to cold (Hochachka and Hayes, 1962); the warmer the environment, the greater the demand for dietary lipids (Gershanovich *et al.*, 1991).

With a change of temperature, the relationships between protein, lipid and carbohydrate anabolism (Romanenko *et al.*, 1991) and catabolism are altered also. During temperature adaptation, somatotropin and prolactin play active roles in the regulation of energy metabolism (resynthesis of ATP and the utilization of macro-ergs: Sautin, 1985; Trenkler and Semenkova, 1990). The metabolism of fish in the early developmental stages is especially sensitive to the effects of temperature (Shekk *et al.*, 1990; Timeyko and Novikov, 1991). At the highest temperatures (30°C and over), the level of total metabolism in many species decreases, respiration and phosphorylation becoming uncoordinated (Johnston *et al.*, 1983; Romanenko *et al.*, 1991). As an attempted counter-measure, heat radiation from the liver and the surface of the fish body increases (Fomovsky, 1981).

At the other extreme, Antarctic fish such as the Notothenidae live at around −1.8°C at which temperature the oxygen concentration in the water is very high. Food is plentiful, and the conditions encourage aerobic metabolism and inhibit the enzymes controlling carbohydrate metabolism (Hemmingsen and Douglas, 1970; Johnston, 1985). The same situation pertains even in the icefish, which possesses no haemoglobin but in which the blood plasma is rich in oxygen.

Ambient temperature is responsible for differences in metabolism, which develop between fish of the same species at different seasons and in waters at different latitudes. The situation is somewhat different in cases where the ambient temperature oscillates. Protracted experiments on several species of fish showed that temperature oscillations of 20% of the average during the day reduced oxygen and nutrient consumption, but accelerated the weight gain (protein), stimulated an increase in the haemoglobin and erythrocytes in the blood, and improved the viability of fish fry (Konstantinov *et al.*, 1989; Konstantinov and Sholokhov, 1990). Many fish live at temperatures that fluctuate during the day; such a condition is clearly beneficial for their vital activity and may be regarded as a case of temperature compensation of their metabolism. Similar results are known from studies on aquatic invertebrates (Galkovskaya and Sushchenya, 1978; Lozina-Lozinsky and Zaar, 1987).

In addition to exerting an influence on enzymatic reaction rates and the character of metabolism, temperature can also be a determining factor in the composition and conformation of biomolecules.

2.1.2.2. Contractile Proteins

Fast and slow muscle fibres contain various isoforms of contractile proteins, which differ in amino acid sequence and functional characteristics. In some species, the expression of these isoforms varies with the acclimation temperature and correlates with adaptive changes in locomotory performance (Johnston, 1993). The ATPase activity of the contractile proteins of some species increases with reduced temperatures, but not if the fish are starving, so protein synthesis is evidently involved in the phenomenon.

Most aspects of the development of fish larvae are sensitive to temperature. It affects not only the rate of development, but also the number of vertebrae and of rays in the fins (Schmidt, 1930; Tåning, 1952), the sequence in which external features and organs appear, the ultrastructure of the muscle and the developmental stage at which hatching occurs (Crockford and Johnston, 1993). In herring larvae 1 day old, the number of inner muscle fibres posterior to the yolk sac is smaller at lower temperatures but, since their average diameter is increased at the same time, the total cross-sectional area does not change (Vieira and Johnston, 1992). The proportion of red fibres to white fibres in the musculature of larval turbot is greater when the larvae are reared at 17° than at 22°C (Calvo and Johnston, 1992). These authors pointed out that the responses of fish larvae to temperature are not fixed, but can change during development, and that the thermal history of the larvae can contribute to subsequent physiological variation in the adults.

Many workers (Ushakov, 1963; Zhirmunsky, 1966; Tsukuda and Ohsawa, 1971; Aleksandrov, 1975; Johnston *et al.*, 1975 and others) have found that the thermal resistance of enzymes (succinic dehydrogenase, aldolase, adenylate kinase, alkaline phosphatase, acetylcholinesterase and protease) and structural proteins (actomyosin, collagen, haemoglobin, serum albumins and globulins) is closely connected with the range of temperature which characterizes the habitat of the species or the temperature of experimental acclimation. According to the hypothesis of Alexandrov (1975), flexibility in protein conformation develops in response to varying ambient temperatures.

Thermal tolerance of isolated muscle tissue and of proteins, such as adenylate kinase and actomyosin, of small and large varieties of Black Sea horse-mackerel has been studied by Altukhov (1962) and Glushankova (1967). Johnston *et al.* (1973) focused on determining the thermoresistance of ATPase in muscle myofibrils of 19 fish species occurring in the North and Mediterranean Seas and the Indian Ocean. Mednikov (1973) has shown that the temperature limits of numerous species depend on the optimum temperature of ribosomal functioning, i.e. on the optimum temperature for protein synthesis.

2.1.2.3. *Collagen*

The tough protein of the connective tissue, binding the blocks of muscle, internal organs and indeed the whole fish together, is collagen. When heated slowly, it suddenly shrinks and the temperature at which this occurs is a measure of its stability. Gustavson (1953) first showed that the shrinkage temperatures of the skin collagens of warm-water fish are higher than in cold-water species. In Alaska pollock, which inhabits relatively cold waters, the connective tissue shrinks below 40°C (Takahashi and Yokoyama, 1954), while that of the mud-skipper, which spends part of its time on land at high temperatures, shrinks at 57°C (Gowri and Joseph, 1968), almost as high as that at which calf skin shrinks (60°C).

Once again we are comparing different species, rather than the same species, from different habitats. Data derived from the same species are very limited. Andreeva (1971) found that the shrinkage temperatures of skin collagen from cod and whiting varied slightly according to the place of capture, warmer habitats leading to higher shrinkage temperatures. However, Lavéty *et al.* (1988) compared the collagens of young turbot reared in open-sea cages with those from artificially warmed effluent water of a nuclear power station (temperature difference 6–10°C) and found no difference in the shrinkage temperatures.

The fishing grounds from which Andreeva (1971) took her cod and whiting were very far apart, and she referred to the respective fish as 'sub-species'. It appears that the collagens of individual species from habitats at different temperatures do not adapt, even in young, rapidly growing fish, but further evidence is needed to establish this.

2.1.2.4. *Lipids*

Ivanov (1929) found a close inverse relationship between the value of lipid unsaturation of plants (determined by iodine value) and the latitude of their habitat. It was found that, with lower environmental temperatures, the iodine numbers of the lipids were greater (i.e. there was more unsaturation). The effect of increasing unsaturation is to lower the temperature at which the lipids change from liquids to a liquid-crystalline state, and so to lower the temperature at which the cells remain flexible.

This phenomenon was later proved valid for poikilothermic animals also, including fish (Kizevetter, 1942; Shkorbatov, 1961; Ackman, 1964; Hilditch and Williams, 1964; Lunde, 1973). The desaturation of lipids in fish and other water animals varies with the climatic–geographical zone, with annual cycles (depending on changing water temperature) and with temperature adaptations which develop under experimental conditions (Hoar and Cottle, 1952; Lewis, 1962; Farkas and Herodek, 1964; Privolnev and Brizinova, 1964).

Polyenoic fatty acids (acids with more than one double bond in the chain) play a leading role where total unsaturation is observed to change. According to current concepts, the structure and functioning of cell membranes are maintained through the agency of a binary film composed of phospholipids and cholesterol. The phospholipids consist of polar, outwardly directed 'heads' and fatty acid 'tails' immersed in the membrane (Fox, 1972; Singer and Nicholson, 1972; Chapman, 1975). In this binary film, there are inclusions of protein molecules.

As the content of polyunsaturated fatty acids in the phospholipids increases, so does the proportion of unsaturated double bonds. The bends in the carbon chains become more numerous and the fatty acid 'tails' become less compactly embedded in the binary lipid layer. With much unsaturation in the phospholipids, the looser structure of the membrane facilitates metabolic activity by enzyme modulation, enhanced ion transport and antioxidative activity. A fall in ambient temperature brings about the elongation and desaturation of the fatty acid 'tails' of the phospholipids. In fish, as in other animals, these processes are engendered by the accelerated synthesis of the most unsaturated fatty acids (Knipprath and Mead, 1968; Walton and Cowey, 1982; Farkas, 1984, Hagar and Hazel, 1985; Hazel and Livermore, 1990). Alternatively, as we shall see later, by desaturation of existing, more saturated fatty acids, such 'compensatory' synthesis and desaturation inhibits phase transition (solidification) and promotes normal functioning of the membrane. With increasing temperature, the reverse process takes place, when the embedding into the binary layer becomes more compact, increasing the viscosity of the membrane and decreasing its metabolic activity. According to present conceptions, the ability of fish to synthesize polyenoic fatty acids is limited; the main way of increasing their proportion is via the food (Sargent and Henderson, 1980; Bell et al., 1986).

The close inverse relationship between ambient temperature and the percentage of polyunsaturated fatty acids in the lipids of poikilothermic animals has been confirmed by many research workers (Johnston and Roots, 1964; Ackman and Eaton, 1966; Knipprath and Mead, 1966; Jangaard et al., 1967; Roots, 1968; Morris and Schneider, 1969; Caldwell and Vernberg, 1970; Kemp and Smith, 1970; Baldwin, 1971; Hazel, 1972; Lynen, 1972; Viviani et al., 1973; Ota and Yamada, 1975; Patton, 1975; Bolgova et al., 1976; Deng et al., 1976; Driedzic et al., 1976; Irving and Watson, 1976; Leslie and Buckley,

1976; Miller *et al.*, 1976; Cossins *et al.*, 1977; Holub *et al.*, 1977; Selivonchick *et al.*, 1977; Cossins and Prosser, 1978; Dobrusin, 1978; Hayashi and Takagi, 1978; Wodtke, 1978; Zwingelstein *et al.*, 1978; Hazel, 1979; Farkas *et al.*, 1980; Pekkarinen, 1980; Shatunovsky, 1980; Kreps, 1981; Lapin and Shatunovsky, 1981; Sidorov, 1983; El Sayed, 1984; Farkas, 1984; Henderson *et al.*, 1984; Bell *et al.*, 1986; Reinhardt and Van Vleet, 1986; Farkas and Roy, 1989; Henderson and Almater, 1989; Ugolev and Kuzmina, 1993). As the temperature decreases, the activities of 5-, 6- and 9-desaturases increase (Christiansen, 1984; Hagar and Hazel, 1985). The most significant changes in the polyenoic acids are in the principal phospholipid fractions, phosphatidyl choline and phosphatidyl ethanolamine.

Among species examined, several are from European seas: horse-mackerel from the north-eastern Atlantic (Dobrusin, 1978), mackerel tuna (El Sayed, 1984), Atlantic mackerel and anchovy from the Spanish sector of the Atlantic (Pozo *et al.*, 1992), sprat from the Adriatic Sea (Viviani *et al.*, 1973), Atlantic herring (Henderson and Almater, 1989), capelin (Henderson *et al.*, 1984), cod from the North Atlantic (Jangaard *et al.*, 1967), viviparous blenny (Pekkarinen, 1980), grey mullet and others (Deng *et al.*, 1976). All these species have demonstrated a substantial increase of, in particular, C20:5 and C22:6 polyunsaturated fatty acids during the cold season as compared with the warm season. Table 1 provides a good illustration. Unfortunately, published data are too few to allow a meaningful comparison to be made between polyunsaturated fatty acids in whole populations of the same species inhabiting waters of markedly different temperatures. E.M. Kreps and his collaborators (Kreps, 1981; Kreps *et al.*, 1977) found a higher content of C20 and C22 fatty acids in the phospholipids of cerebral tissue taken from tropical deep-sea fish living at temperatures lower than those of most of the species which live in the surface waters.

Table 1   Seasonal variation in the principal fatty acids (%) in the muscle of capelin. (After Henderson *et al.*, 1984.)

Fatty acid	January	August	January	August
14:0	4.8	5.9	3.8	7.6
16:0	22.1	25.1	19.3	23.0
16:1	8.5	8.3	6.9	7.8
18:0	1.8	1.2	1.4	1.1
18:1	26.0	28.0	20.8	22.4
18:2	1.4	1.8	1.4	1.7
18:4	1.4	4.6	1.5	4.2
20:1	2.3	1.3	4.1	4.4
20:5	13.8	10.8	16.8	10.7
22:1	2.0	1.0	3.8	4.4
22:6	11.3	6.7	15.4	8.3

Adaptations to change in temperature involve not only the fatty acids in the phospholipids but may affect the proportions of different lipid fractions in the total lipids as well. Some workers (Shatunovsky, 1980; Lapin and Shatunovsky, 1981; van den Thillart and de Bruin, 1981; Sidorov, 1983) claim that a drop in temperature results in an increased proportion of phospholipids in the total lipids. There is a larger concentration of polyunsaturated fatty acids in the phospholipid fraction than in other lipid fractions. However, other workers (Anderson, 1970; Caldwell and Vernberg, 1970; Selivonchick and Roots, 1976; Wodtke, 1978; Hazel, 1979) have not found this effect, and Knipprath and Mead (1966) found that the phospholipid content of goldfish actually decreased during adaptation to cold.

Within each of two Black Sea species, anchovy (warm water) and sprat (cold water), both the concentrations and absolute amounts of phospholipids fluctuate within similar limits, but do not change during the annual cycles in the same tissues. This contrasts with, for example, the considerable differences between the phospholipid contents of red and white muscle or between that of either of them and liver (Shchepkina, 1980a; Shchepkin and Minyuk, 1987). The content of polyenoic acids in the phospholipids of anchovy is higher than that in the sprat (Yuneva, 1990); possible explanations will be given in Chapter 3.

Studies by Johnston and Roots (1964), Roots (1968) and Kreps (1981) have revealed an increased ratio between the plasmalogenic and diacyl forms of phosphatidyl ethanolamine in oceanic fish from low-temperature waters. During cold adaptation, the ratios between the main phospholipid fractions alter: the relative proportion of phosphatidyl choline decrease and phosphatidyl ethanolamine, phosphatidyl serine and sphingomyelin, all of which contain large amounts of polyenoic acids, increase (Caldwell and Vernberg, 1970; Miller et al., 1976; Wodke, 1978; Hazel, 1979; Brichon et al., 1980; van den Thillart and de Bruin, 1981; Zabelinsky and Shukolyukova, 1989).

Cholesterol, another structural lipid capable of influencing membrane 'hardness', also plays an active role in temperature adaptations (Kreps, 1981; Chebotareva, 1983; Chebotareva and Dityalev, 1988; Sautin, 1989). Interposed between phospholipid molecules, it regulates the vital capacity of the membrane.

2.1.2.5. *Glycoproteins and Other Substances*

As to other organic substances in this context, special mention must be made of glycoproteins, which prevent blood from freezing in Arctic and Antarctic fish (De Vries, 1970; Hochachka and Somero, 1973). Glycoprotein, composed of repeated subunits of alanine and threonine bound to a disaccharide derivative of galactose, is an antifreeze agent (De Vries, 1971; Shier et al., 1972). It impedes crystal formation in blood and lowers the freezing point to $-1.8°C$, the lowest temperature possible for fish inhabiting 'normal' sea water. In some examples, the freezing point of the plasma can be reduced to as low as $-2.07°C$ (DeVries and Wohlschlag, 1969; De Vries and Eastman, 1981: notothenids). Pentoses also play a significant role as antifreezes (Clarke, 1985; Johnston, 1985).

Earlier writers observed increases in other constituents of the blood of fish transferred to low ambient temperatures, such changes helping to protect the fish from freezing damage. Eliassen *et al.* (1960) noted that lowering the water temperature from 10° to $-1.5°C$ caused a rise in total osmolarity, notably in the chloride concentration of the plasma in three species of marine fish. The effect in the bullhead was to lower the freezing point of its plasma from $-0.6°$ to $-0.9°C$, which would bestow only partial protection. When killifish were acclimatized to low temperatures, the serum was found by Umminger (1968) to undergo an increase in total nitrogen, calcium, cholesterol and, particularly, glucose, the latter increasing by 440% when the fish were cooled from 20° to $-1°$ C. These changes could represent adaptive protection against freezing, but, equally, the increase in ionic concentration could represent the breakdown of the osmoregulatory system (Woodhead and Woodhead, 1965a,b; Stanley and Colby, 1971) and any increase in blood glucose might be a 'stress' response to the low temperature (Love, 1980, Table 15).

2.1.3. Temperature Preference

Within the extremes of temperature tolerated by any fish species lies a smaller range of 'preferred temperatures', the range chosen by fish which are free to swim through a graded series. The fish probably choose temperatures at which they function best physiologically (Ivlev, 1963; Umminger and Gist, 1973), but their preference varies with acclimation, and changes seasonally according to the environmental temperature (Cherry *et al.*, 1975; Reynolds and Casterlin, 1980). The importance of this characteristic lies in the fact that fish at their temperature optima are less stressed by extraneous factors (reviewed by Love, 1980, pp.322–323).

2.2. SALINITY

2.2.1. Mineral balance

The mineral concentrations in the blood plasma of the 'primitive' cyclostomes (hagfish and lampreys) are close to those of the surrounding sea water (Robertson, 1954; McFarland and Munz, 1958; Bellamy and Chester Jones, 1961). More evolutionarily advanced species maintain relatively lower internal concentrations, presumably because the tissues function best under such conditions (Baldwin, 1948). Freshwater fish maintain internal ionic concentrations higher than in the external environment but usually less than in marine fish (Robertson, 1954). Because of the concentration gradients between the external medium and the blood and tissue fluids of both kinds of fish, there is a tendency for freshwater fish to absorb water by osmosis, becoming waterlogged, while marine fish tend to become dehydrated by a corresponding process. The mechanisms whereby the fish maintain the desired levels of salts within their bodies are known collectively as osmoregulation, described in detail by Conte (1969).

Marine teleosts survive by drinking sea water, subsequently excreting the salt, while marine elasmobranchs retain relatively high concentrations of urea, sodium, chloride, trimethylamine oxide, betaine and similar substances in their blood (reviewed by Love, 1970, p.139), thus increasing its osmotic pressure. Freshwater fish deal with the opposite situation by absorbing and concentrating the very dilute minerals from the water, retaining as much of them as possible while removing excess water as urine. The plains killifish can live in fresh or salt water ('euryhaline'); Stanley and Fleming (1967) showed that the daily output of urine when this species was kept in salt water was 31 ml kg^{-1} day^{-1}, but in fresh water it rose to 304 ml kg^{-1} day^{-1}.

The major ions in different organs and body fluids of euryhaline fish have been studied by a number of authors. The concentrations of sodium, potassium, magnesium and chloride were usually more concentrated in fish taken from sea water than in those from fresh water, the effect being shown in blood, kidney, liver, various secondary muscles and urine. The trend was less clear in the case of swimming muscle, as were the values for calcium (reviewed by Love, 1970, Table 30). All the ions mentioned above were much more concentrated in the urine of the fish from the sea, urine being one channel by which these salts are excreted. A fish with remarkable ability to control its internal milieu is the tilapia, in which the total sodium in the body increases by only 30% when it is transferred from fresh water to double-strength sea water (Potts et al., 1967).

Most species of fish can tolerate only small changes in external salinity, when they are said to be stenohaline, and they will swell up and die if transferred from sea water to fresh water. If the transfer is in the other direction, they die because the large influx of salt exceeds the level at which the tissues can function. However, Hulet et al. (1967) found that, in the presence of increased concentrations of calcium, permanently marine fish could thrive in a much-diluted medium.

The kidneys are not the only channel through which excess salts are eliminated from the blood. Motais et al. (1965) reported that, when flounders, which are euryhaline, were transferred from sea water to fresh water, there was an immediate reduction in the elimination of sodium and chloride by a mechanism located in the gills. The painted comber, a species which dies if the salinity is changed, was found not to possess this mechanism. The function of the gills is not only to exchange dissolved gases with those of the surroundings, but also to clear nitrogenous waste and maintain the acid–base and mineral balance (Maetz, 1971). The passage of individual ions across the gill membranes is extremely complex: the fish does not simply excrete sea salt but treats the different ions independently. Pickford *et al.* (1969) showed that, when killifish were transferred from sea water to fresh water, each serum ion changed in a different proportion, with potassium showing the greatest change. The internal sodium and chloride ions of marine fish are constantly being exchanged with the same ions in the surrounding water by simple diffusion (Maetz, 1971), but sodium is also actively extruded by a mechanism which uses energy and apparently involves the simultaneous absorption of an equivalent amount of potassium from the medium. Flounders kept in sea water from which the potassium has been removed will steadily accumulate sodium in their bodies (Maetz, 1973). In fresh water, flounders appear to acquire sodium from the very dilute solution, through their gills, by exchanging it for hydrogen, ammonium and bicarbonate ions (Maetz, 1971). These and other mechanisms differ between species but the finer details are beyond the scope of this review.

Much of this electrolyte transfer takes place in specialized cells, abundant in the gill epithelium of marine teleosts, known as chloride cells because they resemble the cells of the stomach lining that secrete hydrochloric acid (Maetz, 1974). The change in gill function from freshwater to seawater type is quite slow in some species, since it involves the synthesis of chloride cells. The increase and decrease in the numbers of chloride cells in the gills of Japanese eel after transfer from fresh water to sea water, and vice versa, are shown in Figure 2. The long-jawed mudsucker is an intertidal species in which chloride cells are found in the skin of the jaw as well as in the gills. These chloride cells increase in size, rather than in number, with increasing salinity (Yoshikawa *et al.*, 1993).

The adaptation of salmon smolts entering the sea from the river poses problems, as there are frequent changes in the salinity during passage down the estuaries. The smolts may enter areas of fresh or salt water alternately in rapid succession and, until recently, the reaction of their osmotic apparatus to these fast, reversible systems was poorly understood. Potts *et al.* (1970) identified both rapid and slow components in the adaptation of salmon smolts to salinity changes. More recently, Chernitsky *et al.* (1993) found that, on transfer to full sea water, the internal sodium concentration of Atlantic salmon smolts rose to high values within 1 h, but then returned to control levels as the fish adapted. Smolts transferred from sea water to fresh water adjusted the concentrations of all ions in their blood to normal values within 1.5 h. The rapid restoration of sodium levels was ascribed to an 'emergency system' of extrusion or seques-tration, according to external salinity. These authors were aware of the longer

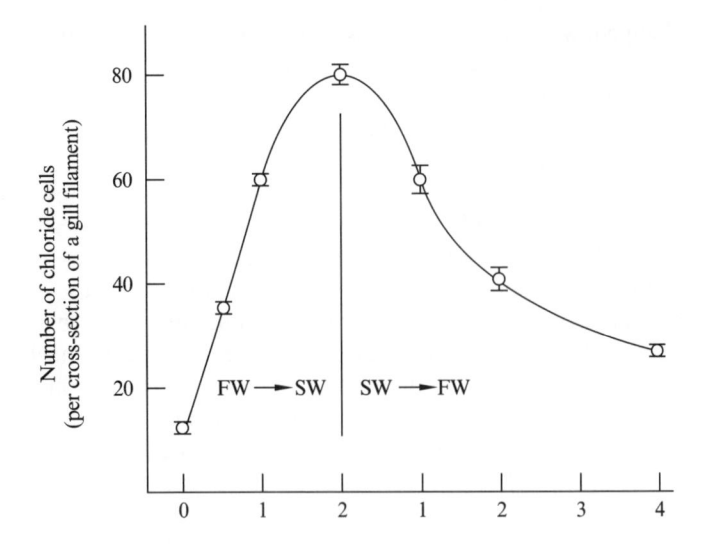

Figure 2   Changes in the number of chloride cells in a section of gill filament when *Anguilla japonica* are transferred from fresh water (FW) to sea water (SW) and vice versa. (After Utida and Hirano, 1973.)

term adaptation of this species described by earlier writers, but did not observe it in their own work, concluding that their results differed from the others because of differences in the condition of the experimental animals.

2.2.2. Energy Requirements

Energy is needed to acquire or excrete salts across a membrane against a concentration gradient. In fish the energy is supplied by ATP (adenosine triphosphate), and the enzyme which releases it (ATPase) increases its activity when fish are transferred from fresh water to salt (several authors listed by Love, 1980, Table 18). The supply of ATP is replenished by the metabolism of liver glycogen. Transfer of rainbow trout to diluted sea water (Soengas *et al.*, 1991) or full-strength sea water (Soengas *et al.*, 1993) has been shown to result in an increase in glycogen phosphorylase and a decrease in glycogen synthetase, causing the level of liver glycogen to fall and of blood glucose to rise. This is clear evidence of the mobilization of energy to satisfy the increased osmotic effort.

It has been calculated that, at a salinity of 30‰, nearly one-third of the energy consumed by a fish is used for osmoregulation (Farmer and Beamish, 1969; Rao, 1969). It follows that an excess of any other activity, or a situation where energy is consumed is likely to affect the ability of a fish to osmoregulate. This proves to be the case, and is immensely important in biochemical ecology. The ability of coho salmon to osmoregulate is reduced more, and the symptoms of stress increase, when they are artificially stressed in sea water, rather than fresh water (Avella *et al.*, 1991). As a corollary, the mullet and some other species, placed in a saline solution isotonic with its own internal fluids, uses minimum energy for osmoregulation (Figure 3). The turbot requires least oxygen at a salinity of 8‰ (Waller, 1992), and mud skippers survive continuous immersion in water of intermediate salinity, soon dying in either fresh or sea water (Ip *et al.*, 1991). In contrast, the metabolic rate of the sheepshead minnow is at its highest in ambient salinities of 15–50‰, being lower at higher or lower salinities (Nordlie *et al.*, 1991), so it is not possible to generalize for all species.

The consequences of these phenomena are extensive. Rainbow trout grow better in brackish water compared with those in fresh water; this suggests a useful modification to culture practice (Teskeredzic *et al.*, 1989). The salinity which permits optimum growth rate in this species was identified as 18‰ by Tsintsadze (1991). Brown trout are more resistant to the toxic effects of acid water and aluminium when the water is slightly saline (Dietrich *et al.*, 1989).

Pérez-Pinzón and Lutz (1991) studied the effect of salinity on swimming performance in juvenile snook. While at rest, the snook were able to osmoregulate in any salinity up to full-strength sea water, but the effect of salinity became increasingly damaging during swimming. In fresh water, the 'fitness' of the swimming fish was reduced, as seen in reduced aerobic scope, much increased muscle lactate and increased haematocrit (packed red blood cell

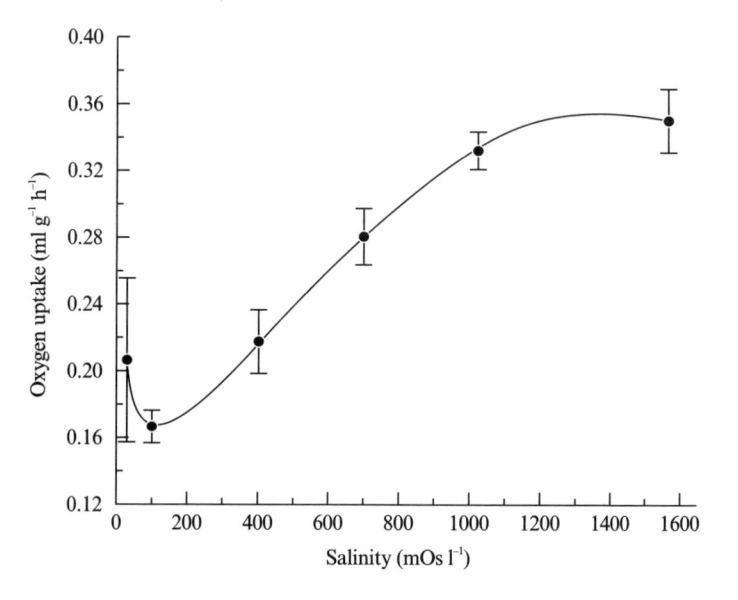

Figure 3   The effect of salinity on the metabolic rate of mullet. (After Nordlie and Leffler, 1975.)

volume). They were also less able to cope with handling stress. In isotonic sea water, they had the highest aerobic scope, lowest maintenance costs and no increase in lactate at the highest swimming speeds. In full-strength sea water, the lactate again increased. These observations are of considerable ecological significance. The authors stated that the salinity of the inshore waters around Florida is steadily increasing because of the diversion of much of the fresh water to use for human consumption, and there has been a dramatic decline in the snook population, perhaps as a consequence.

The alevins of rainbow trout thrive best in fresh water. Raising the salinity just to 5‰ causes a rise in lactate under fully aerobic conditions and markedly decreases the concentration of blood glucose (Krumschnabel and Lackner, 1993), so the advantage of using isotonic saline varies according to developmental stage.

2.2.3. Salinity and Density

The density of sea water increases according to its salinity but the resistance to swimming ('drag') seems to be little affected, at least as regards snook (Pérez-Pinzón and Lutz, 1991). However, the buoyancy of a fish has to be adjusted as the salinity changes. Gee and Holst (1992) found that two species of stickleback maintained their buoyancy by altering the volume of the swim-bladder when placed in various salinities. When they were abruptly transferred from fresh water to 10‰ salinity, it took 96 h for them to reach buoyancy equilibrium and, in the mean time, the fish used hydrodynamic forces to provide the necessary lift, either by swimming or by passing undulations up or down the pectoral fins. These authors then kept sticklebacks in a 'Percoll' solution, where the density can be altered without changing the tonicity from that of fresh water. The results showed that the fish were responding to density changes rather than to salinity as such. It was pointed out that, since both species encountered variations in salinity in their natural habitat, the ability to respond was adaptive.

2.2.4. Hormones

We have seen that the transfer of fish to a medium of different salinity often results in a change in the number of chloride cells but the agents effecting the change have not been mentioned. Acclimation to different salinities affects the output of hormones, which independently influence metabolism (Foskett *et al.*, 1983). The total rates of metabolism are not *per se* reliable estimates of osmoregulation in different salinities, but are merely 'an index of the cost of living in these environments' (Pérez-Pinzón and Lutz, 1991). The hormone picture is likely to be complex, depending, for example, on whether the change in salinity is stressful or is accomplished gradually. Thus adrenaline, a 'stress' hormone, has been shown to inhibit the excretion of sodium and chloride in rainbow trout held in sea water (Girard, 1976).

Chronic stress was shown by Avella *et al.* (1991) to cause a gradual increase in the concentration of circulating prolactin and a more rapid increase in cortisol in juvenile coho salmon, independently of salinity. Acute stress, on the other hand, raised only the cortisol level and did not affect prolactin. No clear correlation was found between the two hormones, which is not surprising since prolactin is the important hormone in the adaptation of fish transferred from salt water to fresh, and cortisol in fish transferred from fresh water to salt (reviewed by Johnson, 1973). The effects of these hormones seem to differ between different species, so the picture is complex. However, Johnson concluded that the prolactin may act at the level of osmotic permeability while cortisol governed the active transport of minerals from the gills. Cortisol is also thought to increase the inflow of water across gill membranes (Ogawa, 1975).

Osmotic stress provokes changes in the level of prolactin, cortisol and adrenaline (several authors, reviewed by Mancera *et al.,* 1993). In the gilthead sea bream, the level of cortisol decreased strongly once the fish had adapted to the sea water; its work was done. Interestingly, sockeye salmon which adapted successfully to a transfer from fresh to sea water showed a temporary large increase in cortisol, plasma ions rising strongly but then settling down to an 'acceptable' level. In salmon that failed to adapt, the cortisol remained elevated, ionic concentrations also remained elevated and the fish became dehydrated and died (Franklin *et al.*, 1992). Cortisol has a wide-reaching activity, influencing several osmoregulatory organs, such as gills, guts, urinary bladder and kidneys. However, Mancera *et al.* (1994) still feel that its exact osmoregulatory role is not clear and seems to depend on the species studied. In some species, as already noted, it stimulates the chloride cells and increases branchial ATPase activity, but in other species it controls the adaptation to fresh water by acting synergistically with prolactin.

Pituitary activity is absolutely central to osmoregulation, and hypophysectomy destroys the ability of fish to adapt to a change in salinity. The prolactin is synthesized in, and is secreted by, the pituitary, which also secretes adrenocorticotrophic hormone (ACTH), which, in turn, stimulates the adrenals to produce cortisol. The level of ACTH in the plasma is therefore raised when fish are in sea water (Nichols and Fleming, 1990). The pituitary also secretes growth hormone into the blood plasma in sea water (Yada and Hirano, 1992: rainbow trout) but its role is not clear in the present context.

In fresh water, salmon change from parr to smolt in preparation for the marine environment and this process involves increases in thyroid hormones (Boeuf, 1987). The thyroid activity of killifish in sea water is greater than that in fresh (McNabb and Pickford, 1970). However, in two species of mudskipper, *Periophthalmus chrysospilos* and *Boleophthalmus boddaerti*, the thyroxine (T4) was involved only in enabling the fish to cope with terrestrial stress and not in osmoregulation in waters of different salinities (Lee and Ip, 1987). These authors found, on the other hand, that the other thyroid hormone, 3,5,3'-tri-iodo-L-thyronine (T3) played a more significant role in osmoregulation under the various aquatic conditions in the latter of the two species. The authors concluded that, since the habitats of the two species differed considerably, they possessed different control mechanisms to deal with changes in the environment. Parker and Specker (1990) measured the levels of both thyroid hormones in whole larval and juvenile striped bass reared for 10 days in fresh water, sea water and intermediate salinity. They concluded that these hormones 'may mediate the beneficial effects of salinity on larval striped bass growth and survival'.

There does appear to be some connection between the hormones of the thyroid gland and the salinity of the medium. However, it is far from simple, and the situation has changed little since Henderson and Chester Jones (1974) concluded that the precise connection between thyroxine and osmoregulation was unclear, even suggesting that the hormone may influence the salinity preferred by the fish.

2.2.5. Salinity and Lipid Composition

Freshwater and marine fish have different lipid compositions, the main difference being that marine fish contain more of the C20 and C22 polyunsaturated fatty acids (reviewed by Love, 1970). Both Kelly *et al.* (1958) and Stansby (1967) considered that the differences were largely dietary in origin, although it was known that both groups of fish were able to synthesize some polyunsaturated fatty acids from non-fatty precursors. Earlier, Lovern (1934) had observed that, while lipids of the parr stage of Atlantic salmon were typical of freshwater fish, they changed to the type peculiar to adult (marine) salmon when the parr changed to the smolt form while still in the river, although some characteristics of freshwater fish lipids remained. Lipid chemistry advanced, and Meister *et al.* (1973) found that the phosphatidyl choline and phosphatidyl ethanolamine of the gills became enriched in docosahexaenoic acid (C22:6 ω3) when eels transferred from river to sea. These authors associated the phenomenon with the increase in ATPase activity of the gill membranes.

More recent work emphasizes the synthesis by the fish of polyunsaturated fatty acids in the marine environment. Borlongan and Benitez (1992) maintained groups of milkfish (which can tolerate salinities from 0 to 100‰) in fresh water or sea water on the same diet, and while the two groups showed no differences in their total lipid contents, the proportion of phospholipids in the lipids of various organs was the greater in fish from sea water. The organs of fish in fresh water contained lipids with higher proportions of neutral lipids.

There were marked differences in the fatty acid patterns of the lipids of gills and kidneys, those from seawater fish possessing more ω-3 in proportion to ω-6 fatty acids and a higher proportion of total polyunsaturated fatty acids. They suggested that the role of polyunsaturates in membrane permeability and plasticity might account for the observation, and pointed out that the ω-3 structure allows a greater degree of unsaturation than do the ω-6 or ω-9 series.

Studies on eels (Hansen, 1987) and rainbow trout (Hansen *et al.*, 1995) show a close connection between salt transport and the metabolism of phospholipids, particularly in the gills. These authors added radioactive precursors to the water and measured their incorporation into different fractions of phospholipids in gills, oesophagus and intestine under different regimes of salinity. Broadly speaking, the phospholipids of the fish in fresh water were dominated by the (saturated) phosphatidyl choline, whereas in sea water they were dominated by the mono-unsaturated phosphatidyl ethanolamine. It was suggested that the incorporation of the latter might modify the cell membranes to change their permeability to ions. According to Hansen (1987), it is known that only the mono-unsaturated fatty acids are involved in the proliferation of mitochondria, so that the seawater pattern of phospholipids in eels might represent the proliferation of mitochondria; chloride cells are rich in them. The absence of phosphatidyl ethanolamine domination in the phospholipids of fish in fresh water indicates that the fish do not face such an osmotic challenge under these conditions (Hansen *et al.*, 1995: rainbow trout). As the latter authors stated, 'The modification of membrane proteins by changes in the surrounding lipid moiety is a neat way of achieving biological adaptation'.

These more recent findings show that, although some differences between the lipids of fish from fresh and salt water may be largely of dietary origin, the lipids of tissues involved in salt exchange are made different (by synthesis) to assist osmoregulation. Phosphatidyl inositol appears also to be important in the marine environment.

2.2.6. Nitrogenous Compounds

It was stated at the beginning of this section that elasmobranchs guard against dehydration in sea water by retaining urea, trimethylamine oxide and other nitrogenous compounds in their blood. Urea is synthesized as the end product of nitrogenous excretion in elasmobranchs, whereas teleosts excrete mostly ammonia. Elasmobranchs have little or no active ion transport at the gills, but have a 'rectal gland', which is found only in cartilaginous fish, and which excretes some of the excess salts. Further removal of salts is carried out by the kidney.

The muscle and blood of various inshore invertebrates, such as crabs, lobsters (Duchâteau *et al.*, 1959; Florkin and Schoffeniels, 1965) and mysids (Moffat, 1996) become enriched with free amino acids when the animals are transferred from fresh or brackish water to full-strength sea water. There is presumably an osmotic advantage in this. The situation in fish, however, is far from clear. Jones (1959) suggested that variations in the free amino acids and taurine in lemon sole and Atlantic cod might be governed by the salinity of the water. On the other hand, the Atlantic salmon shows no such relationship, the level of free amino acids in the blood rising after the fish enter the rivers to spawn (Cowey *et al.*, 1962). This finding is probably linked to the spawning: muscle proteins are being broken down and the amino acids transported to the developing gonads. The amino acid content of killifish serum has also been reported to decrease when the fish are transferred from fresh water to sea (Pickford *et al.*, 1969) when spawning was not involved. Boyd *et al.* (1977) reported an increase in the free amino acid concentrations of two species of fish – but they were both elasmobranchs, which draw on such compounds for osmoregulation anyway. Love (1980, Table 19), reviewing the topic, listed fish species in which the concentrations of free amino acids appear to be influenced by the external salinity, but in all cases but one the tissues examined did not include blood. The exception was the hagfish, which, as mentioned earlier, does not osmoregulate. The role of free amino acids in osmoregulation is still unclear.

Trimethylamine oxide (TMO) is one of the compounds retained by elasmobranchs to assist their osmotic balance (reviewed by Love, 1970). This compound is also used by eels when they are transferred from fresh water to salt. The fish were not fed, but TMO levels increased for about 24 h. The compound is probably synthesized from choline via trimethylamine, and synthesis of the enzymes responsible for the system appears to begin immediately after transfer to the sea water (Daikoku and Sakaguchi, 1990).

Significant studies by workers from the former Soviet Union on the impact of salinity on fish metabolism include Zaks and Sokalova (1961), Natochin *et al.* (1975a), Romanenko *et al.* (1982), Krayushkina (1983), Klyashtorin and Smirnov (1990) and Varnavsky (1990a, b).

2.3. OXYGEN LEVEL

The ambient oxygen level is the most important factor governing modes of metabolism in aquatic animals. The extent to which the dissolved oxygen is accessible to the tissues of fish determines the intensity of aerobic catabolism of proteins, lipids, and carbohydrates and their sub-components – amino acids, fatty and other organic acids, and glucose. It also governs the oxidative phosphorylation that supports the resynthesis of adenosine triphosphate (ATP), the source of 'instant energy' to, for example, the swimming muscles.

When the organism is in an environment deficient in oxygen, the processes of anaerobic catabolism, including glycolysis, the pentose phosphate shunt and the utilization of deaminated amino acids (Hochachka *et al.*, 1973), play an active role. Evaluating the intensity and efficiency of these pathways of energy metabolism relative to aerobic pathways, and their interrelationships under different oxygen regimes, is essential to the study of biochemical and physiological ecology. Another important task is to determine the total content of substances which yield energy, for example, high-energy phosphates which contain adenyl nucleotides (primarily ATP and creatin phosphate), non-esterified ('free') fatty acids, triacylglycerols, glucose, glycogen and free amino acids. Compounds which transport oxygen (haemoglobin and myoglobin) are also significant in this connection. The sum total of these substances indicates the energy capacity of the organism.

Changes in certain morphological features run parallel to and supplement the information obtained from biochemical and physiological studies. They include among other things the numbers and sizes of mitochondria (the main 'energy factories' of tissues), features of myofibril structure, localization of reserve (as distinct from 'structural') lipids, the distribution of glycogen granules in liver and other tissues, the number of erythrocytes (indicating the oxygen capacity of the blood) and the sizes of endocrine follicles, which perhaps indicate the potential endocrine output.

The intensity of metabolism has been studied in a variety of Black Sea and Azov Sea fish (Kovalevskaya, 1956; Alekseeva, 1959, 1978; Ivlev, 1964; Belokopytin, 1968, 1978, 1990, 1993; Skazkina, 1972, 1975; Klyashtorin, 1982; Klovach, 1983; Shekk, 1983). This sort of research takes the form of measuring the oxygen uptake by the entire organism. For the most part, such work has been done under normal oxygen saturation (100–80%). Quantitative relationships between the intensity of oxygen uptake and the body weight, temperature, locomotory activity and the 'group effect' have all been measured, and the results are widely available in the literature.

Insofar as these data are solely physiological, we examine the findings only in general outline. The rate of oxygen uptake (Q) in fish depends on its body weight (W), described by the equation:

$$Q = aW^k \qquad (1)$$

where coefficients a and k are within the range 0.28–0.97 and 0.73–0.80, respectively (Belokopytin, 1993). Dependence of Q upon temperature is generally governed by the Van't Hoff law:

$$Q_{10} = \left(\frac{Q_2}{Q_1}\right) \frac{10}{T_2 - T_1} \qquad (2)$$

where Q_1 and Q_2 are the rates of oxygen uptake at temperatures T_1 and T_2. The average deduced from this equation is 2.2 (Belokopytin and Shulman, 1987), a value close to that given by Vinberg (1983) for fish in general. The rate of oxygen uptake depends on the swimming rate of a fish (V) as in the equation

$$Q = qb^v \qquad (3)$$

where q and b are coefficients (Belokopytin, 1990, 1993). Alexeeva (1959) found that the rate of oxygen uptake per fish is 30% less in a group compared with that of a solitary fish. Critical and threshold oxygen concentrations, the oxygen capacity of the blood and total oxygen demand have all been determined by Yarzhombek and Klyashtorin (1974), Klyashtorin (1982) and Soldatov (1993). The role of erythrocytes in oxygen transport for the tissues was studied by Soldatov and Maslova (1989) and Soldatov (1996). Most studies on fish energy metabolism have been carried out at normal oxygen levels. However, oxygen insufficiency is a fact of life with a great significance for all aquatic organisms, and it influences the level and character of their energy metabolism.

The problem of hypoxia (oxygen deficiency) has often been acute in the southern European seas (Orel *et al.*, 1986, 1989; Zaitsev, 1992). The cause is usually excessive growth ('blooms') of phytoplankton in the Sea of Azov, the north-western Black Sea, the Gulf of Venice and the Gulf of Trieste, which leads to the intensive oxidation of organic matter. In recent years, the phenomenon ('eutrophication') has been exacerbated through human agency in these shallow waters. The discharge of insufficiently treated sewage rich in nitrite, nitrate, phosphate, sulphate, etc. and run-off of excess fertilizer from agricultural land results in excessive production of microalgae. Pesticides, heavy metals and hydrocarbons can also cause oxygen deficiency by impeding its accessibility to the tissues of aquatic animals. This can lead ultimately to mass mortalities of invertebrates and fish.

To add to the list of these calamities, we should mention an important problem in the Black Sea: the existence of a very large deep-water zone rich in hydrogen sulphide. The oxygenated zone is restricted to a water layer from 0 to 150 m deep, extending over the entire sea, and Vinogradov (1990) has shown that living organisms are concentrated mostly at the interface between the oxygenated and the deeper hydrogen sulphide zones – the so-called 'redox zone'.

These phenomena are aspects of a world-wide occurrence of areas deficient in oxygen, which have been much investigated (Schmidt-Nielsen, 1975; Douglas *et al.*, 1976; Kukharev *et al.*, 1988; Vinogradov *et al.*, 1992a, b). Intensive accumulation of phytoplankton and detritus is particularly common in tropical and subtropical waters of the ocean: the Gulf of California and Arabian Sea are typical examples. The result is oxygen depletion in vast water masses at between 100 and 900 m depths, concentrations dropping to values as low as 0.1 ml l^{-1}. Surprisingly, despite such extreme depletion of oxygen, life can be abundant and diverse in these waters; zooplankton, pelagic fish and squid develop actively and yield high biomasses. As recently discovered, life can also be abundant around areas of the mid-ocean ridges where there are hot springs containing high levels of hydrogen sulphide, methane and heavy metals (Grassle, 1986; Gal'chenko *et al.*, 1988; Tunnicliffe, 1991; Jannasch, 1995; Fisher, 1996; Gebruk *et al.*, 1997). There is no space here to discuss these special biotopes that are based on chemosynthetic bacterial production, and which are inhabited by certain species of fish as well as several phyla of invertebrates.

We should note that fish do need some oxygen, even if the concentration is low. Further studies of fish inhabiting the Black Sea and similar hypoxic habitats might contribute to solving regional problems and yield a better understanding of the paradox. Experiments imposing a sudden hypoxia lasting 1–3 h on Black Sea fish (annular bream, scorpion fish and horse-mackerel: Stolbov *et al.*, 1995) show a continuous decrease in oxygen consumption by fish in water where the dissolved oxygen content decreases from 8.6 to 1.7 mg l⁻¹ (Figure 4). It is reasonable to assume that the latter oxygen level is insufficient to allow aerobic

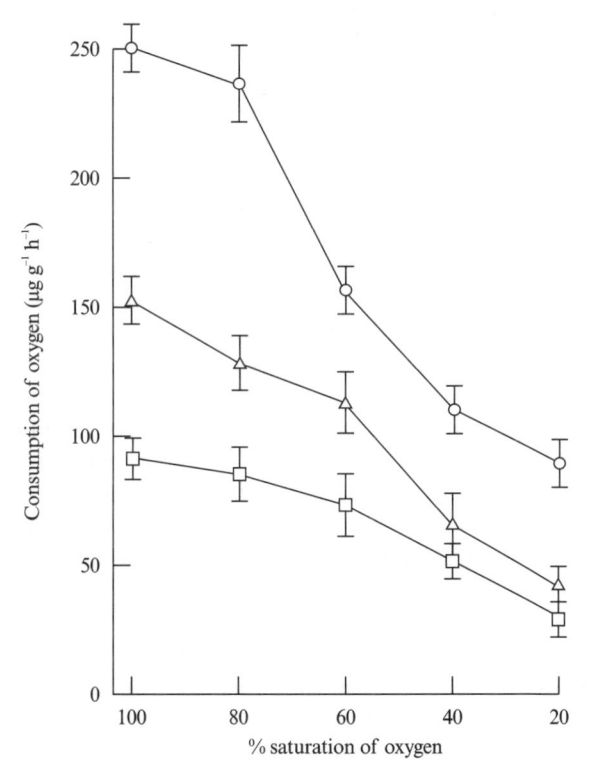

Figure 4   Oxygen consumption by Black Sea species during short-term hypoxia: ○, Annular gilthead (annular bream); △, sea scorpion; □, horse-mackerel. (After Stolbov *et al.*, 1995.)

metabolism to take place in the tissues. Note that aerobic metabolism normally predominates over anaerobic in fish held at a wide range of temperatures, except for the extremes (Hochachka and Somero, 1973; Romanenko *et al.*, 1991).

As stated above, the influence of glycolysis and the pentose phosphate shunt have been much studied in relation to hypoxia. But would the modest stock of carbohydrates be sufficient to support energy requirements in fish suffering continuous oxygen deficiency? Lipids will not be catabolized in the required amount under the hypoxic conditions that may persist for weeks or months in the oceans, and are permanent in some pelagic tropical zones and in the redox zone of the Black Sea. The stock of glycogen, the basic reserve of carbohydrate energy, is usually not more than 1% of the total body weight (4–5% of dry weight). This is sufficient to provide only a short-term burst of energy (Black, 1958; Shulman, 1974; Morozova *et al.*, 1978a) but not long-term or continuous energy metabolism.

Interestingly, it has been stated a priori (Kukharev *et al.*, 1988) that in the hypoxic water of tropical seas it is glycolysis that supports energy metabolism in copepods, squid and even fish. However, our belief is that, like other pelagic animals, fish take quite another pathway to maintain their energy metabolism under oxygen deficiency. Our experiments showed that, when oxygen deficiency becomes more severe, the lessening oxygen uptake by the fish is accompanied by increasing excretion of ammonia, the major end product of protein catabolism (Figure 5). Ammonia nitrogen is known to make up to 80–90% of the total nitrogen excreted by fish, the remainder being urea (Stroganov, 1962; Mann, 1965; Waarde, 1983). It appears that protein catabolism is intensified under hypoxic conditions. This is surprising, for the traditional viewpoint is that the process requires oxygen at the final stages of amino acid deamination, at least in the overwhelming majority of obligate aerobes (Schmidt-Nielsen, 1975; Prosser, 1979).

It therefore follows that the importance of proteins in supporting total energy metabolism of fish increases during short-term oxygen deficiency. In estimating the share of total energy metabolism contributed by protein, the ammonia quotient (AQ) may be used, that is, the ratio between the oxygen consumed to nitrogen excreted (O/N). This coefficient was originally introduced into hydrobiological studies by Stroganov (1956), and has been widely used since then (Prosser and Brown, 1962; Mathur, 1967; Kutty, 1968, 1972; and others). With AQ > 30, lipid–carbohydrate substrates dominate energy metabolism. When AQ values range from 20 to 30, a mixture of protein and non-protein substrates are used, and an AQ < 20 indicates that protein is the main source of energy in metabolism. An AQ value of 8.67 corresponds to 100% of the oxygen having been utilized by protein, and, with values less than this, a proportion of the proteins is catabolized anaerobically. It has been demonstrated experimentally (Stolbov *et al.*, 1995) that, with the onset of sudden oxygen deficiency, the ammonia coefficient steadily decreases in three

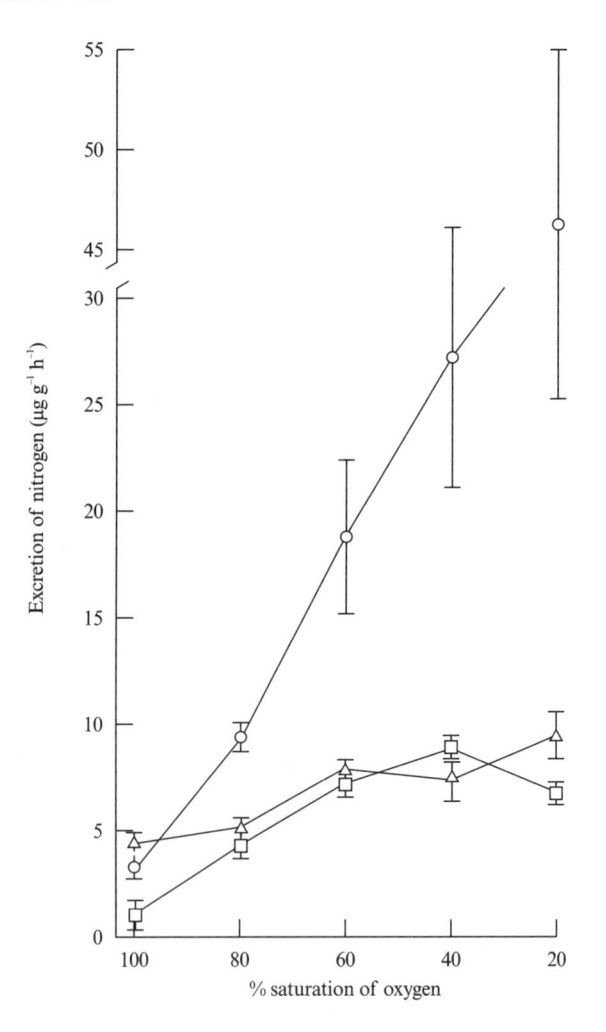

Figure 5   Nitrogen excretion by the same three species, as shown in figure 4, during short-term hypoxia.

species of Black Sea fish. A mixed lipid–carbohydrate pattern of energy substrate utilization has been observed in ambient water 100% saturated with oxygen, which changes to lipid–protein catabolism at 80% saturation and solely to protein catabolism at 60% saturation. With 40–20% saturation, the protein breakdown becomes anaerobic.

These data provide convincing evidence of the active, if not the leading, role that proteins play in maintaining energy metabolism in fish during hypoxic conditions. It is what one should expect, because, unlike glycogen, protein contributes the greatest share of organic material to the body, as much as 80%

of the dry weight and 20% of the fresh weight. It continually replenishes itself during normal living and yields a substantial portion of energy during catabolism. Mathur (1967), Kutty (1968, 1972), Kutty and Mohamed (1975), van den Thillart and Kesbeke (1978), Mohamed and Kutty (1983a, 1983b, 1986) and Solomatina et al. (1989) have demonstrated the leading role played by proteins in energy metabolism (and also their anaerobic utilization) during oxygen deficiency of the freshwater species: common rasbora, St Peter's fish, Rhinomugil corsula (an Indian species), catfish and goldfish. These results have been obtained from calculating the ammonia coefficient and the respiratory quotient (RQ), the latter being the ratio of carbon dioxide excreted to oxygen consumed. The determination of RQ in whole fish is a difficult task because of problems in measuring the excreted carbon dioxide: some of it is bound by the carbonate buffers that occur in sea water.

As stated above, at 40–20% saturation of oxygen, a considerable proportion of protein is catabolized anaerobically. Figure 6 gives the estimated percentage of oxygen used to oxidize protein in Black Sea fish during a sudden oxygen stress, and the percentage of proteins involved in anaerobic metabolism. In horse-mackerel, the anaerobic portion of protein is 35–45%, in scorpion fish 60% and in annular bream up to 80%. Interestingly, these bottom scorpion fish change to anaerobic metabolism of proteins at a lower level of dissolved oxygen than do the (near-bottom) annular bream and pelagic horse-mackerel which normally inhabit more aerated waters.

Experimental results from three species of Black Sea fish exposed in the autumn of 1991 to sudden oxygen deficiency (Stolbov et al., 1995) agree with the data from similar experiments on the same species in spring–early summer 1992 (Stolbov et al., 1997). The only discrepancy was the lower value (30–40%) for the aerobic metabolism of protein. Thus in different seasons and in different physiological states (post-spawning rest in autumn and intensive sexual development in spring) fish maintain the same mode of metabolism of biochemical substrates.

A series of experiments intended to imitate the situation in nature (with a longer period of oxygen deficiency) were carried out on Black Sea scorpion fish by Stolbov et al. (1997), the fish being placed in 16–12% aerated water (1.0–1.4 mg l^{-1} dissolved oxygen). On reaching steady hypoxic conditions, the fish consumed oxygen and excreted nitrogen in the manner described in the earlier experiments using sudden oxygen stress. Consumption of oxygen stabilized but nitrogen excretion showed a continuous increase, the ammonia coefficient dropping to the dramatic value of 8.67, or even further. The experimental long-term hypoxic conditions with scorpion fish differed from natural conditions; in nature, fish probably adapt to increasing deficiency of oxygen more gradually. However, in fish that live constantly in hypoxic zones, the only possible mechanism for supplying energy is that described above.

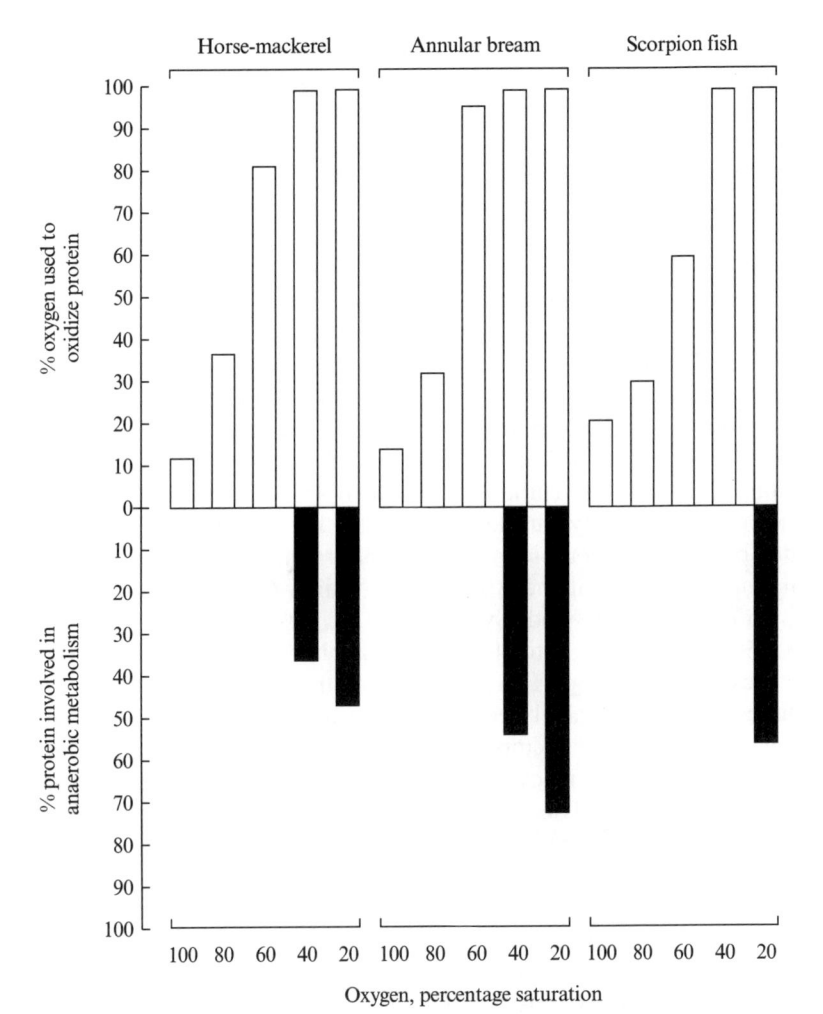

Figure 6   Percentage of oxygen used for the oxidation of protein (upper half) and the quantity of protein involved in anaerobic metabolism during short-term hypoxia (lower half).

The anaerobic mode of protein utilization is entirely possible in theory and in practice. Oxygen is not required for protein and nitrogen catabolism until the final stages of amino acid deamination have been reached. Complete anaerobic catabolism of proteins and nitrogen compounds (to the point where the final products CO_2, H_2O and NH_3 appear) has been known for a long time in prokaryotic organisms, but in eukaryotes only in parasitic worms, which are obligate anaerobes (von Brand, 1946). However, in recent decades, anaerobic metabolism of proteins has been found in some aquatic organisms – facultative anaerobes inhabiting littoral, soft-bottom deposits and dried-up water bodies. It has also been found in benthic and pelagic organisms occurring in hypoxic surroundings. Our survey of available data on the ammonia coefficient yielded by aquatic invertebrates and fish has shown (Shulman *et al.*, 1993a, b) that, under certain conditions, anaerobic catabolism of proteins takes place in coelenterates (*Polaria infeseus, Aglantha digitale*), crustaceans (*Neomysis integer, N. rayse, Acartia clausi, Centropages typicus, Phronima sedentaria, Palaemonetes varians*), cephalopods (*Loligo forbesi, Octopus vulgaris*) an echinoderm (*Himerometra magnipinna*), a chaetognath (*Sagitta hispida*), tunicates (*Salpa fusiformis, Thalia democratica*) and fish (*Channa punctatus, Carassius auratus, Oreochromis rendalli*), that is, in all principal groups of aquatic organisms (Conover and Corner, 1968; Reeve and MacKinley, 1970; Mayzaud and Dallot, 1973; Ferguson and Raymont, 1974; Ikeda, 1974; Caulton, 1978a; van den Thillart and Kesbeke, 1978; Smith, 1982; Mohamed and Kutty, 1983a, b; Ryabushko, 1985). This list has recently been supplemented by data on anaerobic metabolism of proteins in two species of squid from tropical waters of the Atlantic and Indian Oceans (Shulman *et al.*, 1992), a Black Sea copepod (Svetlichny *et al.*, 1994) and a shrimp (Stolbov *et al.*, 1997). The share of total protein catabolism contributed by anaerobic processes varied from 2–3% to 70–90% in all the forms examined.

These data have been analysed by using the ammonia coefficients of aquatic animals and defining the factors which induce anaerobic catabolism. The first of these is oxygen deficiency, that is, low dissolved oxygen concentration in the surrounding water. This situation can be found in mesopelagic waters of the oceans, in eutrophicated areas of semi-closed seas plentifully supplied by rivers (most of the European seas), in waters adjacent to areas contaminated by hydrogen sulphide (Black Sea), soft-bottom deposits and in hydrothermal zones. The phenomenon is widespread in freshwater bodies, affecting aquatic animals from rivers, lakes and ponds with a high level of eutrophication, low or zero flow rates, and ice cover during the winter.

Secondly, there is so-called functional or inner oxygen deficiency (Lukyanenko, 1987), which plays an important part in anaerobic oxygen deficiency. This deficiency is caused by intensive oxygen uptake during active muscle performance. High locomotory activity consumes all the oxygen stored in the tissues, so that energy cannot then be produced in sufficient quantities. When muscular activity takes the form of powerful bursts of swimming, anaerobic glycolysis provides most of the energy. However, stored carbohydrates are inadequate to support such performance over longer periods, and the activity is then assisted by aerobically catabolized free fatty acids, triacyl-glycerols, amino acids and proteins in addition to any anaerobic protein catabolism that there might be. Detailed examples of such deficiency have been demonstrated by Dabrowsky (1986) in coregonid and salmonid fishes. A similar effect on the metabolism of fish is produced by stress, which increases the energy requirements regardless of the level of oxygen available. In this case, the endocrine glands may be of vital importance, triggering anaerobic catabolism of proteins (Selye, 1973; Meyerson, 1981). Overfeeding also results in an internal hypoxic state found in predatory fish and squid. Predators are accustomed to feed to excess, and when they cannot assimilate all ingested protein aerobically, as much as 30–70% is metabolized anaerobically (Sukumaran and Kutty, 1977; Shulman et al., 1992).

An important cause of environmental hypoxia is pollution by man. Among the most hazardous pollutants are nitrites, nitrates, heavy metals, pesticides and hydrocarbons. Toxic products may also be created naturally, as a result of outbreaks of blue–green algae (Cyanophyta). The subsequent toxin results from their metabolism and decay. Such toxicants can induce a sharp decrease in the haemoglobin concentration of the blood of sturgeon from 7–9% to 1.1–1.2% and in the red cell count from 2 to 0.65 million ml^{-3} (Lukyanenko, 1987). The impact of these substances on invertebrates and fish manifests itself in different ways, although their basic action is generally confined to inhibiting the supply of oxygen to the tissues. This impedes processes that require oxygen and leads to the accumulation of surplus peroxides, free radicals that damage the phospholipid proteins of cell membranes. In all probability it is this mechanism that accounts for disintegration of muscle fibres in fish from many water bodies, notably in sturgeons from the Volga–Caspian basin (Evgenyeva et al., 1989; Evgenyeva and Kocherezhkina, 1994; Lukyanenko et al., 1991; Altufiev et al., 1992; Nemova et al., 1992; Shulman et al., 1993a; Yurovitsky and Sidorov, 1993). Histological examination of the ravaged muscle, especially white muscle, shows disintegration of the sarcoplasmic reticulum and myofibrils, and the degeneration of collagenous structures. Structural changes are also seen in the liver and some other organs.

Toxicants that impede the access of oxygen to the tissues of sturgeon appear to disrupt the structure of polyenoic acids, which then give rise to free radicals. In addition, the antioxidative activity of the membrane decreases, perhaps again because of the action of (damaged) polyenoic acids, and cell membranes are destroyed, their protein components being subjected to extensive catabolism. A possible scheme for the catabolic 'collapse' is shown in Figure 7. In addition, the amino acids released from the proteins may become partially involved in gluconeogenesis, leading to glycogenolysis and glycolysis. This may result in an accumulation of lactate, hence a fall in pH which could per se destroy muscular tissue. Observations appear to show that, during muscle destruction, the peroxidase level, activity of lysosomal and non-lysosomal enzymes (including cathepsin D and calpain I and II) and collagenase increase sharply (Nemova et al., 1992; Yurovitsky and Sidorov, 1993). Some protein fractions in the muscle can disappear completely (Bal et al., 1989).

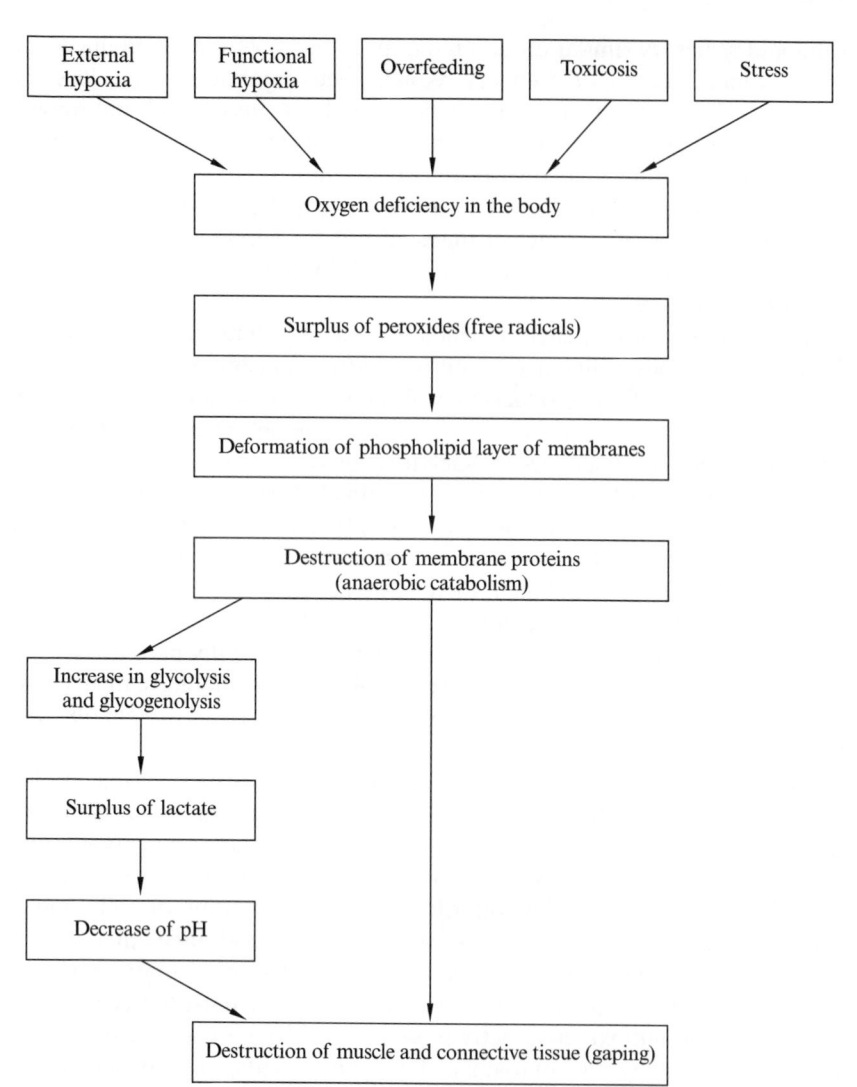

Figure 7   Causes of hypoxia and its consequences in the tissues.

This 'steep' anaerobic catabolism of proteins can be triggered not only by
toxins but also by the other factors mentioned above, including ambient
oxygen deficiency, intensive muscle performance, consuming proteins in
excess and during stress. What appears to be a similar picture is observed in
cod from the North Atlantic, in which heavy feeding following winter starvation results in a lowered pH after death and a weakening of connective tissue
(Love, 1988). In that case, however, there is no evidence of actual catabolism,

as the weakened connective tissue regains its original strength when the pH is raised to neutral again (Love *et al.*, 1972). The collapse of proteins in sturgeons through anaerobic catabolism, although involving most of the muscle tissue, is less marked in red muscle, which enables the fish to maintain continuous swimming, as distinct from powerful surges. Fish exhibiting the destruction of white muscle therefore apparently display normal cruising activity, so much so that they cannot be visually distinguished from non-affected stock.

Dramatic destruction of muscular tissue in fish is not necessarily pathological, but can be part of the natural life cycle. Pacific salmon (*Oncorhynchus* species) degenerate in this way during their spawning migration (Idler and Bitners, 1958; Mommsen *et al.*, 1980; Ando, 1986), when the proteins provide material for gametogenesis and energy (through gluconeogenesis) for the long swim upstream. The destruction does not involve red muscle, so that the males which eventually reach the spawning ground are still able to fight competitively with one another (Yuneva *et al.*, 1987). The smell of ammonia given off by squid caught in the hypoxic zone of the Arabian Sea (Kukharev *et al.*, 1988) may again indicate the destruction of protein. Dr. G. Brizzi (personal communication) found that spiny lobsters fished from the Bay of Trieste during a pronounced plankton bloom, fatal to fish, also smelt of ammonia.

So far, protein catabolism in hypoxic conditions has been discussed using pelagic and benthic marine animals as examples. There is, however, another group of organisms for which glycogen is the basic source of energy when hypoxic conditions develop. In particular, bivalve molluscs show this effect (Hochachka and Somero, 1973; Bayne, 1975; Goromosova and Shapiro, 1984; Lushchak, 1994). These accumulate substantial deposits of glycogen, which can amount to 30% of the dry weight and allows them to sustain normal life during oxygen deficiency, including that induced by heavy metal pollution. Goromosova and Shapiro (1984) have identified high levels of glycogen in Black Sea mussels (*Mytilus galloprovincialis*). Their results, as well as those obtained by Tamozhnyaya and Goromosova (1985) and Antsupova *et al.* (1989), show the important part that proteins also play (via gluconeogenesis) in sustaining energy metabolism during hypoxia. A marked increase in the activities of alanine- and aspartate aminotransferase (enzymes of nitrogen metabolism), an accumulation of malate (an intermediate of amino acid catabolism), and the excretion of the free amino acids alanine and glutamate all provide supporting evidence for this. Similar data were obtained from mussels (de Zwaan and Zandee, 1972; Bayne *et al.*, 1976; Aristarkhov *et al.*, 1988).

The fate of protein catabolites in hypoxia and anoxia provides evidence that the phenomenon is widespread or even universal. Indeed, the principal biochemical mechanism of protein catabolism was originally discovered under hypoxic conditions in molluscs (Hochachka and Mustafa, 1972;

Hochachka *et al.*, 1973; Driedzic and Hochachka, 1975). By now, many workers have identified the pathways by which the products of protein catabolism are utilized, both in normal environments and also during pollution (Johnston, 1975, 1983; Walker and Johansen, 1977; Hughes and Johnston, 1978; van den Thillart and Kesbeke, 1978; Waarde, 1983; Driedzic *et al.*, 1985; Grubinko, 1991; Kurant and Arsan, 1991; Konovets *et al.*, 1994; Grubinko and Arsan, 1995). The pool of free amino acids is therefore an important source of energy during hypoxia.

The complex sequence of operations involved in anaerobic utilization of proteins and carbohydrates in the mitochondria of fish muscle has been described by Owen and Hochachka (1974). Savina (1992) has given a detailed analysis of this scheme. As in molluscs subjected to oxygen deficiency, the major part in the metabolism of amino acids in fish is played by alanine- and aspartate amino acid transferases. The combined use of proteins and carbohydrates provides efficient resynthesis of ATP, which then readily provides the energy for swimming. In molluscs under normal conditions and in fish during oxygen deficiency, succinate and alanine are important intermediate products of amino acid metabolism, which, unlike lactate, are not toxic to the organism. This pathway is open not only to molluscs and fish, but also to diving mammals and to terrestrial mammals during intensive muscular work (Kondrashova and Chagovets, 1971).

The facts presented in this section raise some questions. First, is it valid to claim 'oxygen deficiency' in aquatic organisms whenever a low oxygen concentration (less than 0.1 ml l^{-1}) is found in the environment? Is it possible that 'hypoxia' in human terms is a normal condition for other forms of life? Karpevich (1975), who proposed the concept of concealed adaptation potential supports this idea, which greatly exceeds the limits set by our traditional notions. The concept may provide an answer to the question 'Why are hypoxic forms of life so widespread and obviously thriving in nature'? Secondly, is anaerobic energy always related to protein catabolism, or does it more often relate to those nitrogenous non-protein substances that are so important to the organism, e.g. dipeptides, polypeptides, free amino acids and nucleotides? After all, analysis of the tissues of fish and aquatic invertebrates shows that as much as one-third of the nitrogenous products falls into the category of non-protein components. Similarly, when nitrogenous products enter the organism with food, they are not always solely in the form of proteins. It would therefore be more correct to speak about the catabolism of 'nitrogenous substances' rather than proteins in anoxia and hypoxia. The relative contributions of protein and non-protein nitrogen also need to be ascertained. Thirdly, having demonstrated the importance of anaerobic processes in sustaining energy metabolism, we can begin to dispute the reliability of an index such as total oxygen uptake by these organisms. Energy is produced anaerobically not only by glycolysis but also by nitrogenous catabolism. We feel that the contribution made by protein and nitrogen to the anaerobic pathway of ATP resynthesis has been regularly underestimated.

Stolbov *et al.* (1997) studied the dynamics of the ammonia quotient (AQ) in the early post-embryonic stages of Black Sea turbot and found that, from the 12th to 15th day (stages IV–V), the energy metabolism of the larvae had become mainly anaerobic. Aquatic organisms dwell in hypoxic media in many parts of the world, as mentioned earlier. In this context, it is pertinent to note the tendency of some hydrobiologists to ignore the P/O coefficient, that is, the index of the efficiency of oxidative phosphorylation, which varies widely according to ambient conditions. Neglecting this coefficient, one will fail to grasp the true situation about the amount of oxygen consumed for resynthesizing ATP relative to the amount wasted as heat.

Finally, the productivity of aquatic organisms inhabiting hypoxic water can increase considerably whenever anaerobic metabolism is used in their energy production. Paradoxical though this might seem, it is given further consideration in Chapter 5.

Like proteins, carbohydrates also protect metabolic processes, including the resynthesis of ATP, under hypoxic conditions (Hochachka and Somero, 1973; Murat, 1976; Mohamed and Kutty, 1983a). The remarkable finding of Blazka (1958), that lipids accumulate by means of carbohydrate catabolism in starving and anoxic carp under a crust of ice, is now widely recognized. Further, it has been found that such anoxic carp can also accumulate ethanol (Jürss, 1982; Johnston and Bernard, 1983; Waarde, 1983; van den Thillart and Waarde, 1985; Waversveld *et al.*, 1989). During deficiency of oxygen, wintering carp develop the glycolytic pathways of ATP resynthesis more strongly (van den Thillart and Waarde, 1985; Bilyk, 1989). In general, hypoxic conditions reduce the level of tissue glycogen (and of adenyl nucleotides, ATP and creatine phosphate because of ineffective glycolysis) and raise that of lactate. At the same time, the activity of gluconeogenesis becomes dramatically inhibited (Solomatina *et al.*, 1989; Lushchak, 1994). Oxygen deficiency is stressful, so it increases the level of catecholamines in the blood, leading to increased metabolism of carbohydrates.

It is not yet clear which estimates of the ratio between the levels of protein and of carbohydrate metabolism during hypoxia should be regarded as reliable. It seems likely that the increase in respiratory quotient in freshwater fish to values of 2.5–2.8, as found by Mohamed and Kutty (1983a, 1986), indicates a predominance of protein expenditure over that of carbohydrate. A hypoxic environment shifts the acid–base balance of the fish towards acidosis (Kotsar, 1976), thereby inducing the redistribution of electrolytes, alteration of ion exchange and the activity of Na^+-K^+-Mg^{2+}-ATPases and alkaline phosphatases. It also leads to an increased level of CO_2 in the blood, which enhances the bicarbonate buffer system (Kotsar, 1976). In section 2.1, we described how lipid saturation changes in fish as an adaptation to temperature. However, in the course of an experiment conducted on five species of goby in the Sea of Azov (Shulman, 1972a), we came to the conclusion that saturation might also be influenced by the concentration of dissolved oxygen, or, to be more precise, by the oxybiotic potential of the species.

The iodine numbers obtained from the liver lipids of these gobies were related to the type of bottom (Table 2) where the gobies live. The shallow Sea of Azov warms up considerably in the summer-time throughout its depth, and organic decay and synthesis are intensified in the near-bottom water. This results in a stratification of oxygen concentrations. Different species of Azov gobies are distributed amongst different types of bottom, each ground having a specific oxygen regime. The oxygen concentration in the marine environment is highest in waters washing rocky shores. It is less over a sand and shell bottom and least in areas of silty deposits. The syrman goby inhabits silty bottom grounds and has evolved a series of adaptations to low dissolved oxygen, a reduced total or respiratory metabolism being among the most important (Shulman et al., 1957; Skazkina, 1972). In the syrman goby, the total metabolism is 1.5–2 times less than in the round goby and the highest values were found in the toad goby (Skazkina, 1972). Thus, the unsaturation of liver lipids in the gobies examined relates to their total metabolic activity, indicating also that habitat conditions control peculiarities of respiratory and, possibly, adipose metabolism.

It is possible for water to become supersaturated with oxygen, as under conditions of extreme turbulence. A volume edited by Brubakk et al. (1989) deals with the physical side of this problem and its effect on fish, but biochemical aspects remain to be studied. Water-falls and rapids can cause supersaturation in rivers also, but the best-documented cause is the warming up of ice-cold water already saturated with oxygen: values well over 100% saturation are then found (Wedemeyer et al., 1976). The saturation level must be more than 100% before fish develop symptoms of distress (Nebeker et al., 1976), and salmonids begin to die when it reaches 114% (Nebeker and Brett, 1976). The observable symptom at 115–120% saturation is the formation of gas bubbles (emphysema) under the skin of the fins, in the mouth, head and operculum. When gas bubbles form also inside the sensing chambers of the

Table 2   Iodine value (a measure of unsaturation) in the liver lipids of different species of goby from the Sea of Azov.

	males	females
Neogobius syrman	81.7	80.7
Neogobius melanostomus	128.8	113.7
Neogobius fluviatilis	153.7	145.0
Neogobius ratan	160.0	–
Mesogobius batrachocephalus	160.0	169.0

lateral line organ, the fish become insensitive to pressure waves and appear to become disorientated (Wedemeyer *et al.*, 1976). The incidence of these phenomena is probably rare outside hatcheries and aquaria.

Carbon dioxide oversaturation is important only in fresh water. Not a single example of oversaturation with CO_2 has been reported from the sea, presumably a result of an efficient buffering system. In the ocean, the partial pressure of CO_2 varies within a narrow range, from 1.7×10^{-4} near the surface to 9.9×10^{-4} at 5000 m depth. In enclosed seas, the range is wider. For example, in the Black Sea these values are $3-3.9 \times 10^{-4}$ in near-surface water and $20-25.8 \times 10^{-4}$ in deep water (Alekin, 1966).

There are few data about the influence of changing concentrations of CO_2 on biochemical processes in marine animals. Where oxygen is deficient, carbon dioxide accumulates in considerable amounts in the blood, resulting in a disturbed acid–base balance and impeded oxygen saturation of haemoglobin (Romanenko *et al.*, 1980). It can be assumed that, in this situation, the surplus CO_2 would inhibit aerobic catabolism of proteins and other nitrogenous compounds, and stimulate anaerobiosis. However, studies of the freshwater gudgeon (Stroganov, 1962) indicate that enhanced levels of dissolved carbon dioxide can induce increased excretion of nitrogen.

Carbon dioxide can also have an indirect effect upon marine organisms in that it lowers the pH of the sea water. As industrial pollution of the sea increases, we may expect the CO_2 to exercise an increasing influence on the metabolism of inhabitants of the sea. Evidence of the wide-reaching effects of CO_2 on freshwater fish and on diadromous fish during their life in the river (Karzinkin, 1962; Stroganov, 1962; Romanenko *et al.*, 1980) can help our understanding of similar processes in marine fish. Particular note needs to be taken of the role of carbon dioxide in biosynthesis through carboxylation reactions. The use of radioactive carbon as a tracer has demonstrated the inclusion of dissolved CO_2 into proteins, lipids, glycogen and adenyl nucleotides in fish.

2.4. PRESSURE AND BUOYANCY

Hydrostatic pressure affects marine and freshwater organisms. The general concept is that pressure influences metabolic processes in a manner similar to temperature, particularly by increasing the possibility of, and the rate of, interaction between molecular structures. Hochachka and Somero (1973, 1984) present the theoretical grounds for this concept. An expedition was organized specifically to study the effects of pressure on physiological and biochemical processes in oceanic organisms (Hochachka, 1975). The data obtained from this and other studies (MacDonald, 1975; Siebenaller, 1984; Hennessey and Siebenaller, 1987s) posed questions rather than answered them. The rate of metabolic reactions does not always increase with depth, perhaps as a result of the counter-effect of decreasing temperature or, possibly, oxygen deficiency. On the other hand, as in the situation where temperature decreases with depth, compensatory effects might emerge. Thus, one could expect that fatty acid 'tails' would become more compactly pressed into the phospholipids of cell membranes under pressure. In reality, however, the unsaturation becomes greater and the lipids melt more readily (Patton, 1975; Kreps, 1981), activating the lipid–protein complexes in the cell membranes and increasing the functional activity of the latter.

In deep-sea rat-tails (armoured grenadiers), the activities of fructose diphosphatase and pyruvate kinase increase in high-pressure habitats. In contrast, this activity and some other properties do not change with variable pressure in *Ectreposebastes imus*, which makes extensive vertical migrations. Unlike deep-water and surface species, this particular fish is not influenced by pressure (Hochachka and Somero, 1973). Later experiments (Siebenaller, 1984) showed that the isozyme M_4 of lactate dehydrogenase is unaffected by pressure in deep-water fish (*Sebastolobus altivelis, Antimora rostrata, Coryphaenoides acrolepis* and *Halosauropsis macrochir*), but not in shallow-water fish (*Pagothenia borchgrevinki, Scorpaena guttata, Sebastolobus alascanus*). It seems that some oceanic organisms known to make daily vertical migrations of hundreds of metres (lantern fish, squid, copepods) have evolved compensatory mechanisms that reduce or eliminate effects of pressure. Deep-sea forms differ substantially from those in shallow water in the structural characteristics of their proteins, muscle actin being an example.

Few studies aimed at understanding the effects of pressure on metabolism in fish have been performed in the Black Sea or Sea of Azov because no distinct stratification was found in the distribution of water animals there. The Sea of Azov is saucer-like, its maximum depth being only 15 m, and in the Black Sea the oxygenated water layer extends only to 150 m depth. However, Emeretli (1996) has shown that activity of lactate dehydrogenase in the liver increases 2–10 times in scorpion fish and annular bream placed in a barorespirometer (designed by A. Stolbov) and 'sunk' to a depth of 300 m. This response seems to be peculiar to shallow-water species (Hochachka and Somero, 1984).

Turning now to buoyancy, it is pertinent to refer to the work of Polimanti (1913), carried out in the Bay of Naples. He found that the content of lipid in benthopelagic fish became steadily reduced with increasing depth. This appears to be related to decreasing buoyancy. Conversely, the greater lipid content of pelagic (as distinct from demersal) fish seems to be a feature evolved to ensure greater buoyancy (Aleev, 1963). This is also true for the contents of triacyl-glycerols in the eggs of pelagic and benthopelagic fish. In myctophids, the content of lipids decreases with depth from 14% to 2% (Bailey and Robison, 1986), and in clupeids from 12% to 3% (Childress and Nygaard, 1973). Because the lipid content of herring changes from 1% to 25%

and back during the annual cycle, the buoyancy is much affected, and this affects the distribution and behaviour of the species (Brawn, 1969). Atlantic salmon show the same phenomenon (Pinder and Eales, 1969). Only one species of notothenid, *Pleurogramma antarcticum*, possesses large accumulations of lipid in intermuscular and submuscular sacs. It lives near the surface, while other species live near the sea bottom (De Vries and Eastman, 1978).

The fact that in many benthic and demersal fish species (e.g. gobies, scorpion fish and cod) the main reserve of fat is stored in the liver, rather than muscle or intestine, may be explained by a reduced need for buoyancy (Shulman, 1972a, b). Despite the high content of liver lipids, which can rise to as much as 60% of their fresh weight, the total lipid in the whole animal is relatively low (less than 5%) so that fish such as these have been named 'lean'. Being the main 'metabolic factory', the liver can synthesize and catabolize lipids rapidly and in amounts more than sufficient to compensate for the moderate overall total. In the muscle tissue of 'lean' fish, the lipid concentration is only 0.5–1% of the fresh weight, while in fat pelagic fish it can reach 15–20% or even more. Most of this rich store of lipids consists of triacyl-glycerols, which are simply an energy reserve and not part of the cellular structure. In contrast, a content of 60–70% of phospholipid is found in the small amount of white-muscle lipid of well-fed cod (Love *et al.*, 1975), and values in the region of 90% phospholipid were reported by Ross and Love (1979). Both of these publications provide data to the effect that starvation actually reduces the proportion of muscle phospholipid, which has always been regarded as a purely structural lipid, and so 'inviolate'. However, there is no paradox here, since the starved fish had reached the stage of losing protein from the swimming muscles, so muscle cells were in fact being broken down. The composition of the non-phospholipid fraction was not given in these publications, but Ross (1977) showed that while the white-muscle lipids of Atlantic mackerel, a 'fatty' fish, contained 70% triacyl-glycerols, those from the white muscle of cod, a lean species, contained only 1.5% triacyl-glycerols. We may therefore say that there is no fatty energy store in the muscle of 'lean' species. Fish inhabiting the surface waters regulate their buoyancy with the help of triacyl-glycerols (Lee *et al.*, 1975), but other lipid fractions are more important in this respect in mid-water and deep-water species.

Wax esters are long-chain fatty acids esterified to long-chain fatty alcohols, and contain no glycerol. They occur in large amounts in marine animals, especially copepods (Sargent, 1976). They are important regulators of buoyancy in fish (Nevenzel, 1970). Their density (0.90) is less than that of triacyl-glycerols (0.92) and their concentration in the total lipids is greater in deep-water fish, which migrate vertically, than in fish which remain for most of the time at a particular depth. The advantage of wax esters *vis-à-vis* triacyl-glycerols as buoyancy agents is perhaps that their density is independent of hydrostatic pressure. They may also have an advantage in regulating the membrane and tissue metabolism of deep-water and vertical-migratory fish because of the high unsaturation of their components.

The muscle and bone lipids of the castor oil fish and the coelacanth contain 90% of their lipids as wax esters (Nevenzel *et al.*, 1965, 1966; Sato and Tsuchiya, 1970; Bone, 1972). In some species of the families Gonostomatidae, Myctophidae, Gempylidae and Stromatidae, the contents of wax esters range from 40% to 60%. There is a positive correlation between the wax content and the depth of the habitat of a given fish (Bone and Roberts, 1969; Malins and Wekell, 1969; Nevenzel *et al.*, 1969; Lee *et al.*, 1972, 1975; Phleger and Holtz, 1973; Kayama and Nevenzel, 1974; Kayama and Ikeda, 1975; Phleger *et al.*, 1976; Nevenzel and Menon, 1980; Henderson and Tocher, 1987; Eastman, 1988; Neighbors, 1988; Childress *et al.*, 1990). A similar correlation has been found for copepods (Sargent *et al.*, 1971; Lee *et al.*, 1972, 1975; Sargent, 1978; Reinhardt and van Vleet, 1986). Dietary mesopelagic crustaceans are the main source of waxes for fish, but they can also be synthesized by them (Nevenzel, 1970; Nevenzel and Menon, 1980), the sites for synthesis being muscle, hepatopancreas and gut. The density of waxes depends on the temperature, but pressure changes have little effect. Their presence, therefore, appears to be another useful feature adopted by abyssal fish.

In sharks of the genus *Centrophagus*, the basal lipid that regulates buoyancy is squalene, with a density as low as 0.86 (Corner *et al.*, 1969). Just like the demersal 'lean' fish, these sharks store most of their lipids in the liver, which, however, contributes up to 25% of the total body weight! This liver contains 81% of the total lipids of the fish, and includes 42% of squalene. In the spiny dogfish, the main regulator of buoyancy is diacylglycerol (Malins and Barone, 1970), with a density of 0.908. In some tropical fish such as grenadiers and brotulas, this function is performed by cholesterol accumulated in the swim-bladder (Patton and Thomas, 1971; Phleger and Benson, 1971) and sphingomyelin (Phleger and Holtz, 1973; Phleger, 1975).

One curious fish species appears to achieve near-neutral buoyancy by reducing the amount of protein in its musculature. The jelly cat lives at 300–1000 m depth, mostly north of the Arctic Circle in the Atlantic. Its flesh is extremely soft, and consists of up to 95% water (Love and Lavéty, 1977). In contrast to the muscle cells of severely starved cod or grey sole with the same water content (irregularly shaped remnants of contractile material in connective-tissue sheaths largely filled with liquid: Love, 1980), the cells of this species are circular in cross-section and more-or-less fill each collagenous sheath: they look in fact 'normal'. However, each cell is separated from its neighbour by a relatively large fluid space. The fish has no swim-bladder, and appears to compensate by having spaced-out cells and large quantities of fluid in the musculature. Body fluids are less dense than sea water, and so the jelly-cat should approach neutral buoyancy. Other watery deep-sea fish without swim-bladders combine reduction of the skeleton with high water content of the tissues as a buoyancy mechanism, but there is no evidence of ionic regulation to the same end. Deep-water fish also appear to have a relatively low protein content, and low-molecule protein fractions that have high electrophoretic mobility are also reduced (Vitvitsky, 1977).

2.5. FOOD AND RELATED FACTORS

It is naive to state 'you are what you eat' because much of the ingested material is broken down during digestion to be resynthesized later in a form suitable to the feeder. It is, however, still true that the contents of proteins, lipids, amino- and fatty acids, vitamins and macro- and microelements in the food influence the retention of protein and lipid, the composition of lipids, the metabolism of minerals, the reproductive capacity of the fish and their survival. Evidence for this has been obtained largely from freshwater fish (Karzinkin, 1935, 1952; Ivlev, 1939, 1955; Gerking, 1952, 1966; Stroganov, 1962; Mann, 1965, 1969; Shcherbina, 1973; Ryzkov, 1976; Brett and Groves, 1979; Cowey and Sargent, 1979; Romanenko *et al.*, 1980, 1982, 1991; Lapin and Shatunovsky, 1981; Sorvachev, 1982; Walton and Cowey, 1982; Sidorov, 1983; Sklyarov *et al.*, 1984; Gershanovich *et al.*, 1987, 1991).

The roles of essential food items in metabolism have been demonstrated (Mead *et al.*, 1960; Kayama *et al.*, 1963; Shcherbina, 1973; Sidorov *et al.*, 1977; Cowey and Sargent, 1979; Sorvachev, 1982; Sidorov, 1983; Sklyarov *et al.*, 1984). Naturally, the digestive and tissue enzyme systems must adapt to ensure the assimilation and transformation of foodstuffs preferred by or available to individual species (Korzhuev, 1936; Berman, 1956; Hochachka and Somero, 1973; Shcherbina, 1973; Karpevich, 1976; Sorvachev, 1982; Ugolev and Kuzmina, 1993). Studies made by Kuzmina (1992) showed that digestive enzymes of the prey aided those of the predator. Also, Black Sea scorpion fish, horse-mackerel and pickerel display a close correlation between digestive hydrolysis and feeding pattern: proteases characterize predators, while amylases and sucrases are secreted by herbivores.

Relatively few similar studies have been carried out on marine fish because of problems of keeping them in captivity. However, recent progress in marine fish culture has made it possible to obtain data comparable with those from freshwater and diadromous fish during their period in the river.

The most important results from marine fish show a correlation between the amount of foodstuff available and certain physiological and biochemical features of individual fish and of populations. Data on the accumulation of triacyl-glycerols, which are the principal energy reserves of fish, are very informative.

It is worthwhile to give some examples. As Figures 8 and 9 demonstrate, there is good agreement between the distribution of zooplankton biomass in late summer in the Sea of Azov and the fatness of the Azov anchovy (Shulman, 1972b). In sprat in the Black Sea, the lipid content is greatest in the population which forages adjacent to the estuary of the Danube. This area is distinguished for being rich in nutrients carried down by the river and contains a mass of phyto- and zooplankton. Most convincing are the data on the lipid content of juvenile red mullet from the Adriatic Sea. There is a distinct gradient of food supply, which decreases from the delta of the Po river towards the Strait of Otranto. The fatness of the fry decreases in the same way, from 3.9% at the delta to 0.6% at the Strait. Vivid examples of the impact of food concentration on the lipid content of fish have been demonstrated in the larvae of Californian anchovy (Hakanson, 1989) and their adults (Smith and Eppley, 1982), Pacific sardine (Lasker, 1970) and mid-water fish from the same region (Bailey and Robison, 1986).

The relationships described are not confined to different species from different seas, but have also been found in single species from different seas.

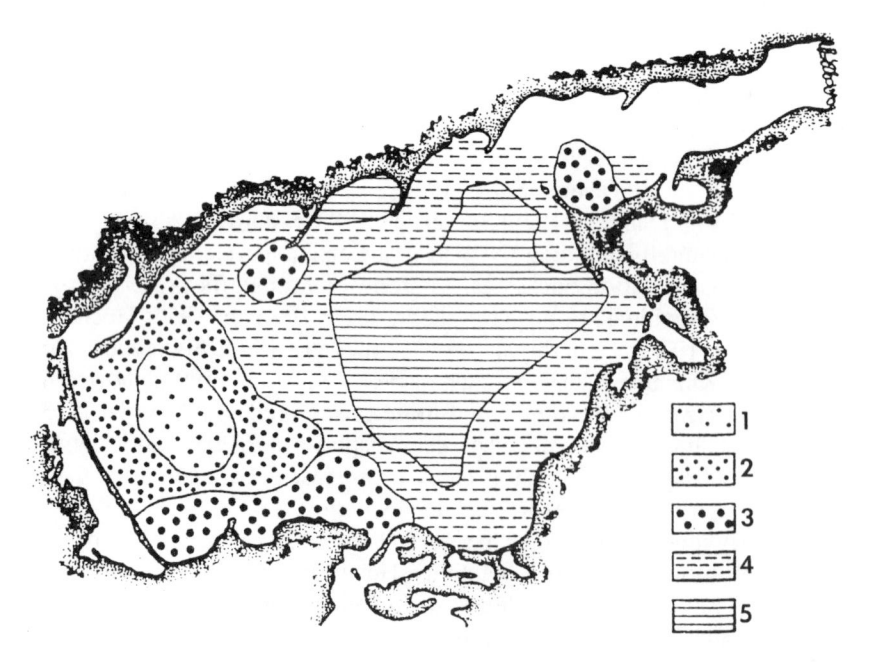

Figure 8   Biomass of zooplankton in the Sea of Azov (mg m⁻³) at the end of August 1957 (Novozhilova, 1960): 1, 1–2 mg; 2, 20–50 mg; 3, 50–100 mg; 4, 100–200 mg; 5, 200–500 mg.

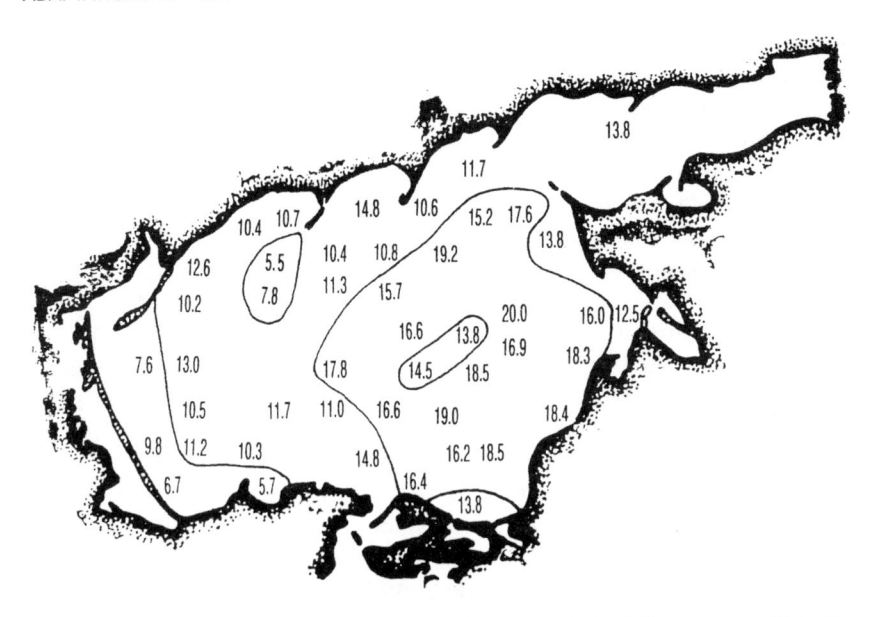

Figure 9   Lipid content of the flesh of anchovy caught in different areas of the Sea of Azov, as a percentage. (After Shulman, 1972a.)

For example, the concentration of zooplankton found in the Sea of Azov (200–500 mg m^{-3}) exceeds that in the Black Sea (100–200 mg m^{-3}), the latter in turn being greater than the averaged estimate (50 mg m^{-3}) known for the Mediterranean (Zenkevich, 1963). The fatness of three races of anchovy, Azov (*Engraulis encrasicholus maeoticus*), Black Sea (*E.e. ponticus*) and Mediterranean (*E.e. mediterraneus*), assessed immediately after completing their foraging periods, was 20–30%, 10–15% and 3–5%, respectively. From this it can be concluded that each sea has a specific trophic capacity that sustains the energy reserves according to the demands of the organisms living there (Shulman, 1972a).

The account of fatness in different races of anchovy is not unique. The degree of fatness in common kilka and red mullet, both warm-water planktivorous fish from the Sea of Azov, is greater than that in mackerel, horse-mackerel and red mullet from similar habitats in the Black Sea. Yet fish from the Mediterranean Sea – mackerel, horse-mackerel, red mullet and pilchard – are less fatty than those from the Black Sea. The actual values of lipid contents of these fish were found to be of the same order of magnitude, sea for sea, as those of the different races of anchovy. By the end of the feeding period, the fatness of common Caspian kilka and the Baltic sprat was as high as 15–20%, a range greater than that found in Black Sea plankton-feeding anchovy and sprat. Placing the seas in order of their nutrient capacity results in the following (decreasing) sequence: Azov, Caspian, Baltic, Black, open part of Mediterranean.

Up to now we have been comparing the responses of fish to different feeding regimes in enclosed, or semi-enclosed, seas which often feature ingress of fresh water from rivers. Similar studies on fish from different fishing grounds in the open ocean are rare. However, the unique properties of Atlantic cod from the Faroe Bank (60°53′N, 08°20′W) have been well documented and compared with cod from many other grounds in the North Atlantic Ocean, Barents Sea and Labrador Sea (Love *et al.*, 1974). The Faroe Bank is said to benefit from a strong upwelling of ocean currents which deposit large quantities of food on it. There is a deep channel between it and the Faroe Islands, so the Faroe Bank cod remain as an isolated stock all their lives.

The results of the unusually rich feeding are dramatic. The cod are uniquely corpulent compared with any others (Figure 10) and their livers are unusually large and creamy, although the actual concentration of lipid in them is not especially high. However, the real interest of these fish lies in the fact that the white swimming muscle actually contains a greater concentration of protein (Figure 11) and the water content of the muscle in many specimens is correspondingly lower – mostly and uniquely below 80%. On other grounds investigated, the water content ranges from 80.0% to 80.9% in non-starving specimens (Love, 1960; Love *et al.*, 1974). Further, although as stated earlier there is virtually no triacyl-glycerol in cod muscle, the total lipid was found to be 0.63% in spring-caught Faroe Bank cod, compared with 0.55% in Aberdeen Bank cod. At the end of the feeding season, these two values rise to 0.78% and 0.67%, respectively (Love *et al.*, 1975). The increase in lipid between the two seasons appeared to be in the

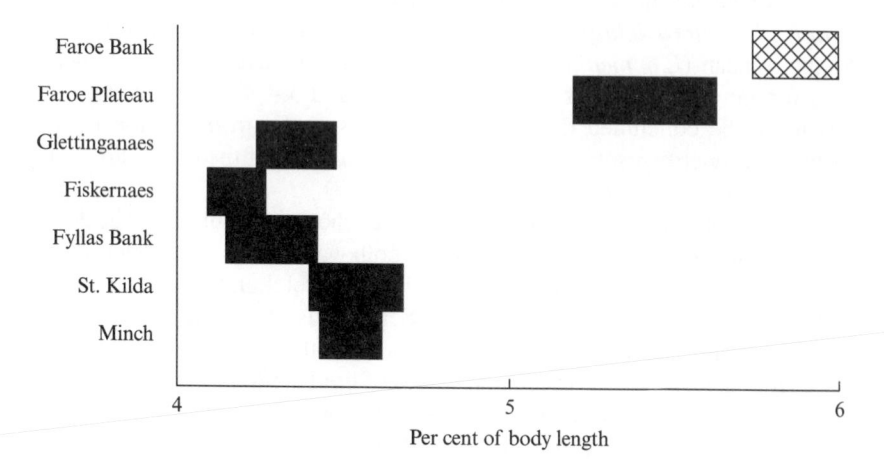

Figure 10   Corpulence of cod caught on various grounds in the spring of 1966, as shown in the maximum diameter of the caudal peduncle expressed as a percentage of the body length. (After Love *et al.*, 1974.) In this figure, and figures 11 and 12 the Faroe Bank stock is demarcated by cross-hatching.

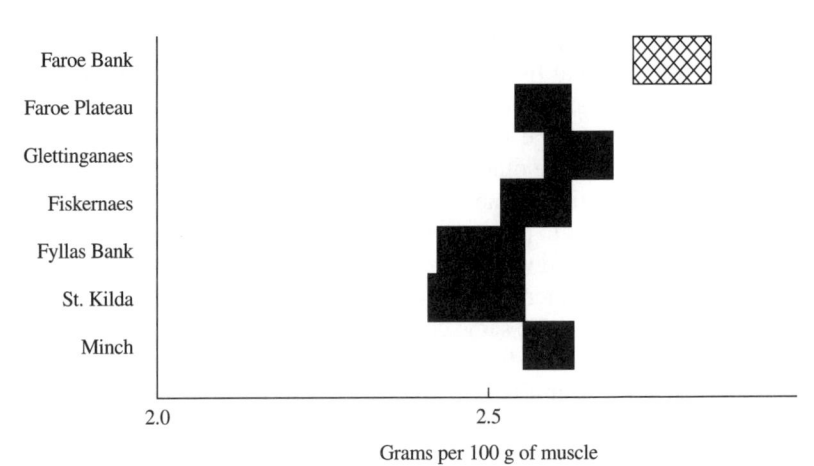

Figure 11   Total protein nitrogen in the musculature of cod caught on various grounds in the spring of 1966. (After Love *et al.*, 1974.)

phospholipid, which suggested that there was a larger concentration of actual contractile cells in the cod from both grounds. The pH of the flesh 24 h after death is lower than in fish from other grounds (Figure 12), signifying a higher glycogen content (Black and Love, 1988). A high value of glycogen in muscle mirrors a high glycogen resource in the liver (Black and Love, 1986). Both of these observations are discussed in more detail in Chapter 4, and shown in Figures 41 and 42.

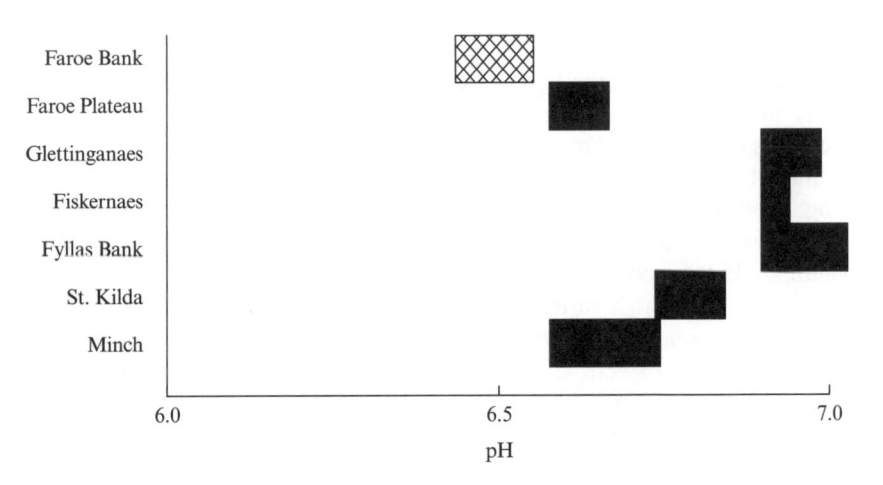

Figure 12   The pH (measured 24 h after death) of the musculature of cod caught on various grounds in the spring of 1966. (After Love *et al.*, 1974.)

Food science is beyond the remit of this review but, to help complete the picture of this unusual fish stock, it should be recorded that Faroe Bank cod spoil less rapidly after catching than some others (Reay, 1957). This is probably because of the relatively low pH of the muscle, which inhibits bacterial growth. On the other hand, if they are frozen at sea, thawed out and filleted, the connective tissue is so weakened by the low pH that the fillets 'gape' and tend to fall to pieces. Also, the slightly higher phospholipid content of the muscle compared with that of cod from other grounds causes the frozen fish to develop a stronger rancid flavour during frozen storage compared with other cod stored under the same conditions (Love, 1975).

A notable feature of the Faroe Bank cod is their colour. Cod, like many other fish species, have the ability to change their skin colour to match that of the sea bottom (see Love, 1970, for colour pictures of cod from different grounds). They receive information about the sea bottom through their eyes – a black cod swimming in a shoal of greenish or yellowish cod is probably blind. The Faroe Bank is unusual in being composed of the crushed remains of shells, and is brilliant white in colour (Love, 1974). Fish of all species caught on the Faroe Bank appear very pale in colour as a consequence of this.

However, a feature of the adaptation of skin colour to the colour of the environment is of unusual interest, and may originate in the fact that the population has been fixed in one locality for a long time. When living Faroe Bank cod were placed in an aquarium with a much darker-skinned group of cod caught on the Aberdeen Bank (off the coast of Scotland), the colour of both groups gradually changed towards the colour of the aquarium, which was intermediate between the two. However, the colours of the two stocks of cod never became identical within the period of the experiment: Faroe Bank cod remained considerably lighter than Aberdeen Bank cod, so that even after 8.5 months a distinction remained (Love, 1974). There are two possible explanations. The first is that the Faroe Bank stock has lived in that locality for so many millennia that their colour matching is genetically established. The second is that, at a certain age, the juveniles establish a certain limited range within which they can alter their colour in response to the surroundings. The range of this stock would not overlap with that of the Aberdeen Bank fish. There is a precedent for the latter explanation. Jordan (1892, quoted by Tåning, 1952) found that the numbers of vertebrae and the numbers of bony rays in the fins of fish varied according to the fishing ground, and that there was a correlation with the temperature of the ground. Schmidt (1930) showed that the numbers of vertebrae and fin rays were fixed very early in life, and envisaged a 'sensitive' period during which the count was influenced by the environment, becoming fixed thereafter. Hempel and Blaxter (1961) found a temperature-sensitive phase 6–7 days after the fertilization of Atlantic herring larvae incubated at 10°C; at this point, transfer of the eggs to an environment at 4°C for 24 h resulted in an increased vertebral count.

There is a lesson to be learned here: if a fish species responds in a certain way to a feature of the environment, the range of its response may be limited according to the environment where it spent its early life. The food of cod on the Faroe Bank consists of organisms that live on or near the sea bottom, such as crustaceans, brittle stars and other echinoderms, and small fish. It is not known how the food supply varies through a whole year, but in our limited experience (Love *et al.*, 1974) we found that the ranges of water contents were almost identical in fish caught on 6 June 1966 and on 13 September 1968. The dates of similar voyages in other years lay between these two extremes.

There is a strong interannual variation in the lipid concentration in the flesh of planktivorous fish such as anchovy, kilka and sprat, which is governed by the varying abundance of the plankton (Shulman, 1972b, 1996; Luts and Rogov, 1978; Luts, 1986).

The qualitative composition of food consumed influences the protein metabolism of the fish (Shcherbina, 1973; Cowey and Sargent, 1979; Ostroumova, 1979, 1983; Sorvachev, 1982; Walton and Cowey, 1982). The importance of essential amino acids in stimulating growth has been well documented in papers on fish farming, as has the catabolism of amino acids and the activity of enzymes of transamination and deamination according to food loading. The role of amino acids is not confined to the biosynthesis of proteins, but extends to that of glyco- and liponeogenesis also.

Although proteins and amino acids in the food influence the metabolism of the fish, they do not alter the amino acid composition of proteins in the body. In contrast, the lipids in the body of the fish are greatly influenced by the dietary lipids. In particular, it is the triacyl-glycerols (the main constituents of reserve energy) which are influenced by diet (Lovern, 1937, 1942, 1964; Kelly *et al.*, 1958; Brockerhoff *et al.*, 1963, 1964; Ackman, 1964, 1967; Ananyev, 1965; Ackman and Eaton, 1966, 1976). Food lipids also influence the structural lipids of the fish.

Lovern (1937, 1942) was the first to ascertain the basic fact that the composition of unsaturated fatty acids of lipids differs distinctly between marine and freshwater fish. Derivatives of the linolenic acid series ($C18:3\omega3$) prevail among the former, while derivatives of linoleic acid ($C18:2\omega6$) prevail in the latter. Accordingly, the bases of polyenoic acids of marine fish are eicosopentaenoic acid ($C20:5\omega3$) and, in particular, docosohexaenoic acid ($C22:6\omega3$), while in freshwater species it is arachidonic acid ($C20:4\omega6$). The fatty acid patterns described above follow those of the algae and invertebrates on which the fish feed, the fatty acid profiles of the prey being transferred from the lowest to the highest links of the food chain. Brockerhoff *et al.* (1963) and later Sargent and Henderson (1980) described every link of the chain in detail.

The factors that determine the specific fatty acid patterns of freshwater and marine algae are beyond our present remit. However, it is worth saying that, as in fish, the patterns can be influenced by the salinity and temperature of the water (Farkas and Herodek, 1964; Akulin *et al.*, 1969; Culkin and Morris, 1970). It is not by mere chance that estuarine species possess an intermediate fatty acid composition (Lovern, 1964). Moreover, by making changes in the experimental living conditions, one can alter the fatty acid composition of captive fish.

But does it follow from this that the fatty acid composition of fish is for the most part a replica of that of the algae? As some authors have explained (Mead *et al.*, 1960; Kayama *et al.*, 1963; Cowey, 1976; Kanazawa *et al.*, 1979; Yamada *et al.*, 1980; Walton and Cowey, 1982; Takahashi *et al.*, 1985; Kayama, 1986), the more or less unchanged transfer of polyenoic acids through the food chain is confined to predatory fish, say, sea bass, which do not possess a developed system for the desaturation–elongation of linoleic and linolenic acids. In phytophagous fish, e.g. guppy, arachidonic, eicosopen-taenoic and docosohexaenoic acids are synthesized *de novo*, so presumably both lower and higher links of the trophic chain in water bodies display a convergence strategy in the biosynthesis of polyenoic fatty acids.

In nature, fish apparently acquire polyunsaturated lipids in one of two ways. The first of them conforms with the concept of Sargent and Henderson (1980), Watanabe (1982) and Henderson *et al.* (1985), that some species of fish do not need to synthesize long-chain polyenoic acids, since they occur in phyto-plankton, which are eaten by zooplankton which in turn are food for fish. Takahashi *et al.* (1985) described the situation as unsaturated fatty acids being transferred from plant organisms to phytoplankton-eating fish to predatory fish.

On the other hand, some fish are able to synthesize long-chain polyenoic fatty acids (Kayama *et al.*, 1963) from shorter carbon chains. Docosohexaenoic acid is laid down in coho salmon in quantities related to the size of the fish, rather than to its availability in the diet (Tinsley *et al.*, 1973). Rainbow trout fed on 18:2 and 18:3 fatty acids can produce 20:3, 22:5 and 22:6 fatty acids in substantial quantities (Owen *et al.*, 1975), but these workers noticed that the capacity of marine flatfish to elongate or desaturate the carbon chains was more limited. They found that 70% of the radioactivity of labelled 18:3 appeared later in the 22:6 fatty acid of rainbow trout, but that turbot converted only 3–15% of labelled precursors into polyunsaturated fatty acids of longer chain length. It was suggested that turbot in the wild probably received adequate polyunsaturated acids in their diet, which the fish therefore did not need to modify. The elongation of the carbon chains and the creation of more double bonds is also only slight in Atlantic cod, another marine teleost, presumably for the same reason (Ross, 1977).

Increasing the degree of unsaturation and chain length of these fatty acids is achieved by enzymes known as desaturases and elongases (Sinnhuber *et al.*, 1972). Their activity is greatly influenced by temperature, as we have seen earlier under 'temperature compensation', even in isolated systems. Schünke and Wodtke (1983) observed a 30-fold increase in the desaturase activity of the membranes from the endoplasmic reticulum of the liver in carp which had suddenly been transferred to water of low temperature after being acclimatized at 30°C. Furthermore, phosphatidyl choline in the microsomes of goldfish liver becomes more saturated as the incubation temperature *in vitro* is raised from 10° to 30 °C (Leslie and Buckley, 1976). The most significant point emerging from these researches is that the saturation or desaturation depends on the temperature at the time of the experiment, rather than that prevailing beforehand. This suggests that fish are quick to adjust the degree of unsaturation of fatty acids appropriately to the temperature, ensuring that their membranes possess the correct physicochemical properties.

Kreps (1981) and his associates (Akulin *et al.*, 1969; Chebotareva, 1983), in their studies on Far East sockeye salmon, found that the fatty acid composition of lipids in the food influences not only that of the triacyl-glycerols of the fish but also the phospholipid fraction. This comprises, firstly, the phosphatidyl choline, phosphatidyl ethanolamine and phosphatidyl serine and, secondly, the sphingomyelin and cardiolipin. All tissues except nervous tissue were found by these workers to be strongly influenced by the composition of the food. However, the fatty acids of the brain have also been found to change according to the diet in European bass (Pagliarani *et al.*, 1986), the proportions of phospholipid to non-phospholipid being unaffected.

The fatty acid profile specific to each food item appears in the tissues of the predator whatever food is consumed (Lasker and Theilacker, 1962; Brenner *et al.*, 1963; De Witt, 1963; Kayama *et al.*, 1963; Lovern, 1964; Saddler *et al.*, 1966; Addison *et al.*, 1968; Ratnayake and Ackman, 1979; Bolgova, 1993; and others). The profound influence of food has been demonstrated in many species of marine fish: California sardine (Lasker and Theilacker, 1962), Pacific anchovy (Hayashi and Takagi, 1978), Atlantic herring (Lovern, 1937, 1964; Ackman and Eaton, 1966, 1976; Henderson and Almater, 1989), Pacific sardine (Takahashi *et al.*, 1985), ocean sunfish (Hooper *et al.*, 1973), Atlantic cod (Addison *et al.*, 1968), Australian tuna (Bishop *et al.*, 1976) and others. For freshwater species, the reader is referred to the comprehensive survey by Sidorov (1983).

Both marine and freshwater fish are often overfed in fish farms, so their own lipids are less likely to follow changes in the dietary lipid pattern. In the natural state, the influence of food on the lipids of the tissues is somewhat blurred by the fact that some of the tissue lipids are generated from carbohydrates and proteins in the diet and so synthesized *de novo*. In his classic work, Lovern (1964) showed that the impact of food on the fatty acid composition of herring lipids was most pronounced during intensive feeding. In winter, at the end of that period, the fatty acid composition became different from that of the zooplankton on which the fish had been feeding earlier.

Wax esters are of great importance to some marine fish. They are obtained from dietary copepods (Henderson *et al.*, 1984; Reinhardt and van Vleet, 1986), and whilst most species convert them to triacyl-glycerols, some accumulate them as long-term stores which are utilized only when their triacyl-glycerols have been expended.

The influence of other food items, e.g. carbohydrates and inorganic substances, on the chemical constitution and metabolism of fish has been less studied (Cowey and Sargent, 1979; Walton and Cowey, 1982). The importance of food in the energy balance of fish populations will be considered in Chapter 5.

Competition for food between populations and between individual fish is an important part of feeding studies. The amount of foodstuff available in a body of water is not necessarily the most reliable indicator of food sufficiency for each 'consumer'. It is helpful to know the actual numbers of consumers. Figure 13 shows the simple relationship between the richness of feeding and the fatness of the anchovy, but a better correlation is given in Figure 14, which takes into account the numbers of fish as well (Shulman, 1972a).

In 1986, a sudden surge in the population of sprats was recorded in the Black Sea. Their average degree of fatness had dropped from 13% to 10% by the end of the feeding period, presumably because of the reduction in the food supply for each fish (Shulman, 1996). However, as regards competition for food, one should bear in mind that the greatest threat does not come from other individuals of the same species, but rather from those of other species. Thus, for Black Sea sprat, the strongest competitor in the early 1980s was the medusa *Aurelia aurita*, excessive numbers of which caused a sharp reduction in the fatness of sprat, from 15.5% in 1981 to 9% in 1983. The ctenophore, *Mnemiopsis leidyi*, a new colonizer introduced into the Black Sea in the late 1980s, has also damaged the food supply of sprat, in which the fatness

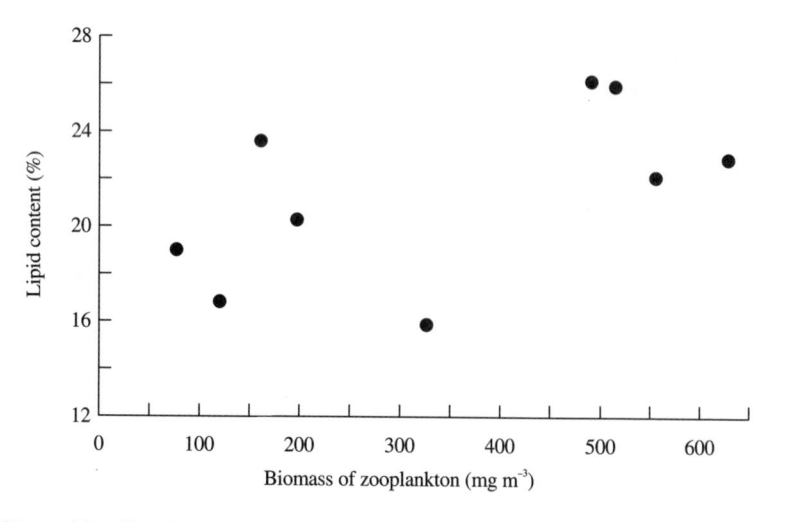

Figure 13   Simple relationship between the fatness of anchovy in October and the biomass of zooplankton in the Sea of Azov (combined data over several years).

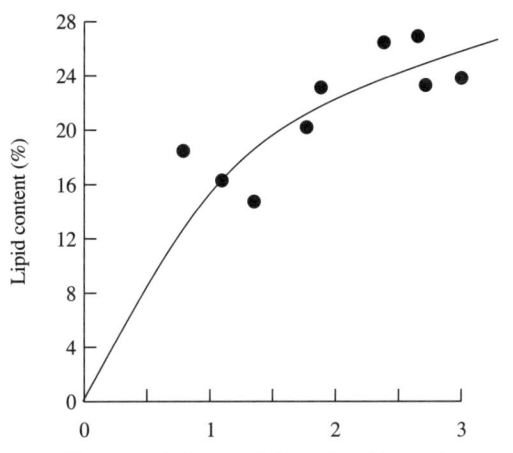

Figure 14 Relationship between the fatness of anchovy in October and the weight of zooplankton food, as grams available to each gram of fish. The zooplankton biomass is computed from the summed concentration (as mg m^{-3} × 300 km^3) the volume of the Sea of Azov, and is applied to the long-term average yield for anchovy.

decreased from 14% in 1988 to 10% in 1989–1990 (Shulman, 1996; Minyuk *et al.*, 1997). *Mnemiopsis* made an even greater impact on the nutritive base of anchovy. In the Sea of Azov, the stock of these fish suffered heavy losses over the period because the ctenophore had grazed out their traditional food items, and their lipid content decreased from 20–30% down to 10–15% (Studenikina *et al.*, 1991).

A peculiar form of interspecific interaction is parasitism, which disturbs the efficiency of food utilization, and reduces the abundance, biomass and production of populations of fish. Shchepkina (1980a,b) reported a considerable decrease of triacyl-glycerols in the liver, and in red and white muscle of Black Sea anchovy infested with the larvae of the nematode *Contracaecum aduncum* (Figure 15). A similar effect, caused by the trematode *Cryptocotyle concavum*, was noted in the round goby. The decrease in triacyl-glycerols varied from 25% to 71%, according to which tissue of either species was analysed.

Calculations show that total lipid production is 20–25% less in infested populations of anchovy and goby than in those clear of the parasites. What is more, infestation weakens fish, rendering them scarcely able to migrate, survive the winter or spawn. This thesis is not indisputable, however. Like predators, parasites regulate the numbers of their victims, rather than wiping them out, by withdrawing the most weakened or unviable hosts from the process of competition for food.

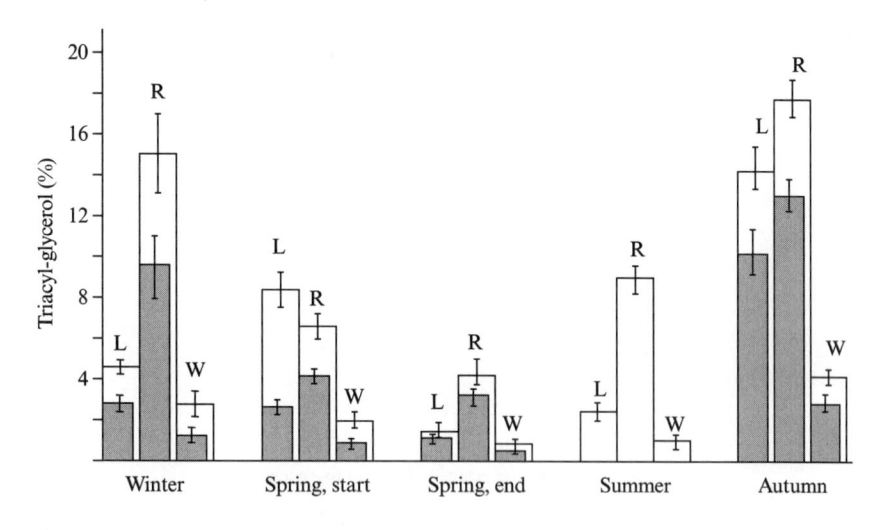

Figure 15   Effect of nematode infestation on the concentration of triacyl-glycerols in anchovy tissues (% wet mass): L, liver; R, red muscle; W, white muscle. Open columns, weakly infested; shaded columns, heavily infested. (After Shchepkina, 1980b.)

Guryanova (1980) and Sidorov (1983) have shown that infestation with helminths (e.g. plerocercoids of the broad tapeworm, *Diphyllobothrium vogeli*) leads to changes in the phospholipid content and composition of tissues of the stickleback (*Pungitius pungitius*) and the burbot. Most pronounced is the drop in phosphatidyl choline and, occasionally, phosphatidyl ethanolamine concentrations, while that of lysophosphatidyl ethanolamine rises. Such a disturbed phospholipid status of the biomembrane usually affects its permeability, permitting parasitic invasion and colonization. Parasites have also been shown to take up (selectively) essential amino acids of the host (Sidorov and Guryanova, 1981). In the developing eggs of Atlantic salmon infected with saprolegniosis, the store of glycogen was found to be depleted, with consequent loss of viable eggs (Timeyko, 1992). Carp infested with ectoparasites reduce their consumption of oxygen, while the content of haemoglobin in the blood and the activity of oxidative enzymes in other tissues decline (Kititsina and Kurovskaya, 1991). On the other hand, good feeding conditions increase the resistance of fish to diseases (Mikryakov, 1978; Mikryakov and Silkina, 1982).

3. Strategies of Adaptation

3.1. ALTERNATIVE STRATEGIES OF BIOLOGICAL PROGRESS

Evolution of species has always been manifested by a tendency to expand, that is, to occupy as many niches in as large an area as possible, increasing the numbers of organisms. This is the essence of biological progress, according to Severtsev (1934). Biological progress can be achieved in two possible ways. One is through intensifying vital processes and enhancing functional activity (the sum of the vital reactions of a living system: Orbeli, 1958; Shmalgausen, 1969) in individual organisms. The other is achieved via the utmost specialization of vital functions, enabling the organism to occupy narrow ecological niches uncongenial to competitors. The first way, Severtsev believed, is adequate for morphophysiological progress, which implies enhanced energy metabolism by the organisms in question. The second way is not necessarily linked to morphophysiological progress, and it sometimes displays every sign of regression and even degradation, as in the case of parasitic forms.

There are two possible life strategies for adapting to the environment, based on alternative modes of metabolism, active and sluggish. The two extremes are part of a continuous series which includes a number of intermediate forms. Examples will be given of fish which use the first, second and intermediate strategies of living.

3.2. DIVERGENCES OF ENERGY METABOLISM

Differences in the motor activity of animals are reflected in the energy used, so it is useful to examine the intensities of oxygen consumption.

Table 3 Rate of oxygen consumption by different Black Sea fish, as ml g^{-1} h^{-1}.

Species	Basal metabolism	Standard metabolism
Anchovy	0.235	0.970
Horse-mackerel	0.240	0.700
Mullet	0.231	0.572
Pickerel	0.212	0.572
Red mullet	0.185	0.247
Whiting	0.218	0.276
Scorpion fish	0.091	0.084

Comprehensive investigations by Belokopytin (1968, 1990, 1993) on fish from the Black Sea provide relevant data (Table 3). The basal and standard metabolisms conform to coefficient 'a', which describes oxygen consumption by fish of conventional weight of 1 g. As coefficient 'K' varies within the narrow range of 0.71–0.90 and is similar in all the fish compared, it is valid simply to use the 'a' coefficient. All estimates are brought to a temperature of 20°C. From the table it is apparent that the level of standard metabolism differs considerably between species. In active ones, such as anchovy, horse-mackerel and mullet, it is several times as high as that of the sluggish scorpion-fish, while pickerel, red mullet and whiting occupy inter-mediate positions. There is also some difference between the rates of basal metabolism.

Similarly, the red muscle of Black Sea horse-mackerel (active) takes up more oxygen than does that of the moderately active pickerel and red mullet (Stolbov, 1990). As stated earlier, it is the red muscle which is responsible for continuous swimming ('cruising'). The red muscle cells of active fish species have more mitochondria, the energy factories of the cell, than those of slow fish. Mitochondria make up 45.5% of the total weight of red muscle in the most active species (Johnston, 1982; Shindo *et al.*, 1986), 25–31% in moder-ately active species (Kryvi, 1977; Kryvi *et al.*, 1980; Johnston, 1981a, b, 1982; Savina, 1992) and 3–5% in slow species (Totland *et al.*, 1981). Furthermore, the mitochondrial content is known to be greater in the tissues of active than of sluggish fish (Savina, 1992).

Active fish have a better developed capillary system in the red muscle to supply oxygen to the mitochondria, and a higher haematocrit (Blaxter *et al.*, 1971). The red muscle tissue also contains more cytochromes (respiratory proteins), and exhibits more cytochrome oxidase activity, which is responsible for transferring electrons in the respiratory chain, more efficient respiration control (oxidative phosphorylation and P/O coefficient) and a greater Atkinson charge, which characterizes energy reserve accumulated in adenyl nucleotides:

$$\frac{ATP + \frac{1}{2}ADP}{ATP + ADP + AMP} \tag{4}$$

(Verzhbinskaya, 1953; Hochachka and Somero, 1977; Johnston, 1981b; Demin et al., 1989; Savina, 1992). Active fish have all these advantages over inactive ones.

Some of the dark colour of the red muscle in any fish is contributed by the muscle pigment myoglobin, which, like cytochrome, aids the transfer of oxygen from the blood to the contractile tissue. While the red muscle of active fish, such as Atlantic mackerel, is darker than that of inactive ones such as plaice, there is also a dynamic relationship between the activity of fish within a single species and the concentration of myoglobin. Love et al. (1977) showed that the red muscle of a migratory stock of Atlantic cod was darker than that of other cod stocks which remain in one locality all their lives. Also that the red muscle of captured cod could be induced to darken (from increased myoglobin) by making them swim for long periods in a flume, while the colour faded in control fish maintained without disturbance in a dark tank.

Between the active and inactive cod just described, there was no measurable difference between the proportions of red to white muscle, but there are large differences between species: the active herring and mackerel contain a larger proportion than the relatively inactive cod or North Sea whiting (Love, 1970). One of the most active genera of all, tuna (*Thunnus* species), possesses an extra band of red muscle near to the vertebral column, in addition to the lateral band under the skin. Salmonids also possess extra red muscle as individual fibres distributed through the white muscle (Webb, 1971). Quantitative estimates were given by Greer-Walker and Pull (1975), who showed that the proportion of red muscle in the musculature of Scombridae was 26.1%, in Clupeidae 19.8%, in Carangidae 18.3%, in Sparidae 15.7%, in Mugilidae 14.5%, in Gadidae 10.6%, in flat fish and Anguillidae 8.8%, in Scorpaenidae 5.5% and in Gobiidae 4.5%, pointing to a steady decline as habitual activity decreases.

There can also be links between the amount of red muscle and the life cycle. Metamorphosis in the eel is accompanied by an increase in the amount of red muscle, interpreted by Lewander et al. (1974) as preparation for spawning migration. The Antarctic fish, *Notothenia rossi*, is pelagic in its early life and possesses a well-marked streak of dark muscle. After 1 or 2 years, however, it sinks and lives among the weeds, using only its pectoral fins for locomotion. Red muscle develops at the bases of these fins, and at the same time the lateral red muscle fades until it can no longer be distinguished from white muscle (Johnston and Walesby, personal communication).

Emeretli (1990) demonstrated a convincing link between the activity of succinic dehydrogenase in the red muscle mitochondria of Black Sea species and their motor activity. This enzyme is one of the most important in the Krebs cycle, which controls the intensity of aerobic energy metabolism.

Quite apart from their greater vascularity, active fish have a higher capacity to transport oxygen. They have increased concentration of erythrocytes, haemoglobin and total blood volume (Egorova, 1968; Putman and Freel, 1978; Rambhasker and Rao, 1987; Tochilina, 1990). The oxygen capacity of the haemoglobin itself is also increased (Klyashtorin, 1982; Soldatov, 1993), and the Bohr effect, the ability to release oxygen bound to haemoglobin at the low pH caused by lactate accumulation, is enhanced (Klyashtorin, 1982; Soldatov, 1983).

All of these features promote the capacity to maintain the levels of energy metabolism characteristic of each species. Tochilina (1990) has provided good examples of the relationships between the haematological index and the habitual motor activity of Black Sea fish (Figure 16). This author recorded that, in the summer, the most active fish possessed the highest haemoglobin concentrations and erythrocyte numbers in their blood, near-bottom pelagic species having less and sluggish bottom fish the least. Active fish also possess greater concentrations of electrolytes in the blood serum (Monin *et al.*, 1989).

It is known that, among the substrates used in ATP resynthesis by oxidative metabolism, lipids possess the greatest energy capacity, yielding ultimately 9.3 cal g^{-1}. Both in this respect and as regards the convenience of storage, lipids are more profitable than proteins or carbohydrates which yield only 4.1 cal g^{-1}. When the respiratory quotient is measured in freshwater fish, the main source of energy in highly active species is the triacyl-glycerol fraction of the lipids, while in less active species it is the proteins and glycogen (Krivobok, 1953; Mathur, 1967; Kutty, 1972; Alikin, 1975; Kutty and Mohamed, 1975; Sukumaran and Kutty, 1977; Mohamed, 1981; Waarde, 1983). These substrates are not, of course, taken up directly, but in the form of their catabolites such as free ('non-esterified') fatty acids, amino acids and glucose. It has not been possible to assess the RQ in marine fish, as pointed out in Chapter 2, section 2.3.

An increased concentration of non-esterified fatty acids in the blood and muscle of active fish (Table 4) clearly indicates a more intensive utilization of lipid substrates than is found in those less active (Shulman *et al.*, 1978; Astakhova, 1983). Enhanced activity of tissue lipases in mobile species provides further evidence for this conclusion (Johnston and Moon, 1981; Johnston and Bernard, 1982). The table also shows that the content of triacyl-glycerol (the energy reserve) in the flesh of the two species correlates positively with their normal swimming activity. The main fatty acids involved in providing energy are oleic acid (monoenoic, 18:1) and palmitic, which is saturated (16:0): Krueger *et al.* (1968); Ackman (1980); Sargent and Henderson (1980) and Yuneva *et al.* (1991). Zabelinsky *et al.* (1995) found that the content of palmitic acid in the gills of active fish such as striped greenling, masked greenling and Pacific herring was up to twice as high as that in snailfish and yellowfin sole, both of which are of low activity.

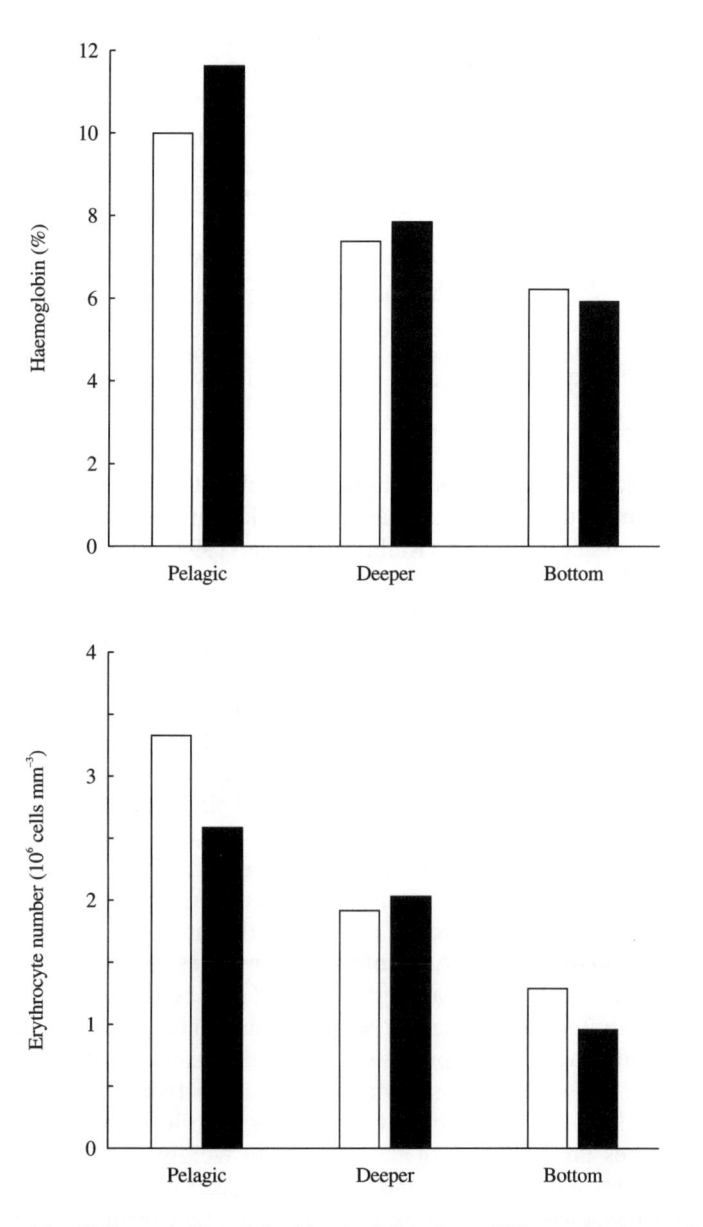

Figure 16   Characteristics of the blood of fish from different depths, surface fish being the most active and those from the sea bottom the most sluggish. White columns, Mediterranean and Atlantic fish. Black columns, Black Sea fish. (After Tochilina, 1990.)

Table 4   Lipid fractions in muscle and blood serum of horse-mackerel (active) and scorpion fish (sluggish) from the Black Sea, as mg % wet weight of tissue. (After Shchepkin, 1972.)

	Phospholipid	Cholesterol	Free fatty acids	Triacyl-glycerols	Cholesterol esters	Total lipids
Horse-mackerel						
Red muscle	2071±145	1379±58	291±32	6170±624	615±67	10496±649
White muscle	1152±60	1026±53	166±22	2190±155	390±57	4924±146
Blood serum			159±87	1269±194		4615±216
Scorpion fish						
Red muscle	1709±115	1418±109	31±7	1084±91	927±43	5169±274
White muscle	1542±110	605±21	22±6	122±39	530±44	2822±66
Blood serum			80±17	453±60		1597±15

In the muscle of active fish such as horse-mackerel and pickerel, the antioxidative enzymes superoxide dismutase and catalase are more active than in the more sluggish scorpion fish (Rudneva-Titova, 1994; Rudneva-Titova and Zherko, 1994).

Differences in energy metabolism are clearly shown, not only between fish of different natural mobility, but also between different types of muscle within each fish. Red muscle continuously engaged in locomotion has a higher functional activity and aerobic metabolism than white muscle. Stolbov (1990) reported that, in Black Sea horse-mackerel, the intensity of oxygen uptake by red muscle was 2.3–4 times greater than that by white, in pickerel 1.2–2.5 times greater and in red mullet 1.5–2.8 times greater, respectively. Moreover, as mentioned earlier, the characteristics of oxidative metabolism are more pronounced in red muscle (Demin *et al.*, 1989; Emeretli, 1990; Savina, 1992). The higher level of aminotransferase activity in red muscle indicates a higher level of nitrogen catabolism than in white muscle (Lyzlova and Serebrennikova, 1983). As Table 4 demonstrates, the energy substrates for the resynthesis of ATP (triacyl-glycerol and non-esterified fatty acids) are more concentrated in red than in white muscle. The greater mass of red muscle found in active fish species ensures their greater mobility by aerobic metabolism. This has been shown for Black Sea fish (Kondratyeva and Astakhova, 1994). Red muscle of the highly mobile horse-mackerel and herring makes up 3.8–6.3% of the total body weight, that of moderately mobile pickerel, annular bream and red mullet makes up 1.2–1.9%, while the figures for the slow-moving whiting and scorpion fish average less than 1%. Astakhova (1983) found that both cardiac and brain indices are greater in active species.

The aerobic pathway via the Krebs cycle yields 36 moles of ATP, while the anaerobic process utilizing glycolysis yields only 2–3 moles. Clearly, fish performing work of great capacity must cultivate oxidative metabolism involving red muscle fibres (Bokdawala and George: 1967a, b; Bilinsky and Jonas, 1970; Greer-Walker, 1970; Itina, 1970; Hamoir *et al.*, 1972; Johnston, 1977, 1982, 1985; Walker *et al.*, 1980).

Electrical activity is greater in red muscle than in white (Matyukhin *et al.*, 1984). This has been confirmed in Black Sea fish by Efimova (1982). Especially active oxidative metabolism is shown by heart muscle (Patton and Trams, 1973), the main substrate being free fatty acids.

Optimum living conditions also give rise to intensification of aerobic metabolism. Within the zone of temperature tolerance, the intensification depends directly upon temperature (Van't Hoff–Arrhenius' law). Many workers have shown that, in the optimum temperature zone, there are increases in oxygen consumption, activities of cytochrome oxidase and succinic dehyrogenase, respiration/phosphorylation ratio (respiratory control) and muscle electrical potential (Hochachka and Somero, 1973, 1977; Wodtke, 1974; Khaskin, 1975; Derkatchev *et al.*, 1976; Walesby and Johnston, 1980; Romanenko *et al.*, 1991).

Most clearly, the intensification of aerobic metabolism is manifested in the seasonal rhythms of vital activity; this topic will be explored further in Chapter 4. Here we stress that the Black Sea fish that are of warm-water distribution spawn during the summer months when the temperatures are highest. Spawning consumes a great deal of energy and in the majority of these fish it lasts for 3 months.

Emeretli (1990) found that, in the summer, the ratio between succinic dehydrogenase and lactic dehydrogenase activity increases sharply. It has already been stated that the SDH activity marks the intensity of aerobic catabolism associated with the Krebs cycle and the activity of LDH that of anaerobic glycolysis. Both processes increase in Black Sea fish during the summer, especially the aerobic catabolism, which is more pronounced in active fish. It is during the spawning season that the number of circulating red blood cells and their content of haemoglobin and methaemoglobin increase (Soldatov and Maslova, 1989; Tochilina, 1990).

Another example illustrating the importance of aerobic metabolism at higher temperatures is related to the dynamics of the adenyl nucleotides (ATP, ADP and AMP) in red muscle. Trusevich (1978) noted that in summer time (temperature 19–22°C) the ATP content of cruising horse-mackerel remained relatively constant, indicating active resynthesis of this compound, probably through the Krebs cycle. At lower water temperatures (11–12°C) in spring and late autumn, the ATP content of the same fish decreased significantly, perhaps through a reduction of aerobic processes. Under such conditions, the fish cannot cruise for the unlimited periods possible in the summer-time, but usually cease swimming after a 2–3 h stint. This agrees well with the data of Shustov *et al.* (1989), which showed that in juvenile lake salmon the vigour of swimming drops considerably when the ambient temperature falls. In contrast, the adenine nucleotides in the muscle of freshwater fish increase in the summer-time (Malyarevskaya *et al.*, 1985). To sum up, we can conclude that high functional activity, as manifested by mobility, is usually serviced by intensified aerobic metabolism.

It is difficult to be certain that all the biochemical characteristics of energy metabolism are increased in more active fish. We can point only to the key steps described above – oxygen uptake, activity of oxidative enzymes, content of substrates which provide energy (ATP, triacyl-glycerols and non-esterified fatty acids), the concentration of haemoglobin and a few others. However, the nature of living organisms is not 'artless and unpretentious', so answers to problems are seldom straightforward. Comparing active and inert species of fish raises questions which blur the original clarity of the subject. The white muscle of animals in general and fish in particular is capable of 'burst' activity, which is necessary in pursuit of or escape from other animals (Black, 1958; Drummond, 1967; Prosser, 1979). These sudden surges of power are characteristic of fish having high natural mobility (e.g. salmon travelling upstream to spawn) and also fish which have little mobility but which lie in wait for prey – pike, scorpion fish, moray eel – and also benthic fish escaping from predators – goby, blenny and some coral-reef fish. In order to perform these feats of power output, the fish concerned must have evolved the appropriate metabolic machinery. In principle, this is based on biochemical substrates that can be mobilized almost instantaneously: creatine phosphate, glycogen and glucose (Johnston, 1982, 1985). In addition, there are metabolic pathways capable of utilizing biochemical substrates immediately: anaerobic glycolysis and the pentose phosphate shunt (Wardle, 1978; Wardle and Videler, 1980; Duthie, 1982; Ogorodnikova and Lebedinskaya, 1984). Although these pathways have nothing like the energy capacity of aerobic catabolism, their quick energy output allows the fish to perform sudden work; fish may therefore be classified as 'stayers' or 'sprinters' (Gatz, 1973; Hudson, 1973).

In horse-mackerel, the high-energy substances and their products are more concentrated in white muscle than in red (Table 5), as shown by Trusevich (1978), Morozova *et al.* (1978b) and Emeretli (1990). The activities of key enzymes of glycogenolysis and glycolysis other than LDH are also greater in white muscle (Serebrennikova, 1981; Malinovskaya, 1988; Serebrennikova *et al.*, 1991;

Table 5   Substances associated with high energy in the muscle of horse-mackerel.

	White muscle	Red muscle
Creatine phosphate (m$_M$ g^{-1} wet tissue)	13.30	5.20
ATP (m$_M$ g^{-1} wet tissue)	5.68	2.28
Lactate (mg % wet tissue)	214.0	132.4
Lactate/glycogen ratio	1.1	0.2
Lactate dehydrogenase (m$_M$ min^{-1} mg^{-1})	4.3848	1.2769

Savina, 1992). Moreover, as these authors and Silkina (1990) note, the white muscle of sluggish fish often exceeds that of highly active species in the characteristics described above. This thesis is supported by the data of Emeretli (1990) on LDH and ATPase activity measured in the muscles of horse-mackerel and scorpion fish (Figure 17). The creatine- and adenylate kinase reactions in the white muscle of sluggish fish appear to proceed at a greater rate than in more

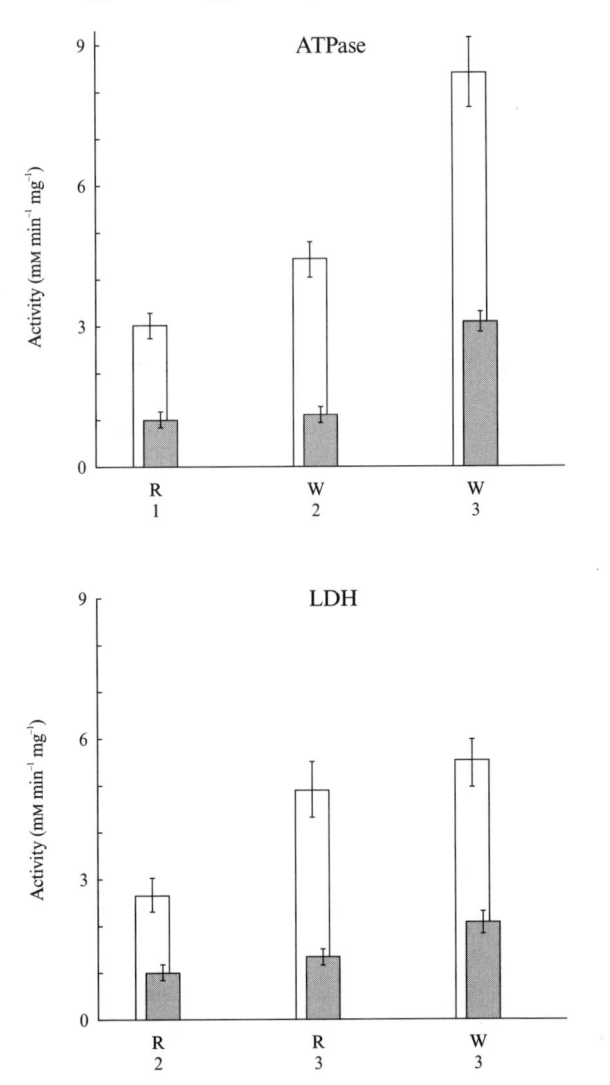

Figure 17   ATPase and LDH activity in scorpion fish (white boxes) and horse-mackerel (shaded): in spring (1); summer (2); and autumn (3). R, red muscle; W, white muscle. (After Emeretli,1990)

active fish (Trusevich and Anninsky, 1987). Similarly, the activities of pentose phosphate shunt enzymes are higher in the heart, and the white and red muscle in these species (Kudryavtseva, 1990). Thus, fish of low mobility compensate their normally small scope for swimming by being capable of bursts of high activity, and hence the ability to compete for existence against highly active forms. Strictly speaking, this adaptation renders it inappropriate to label fish as 'active' and 'sluggish'. Why is a scorpion fish capable of burst speeds of 1.5 m s^{-1} (equivalent to 10 body-lengths) called a fish of low activity? Neither active nor highly mobile fish of similar size would reach such a speed. Therefore, by 'active forms' we define fish that perform much continuous motor work and great total metabolic turnover. Thus, ATPase and phosphocreatine kinase activity is higher in 'sprinters' than in 'stayers' (Johnston and Tota, 1974).

To provide sudden triggering of energy release for bursts of activity, low-mobile species must have evolved mechanisms which differ from those in highly active ones. They involve the rapid mobilization of carbohydrates and intensification of metabolic processes localized in the liver, the main meta-bolic power-house of the organism. According to Khotkevich (1974, 1975), the turnover of lipids and proteins containing ^{14}C tracer is higher in the liver of scorpion fish than in pickerel. Anti-oxidative activity of the liver is more pronounced in goby and scorpion fish than in horse-mackerel (Rudneva-Titova, 1994). The ATP, ADP and AMP contents, as well as the sum of adenyl nucleotides, are greater in the livers of scorpion fish than in that of horse-mackerel (Savina et al., 1993). Fish of low mobility have a greater weight of liver than that found in active species (Silkina, 1990). The basal energy reserve concentrated in the liver of sluggish fish is either comparable with or signifi-cantly greater than that in active fish.

As regards stores of energy, the glycogen content of the liver of Black Sea scorpion fish ranges from 5.4% to 9.1%, but in the active horse-mackerel it is only 0.53–1.09% (Plisetskaya, 1975). The high concentration of liver oil in such an inactive fish as the cod is well known. This oil, composed for the most part of triacyl-glycerols, can form up to 70% of the weight of the liver. The triacyl-glycerol content of the liver oil of scorpion fish is lower than that of horse-mackerel (10.6% against a maximum of 17.6%, respectively, while the non-esterified fatty acid content is higher (1.6–1.9% against 1.2–1.4%). This indicates a greater turnover of lipids in the liver of scorpion fish (Shchepkin, 1979). If we take the absolute weights of the livers into account, the extra reserves of energy in sluggish fish as compared with those of active species become even more evident. That the energy is actually stored in the liver facil-itates its mobilization and makes the metabolism more efficient. The concen-tration of 'transport' proteins, primarily α- and β-globulins (Golovko, 1964; Kulikova, 1967), is higher in the serum of sluggish fish. This facilitates the supply of various substances to the tissues. A more comprehensive discussion of this topic will be given later.

Proteins and nitrogenous products play a more important part in the energy metabolism of sluggish species than of active species. This is demonstrated by the respiratory quotient and the ammonia quotient, which have been measured mostly in freshwater fish (Sukumaran and Kutty, 1977; van den Thillart and Kesbeke, 1978), but later also in marine fish. During cruise-swimming of Black Sea horse-mackerel, 80% of the energy is derived from lipids and 20% from protein substrates (Muravskaya and Belokopytin, 1975). At rest, this proportion became 70:30. Scorpion fish, however, use proteins to cover their basic energy requirements. Interestingly, the efficiency (not intensity) of oxygen utilization in sluggish fish is higher than in active ones (Klyashtorin, 1982; Soldatov, 1993). Fish of moderate natural mobility use mostly proteins and the products of nitrogen catabolism to maintain their energy metabolism. One can regard this as a kind of intermediate strategy of metabolism, which is displayed by marine perciformes, porgies, coral fish and some others. In Mediterranean pickerel during several hours of cruising (50 cm s^{-1} on average) in a flume (Shchepkin et al., 1994), the consumption of muscle proteins was much greater than that of lipids. Eels also use proteins as the principal source of energy substrate during swimming (Boëtius and Boëtius, 1980, 1985), despite the presence of large quantities of lipids. The latter is used as a source of energy and gonad substance during maturation.

Another problem, not yet completely resolved, is the relative roles played by red and white muscle in fish locomotion. Bone (1966) reviewed the distribution of red and white muscle in teleosts, and showed evidence that the white muscle was used for rapid bursts of activity, while the red muscle provided sustained swimming. It should be noted that the distribution of red and white muscle may change during ontogeny. In herring there is a single outer layer of red muscle in the hatched larva. The adult arrangement, where the red muscle is concentrated along the midline of the body, under the skin, develops after the gills and blood circulation have become functional (Batty, 1984).

Red muscle is superior to white according to biochemical parameters concerned with aerobic metabolism (reviewed by Love, 1970, 1980). There is also an intermediate or 'mixed-type' musculature, described by Hochachka and Somero (1977). However, there is a greater total mass of white muscle in the body of a fish compared with red muscle. For example, white muscle comprises 97–100% of the musculature of Black Sea fish of low mobility, 95–96% in moderately active and 88–91% in highly active species. In estimating the swimming performance of a fish, one should regard the organism as a whole, rather than studying individual tissue units. If we do this, it becomes clear that the total content of energy substrates in the white muscle of a fish is much greater than that in the red muscle.[*]

*The content of any substrate calculated on the basis of an entire fish can be called the provision of the organism with the substrate. This term was first introduced by Korzhuev (1964) to denote the total amount of blood in the whole body.

The total consumption of energy substrates during swimming is also significantly greater in white muscle than in red, as demonstrated in the cruising of
Black Sea horse-mackerel (Shulman *et al.*, 1978) and red mullet (Shulman and
Khotkevich, 1977), the experiments being carried out in a flume. The ratio of
white muscle to red is 10:1 in horse-mackerel and 33:1 in red mullet, so the
relationship between absolute amounts of lipid contained in white and red
musculature of horse-mackerel, calculated on the basis of the entire fish, is
approximately 6:1 and in red mullet 7:1. The energy expenditure ratio between
the two types of tissue in the two species during cruising is again 6:1 and 7:1,
respectively. This indicates that the lipid reserve is consumed evenly in both
types of muscle during moderately long periods of swimming.

The thought processes behind the estimation of substrates on the basis of a
whole fish have also been used by Black and Love (1986). They calculated the
liver glycogen in North Sea cod as the weight in the whole liver of a fish
weighing 1000 g, assuming that the average size of the organ was proportional
to the size of the fish. Such values are more informative than concentrations of
glycogen in a given weight of liver tissue, the latter being greatly influenced
by changes in the size of the organ caused by influx or efflux of lipids, carbohydrates and sometimes proteins.

Electron microscopy shows a large number of lipid granules in the white
muscle of certain species (Graf, 1982; Shmerling *et al.*, 1984). This raises the
question of whether white muscle, rather than red, is the more important in
locomotory activity, whether the fish is swimming steadily or in sudden bursts.
It is pertinent to recall the thesis of Braekkan (1956), Drummond (1967) and
Wittenberger (1971, 1973), that red muscle plays a very subsidiary role, or
even no role at all, in the swimming of fish because of its small mass and
because of certain compositional similarities to liver tissue. These authors
envisaged the red muscle solely as a metabolic organ, an 'extra liver', the
function of which was simply to process and supply energy substrates to the
white muscle whenever necessary. The hypothesis appeared to be supported
by the fact that in fish both the absolute and relative weights of the liver are
inversely proportional to the weight of the red musculature.

High-speed photography of cruising fish shows it is not always a smooth or
steady process, but rather a sequence of bursts of activity alternating with gliding
(Webb, 1975; Videler and Weihs, 1982; Rome *et al.*, 1986). The herring is an
example of a fish with such alternating activity, while mackerel is not. White
muscle is clearly of greater importance in this mode of swimming. Further
evidence comes from the fact that the relatively inactive scorpion fish and
whiting undertake long, calm swimming for many kilometres as a daily routine
(Belokopytin, 1993). And red muscle is virtually absent from these species!

It appears that the white musculature of fish is provided with metabolic
mechanisms that allow lipids as well as glycogen to contribute energy.
Experiments using ^{14}C tracer have shown that triacyl-glycerols in white

muscle dissociate into glycerol and non-esterified fatty acids (NEFA) during lipolysis. The NEFA are oxidized to produce energy, while glycerol is converted into glucose and utilized anaerobically (Savina and Plisetskaya, 1976). Labelled triacyl-glycerols are metabolized comparatively actively in the white muscle of pickerel and scorpion fish (Khotkevich, 1974, 1975). Such observations as these presumably do not apply to the white muscle of Atlantic cod, which is virtually free from triacyl-glycerols (Ross, 1977).

   Considering all the evidence, we still incline to believe that red muscle is of greater importance than white in the swimming of fish, particularly in cruising. Much evidence for this has come from electrophysiological studies, when not a single sign of excitation ('spike') has been found in the white muscle of cruising fish (Matyukhin, 1973; Walker *et al.*, 1980; Matyukhin *et al.*, 1984). On the other hand, these spikes were amply demonstrated in red muscle. The electrical excitation measured in Black Sea fish is considerably higher in red muscle than in white (Efimova, 1982). Biochemical data also testify to the leading role of red muscle. Activity of the oxidative enzymes, essential for the production of energy from triacyl-glycerol and fatty acids, is low in white muscle (reviewed by Love, 1970, 1980; Hochachka and Somero, 1973, 1984). The finding that lipolysis does actually take place in white muscle does not contradict the fact that in red muscle it is aerobic catabolism that predominates.

   Could it be that the situation really is the other way round? What if white muscle, not red, is the tissue that contains the reserves of lipid and protein to enable long-distance swimming? In Pacific salmon, during spawning migration, the white muscle[*] is broken down to supply material for the gonads, for swimming and other needs of the fish (Greene, 1919; Pentegov *et al.*, 1928; Kizevetter, 1942; Idler and Clemens, 1953; Idler and Bitners, 1958). At the same time, the red muscle remains intact and allows normal activity to continue (Yuneva *et al.*, 1987). White muscle destruction in sturgeons leaves more of the red muscle intact (Lukyanenko *et al.*, 1991). In all probability, the elements involved in energy metabolism, triacyl-glycerols, NEFA, proteins and amino acids, are transported from the white muscle to the red via the circulatory system.

   A good example of the importance of red muscle in locomotion is found in species of tuna, where red muscle comprises up to 30% of the total muscu-lature and contains a complete set of oxidative enzymes. The oxygen-binding functions of haemoglobin (Rayner and Keenan, 1967; Gordon, 1968; Sharp, 1973; Johnston and Tota, 1974; Brill and Dizon, 1979; Guppy *et al.*, 1979; Hulbert *et al.*, 1979; Gemelli *et al.*, 1980; Graham *et al.*, 1983) operate inde-pendently of pH and temperature, and the tissue is rich in energy substrates

---

[*]Usually known as 'mosaic' muscle in these fish, as the white muscle fibres are interspersed with individual red muscle cells (Walker and Emerson, 1978; Johnston, 1982). This mixed type of muscle is not considered in detail here.

(Hochachka and Somero, 1977). Even glycogen and glucose are used intensively in red muscle of tuna under aerobic conditions (so-called aerobic glycolysis). In contrast, the sluggish scorpion fish and whiting can perform only relatively slow movements using the white muscle.

Some conclusions need to be drawn from what appears to be the contradictory evidence presented above. The divergence of styles of energy metabolism in different species comes about because there is a 'choice' of fundamentally different strategies. The first of these is the maximum intensification of oxidative (aerobic) metabolism, involving lipid substrates of high caloric content which give the fish great scope to perform muscular activity. The second strategy entails rather low total energy expenditure for utilization of non-lipid substrates (proteins, amino acids, glycogen and glucose) and instant access to energy through anaerobic glycolysis. The third, or intermediate, strategy is inherent in fish of moderate activity, and is based mostly on using proteins and the products of their catabolism – amino acids and, perhaps, oligopeptides and polypeptides.

These three features predominate in different groups of fish, but the actual situation is more complicated. For instance, the initial 15 minutes of cruising by the very active horse-mackerel requires an 'igniting charge', that is, glycogen, to be triggered (Morozova et al., 1978a, Figure 18). The next stage, ordinary swimming, may last for hours and involves the intensive uptake of triacyl-glycerols (Shulman et al., 1978; Yuneva et al., 1991), while the content of glycogen in the muscle settles down to steady values. During the third stage, the glycogen content drops again while the concentration of lactate increases. The fish may compensate for the sharply reduced level of glucose in the blood by reducing the supply to the brain, and symptoms of fatigue appear (Morozova et al., 1978b), as shown in Figure 19.

The dynamics of muscle glycogen in cruising horse-mackerel usually correlate with the dynamics of ATP concentration (Trusevich, 1978), as shown in Figure 20. During the first and last stages of swimming, when glucose supplies the main energy uptake, the resynthesis of ATP is weak and its level fails to keep a stable value. In the course of prolonged swimming, the level is restored and maintained because it originates in lipids which yield 36 cal during ATP resynthesis; glycogen yields 3 cal only. Thus carbohydrates, which are easily mobilized, while supplying energy, also perform an important regulatory function by initiating and terminating the swimming of fish. In general terms, very active fish use protein substrates when at rest, lipids when swimming and carbohydrates when attacking prey or escaping predators. Moderately active and sluggish fish depend on protein substrates when swimming strongly. Carbohydrate substrates are used in urgent starts or cessation of swimming in all three groups. The salmonidae possess an interesting system, migrating upriver using lipid reserves for energy, but jumping the waterfalls and rapids using their carbohydrate substrates (Gatz, 1973;

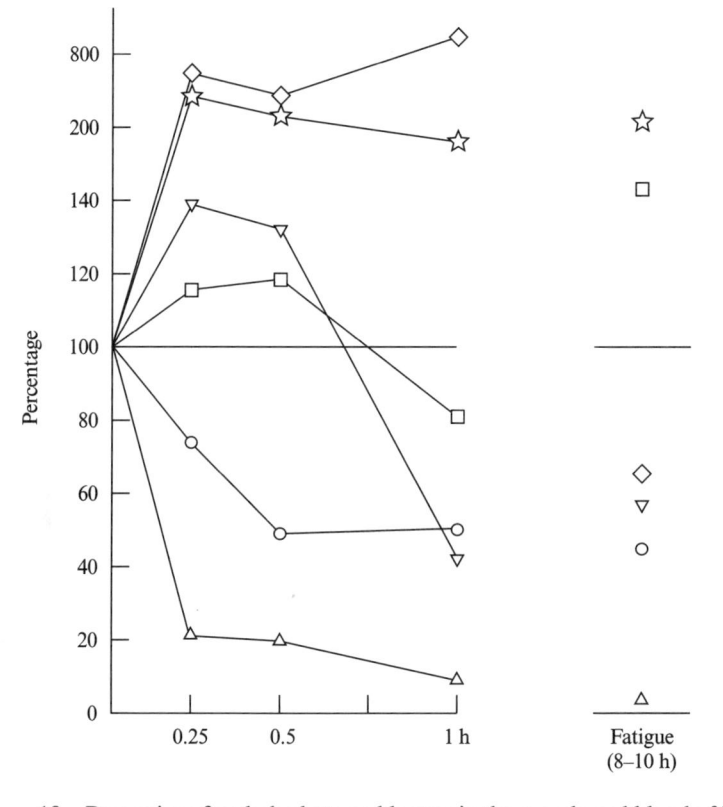

Figure 18 Dynamics of carbohydrate and lactate in the muscle and blood of horse-mackerel during cruise swimming. The 'fatigue' point lies anywhere between 8 and 10 h after the start. The starting level is assumed to be 100% and the curves show the percentage increase or decrease. ○, glycogen in white muscle; △, glycogen in red muscle; □, lactate in white muscle; ▽, lactate in red muscle; ◇, glucose in blood; ☆, lactate in blood.

Videler and Weihs, 1982). Dunn and Johnston (1986) observed that the first short burst of swimming of Antarctic fish is fuelled by creatine phosphate. Each energy substrate in fish has specific functions that allow diversity in locomotory activity.

Of the three basic substances used in the resynthesis of ATP, the proteins predominate by weight. Many groups of animals (e.g. coelenterates, free-living worms, molluscs, echinoderms, tunicata) obtain most or all of their energy from protein. Lipids and glycogen provide energy for the more physically active species (e.g. many planktonic crustaceans, insects, fish and birds). Unlike 'high-speed' fish, the less active animals are content to use mostly proteins and amino acids as their energy substrates. For this reason, the moderate swimmers have recourse to using nitrogenous

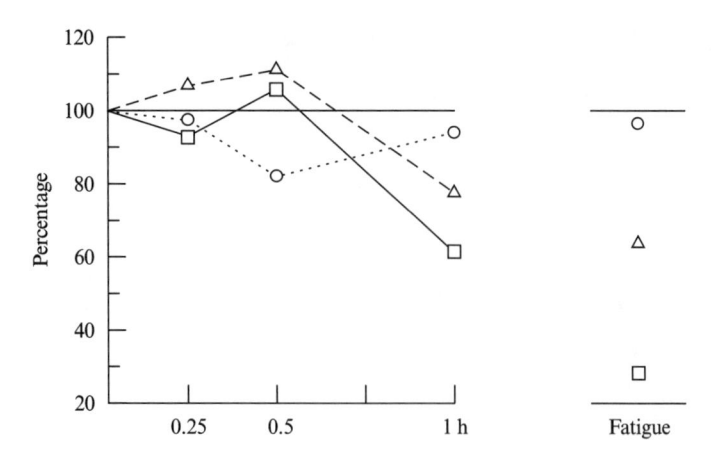

Figure 19   Dynamics of carbohydrate in brain and heart of cruising horse-mackerel. ○, glycogen in brain; △, glucose in brain; □, glycogen in heart.

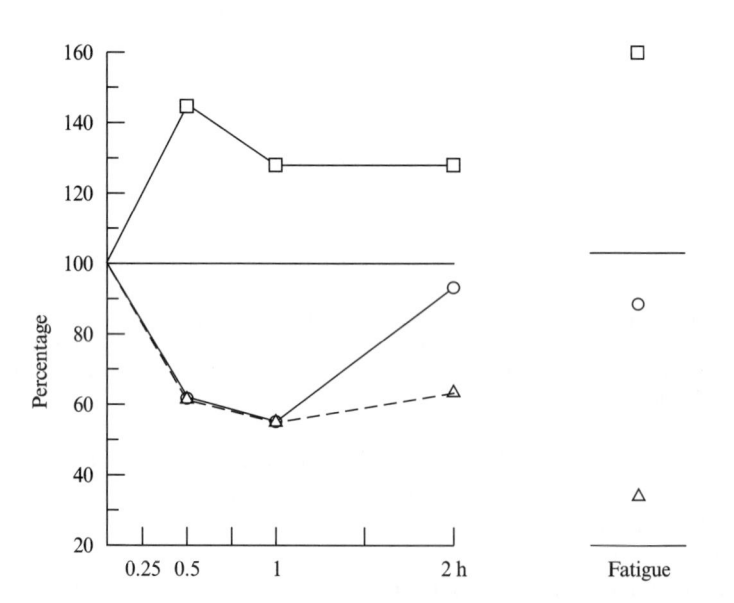

Figure 20   Dynamics of high-energy phosphates in red muscle of cruising horse-mackerel. ○, ATP; △, creatine phosphate; □, inorganic phosphate released.

compounds when hypoxic conditions pertain in the water, as already noted by Kreps (1981). Surely, having once 'invented' such a pattern of functioning, nature will not permanently put it aside but will draw on it whenever the need arises.

3.3. DIVERGENCES OF PLASTIC METABOLISM

Fish species that differ in their physical activity also differ considerably in their level of plastic metabolism. Plastic metabolism is the metabolism of the whole organism, the sum of all activities, including the type of food consumed, the transformation of its basic constituents in the organism and the catabolic products excreted.

Food consumption and transformation rate in horse-mackerel (active) and whiting (sluggish) are shown in Table 6 (Shulman and Urdenko, 1989). The highly mobile fish is superior to the more sluggish one in all the characteristics tabulated. The plastic catabolism can be assessed from the rate of nitrogen excretion. Stolbov *et al.* (1995) and Muravskaya (1978) found that in the highly mobile horse-mackerel, the excretion of nitrogen in standard metabolism is three times that of the less-active scorpion fish. The intensity of transamination (Lyzlova and Serebrennikova, 1983; Savina, 1992) and labelled precursor (^{14}C-acetate) incorporation into muscle proteins and lipids (Khotkevich, 1974) is also higher in mobile species.

Lipids are of special concern in comparing plastic metabolism between the two groups of fish. Triacyl-glycerols, cholesterol ethers and non-esterified fatty acids, which are the direct sources of energy, have already been discussed in the previous section. We now turn to phospholipids and cholesterol, which are essential to the structure of cell membranes, and to polyunsaturated fatty acids, which determine to a large extent the functional activity of these membranes.

In Chapter 2, section 2.1, the importance of lipids in temperature adaptation was noted, and the versatile role of phospholipids, cholesterol and polyenoic

Table 6 Consumption and rate of transformation of food in active and sluggish fish. (After Shulman and Urdenko, 1989.)

	Horse-mackerel	Whiting
Consumption* (mg g^{-1} day^{-1})		
Wet substance	52	13.0
Protein	15	2.0
Lipid	4	0.6
Minerals	3	0.3
Glycogen	1	0.1
Weight metabolized[†] in 24 h (mg g^{-1})		
Wet substances	39.7	3.1
Protein	16.7	2.2

* Estimates of maximum food consumption during July in horse-mackerel and during February in whiting.
[†] Calculated from the average annual values for the time span required to metabolize the weight of food equal to the weight of the fish.

acids in supporting metabolism was mentioned. These compounds are connected with the modulation of enzymatic activity, ion transport and protection against peroxidation (Boldyrev, 1979; Sargent and Henderson, 1980; Pertseva, 1981; Boldyrev and Prokofieva, 1985; Bell *et al.*, 1986; Kreps *et al.*, 1987). They also provide the base that secures the functional activity of tissues.

Evidence to support this thesis can be found in a paper by Shchepkin (1979), whose data on Black Sea horse-mackerel and scorpion fish showed that: (1) red muscle tissue, which is more functionally active than white, also contains more phospholipids; (2) red muscle of the highly mobile horse-mackerel contains more phospholipids than that of scorpion fish; and (3) white muscle of scorpion fish, which serves to perform sudden bursts of activity, is richer in phospholipids than is the white muscle of horse-mackerel (Table 4). Shulman *et al.* (1990) found that liver tissue, which has the highest metabolic activity, also has the greatest content of phospholipids – much greater than in red muscle. Interestingly, there is more phospholipid present in the liver of scorpion fish than in those of all other species examined, both highly and moderately mobile species, with the sole exception of horse-mackerel. This finding, together with the high content of phospholipids in white muscle, is a telling example of functional compensation provided by plastic- and energy metabolism in less mobile fish. It is curious that the content of phospholipid in the liver and functional activity are greatest in the extremes, in fish of both high and of low mobility. In the former, the liver makes possible the long-term locomotory activity of red muscle, while in the latter it operates the surge of activity ('burst') in white muscle. The intermediate, moderately active species have considerably lower contents of phospholipids in their livers.

The relationship between functional activity and the polyenoic acids is even more evident. The latter regulate functions such as enzyme modulation and ion transport, as stated above (Sargent and Henderson, 1980; Bell *et al.*, 1986). They therefore perform a universal role in adjusting the organism to a specific level of functional activity. The polyenoic acids (especially $C20:4\omega6$ and $C20:5\omega3$) are also important in the synthesis of such substances as the prostaglandins, leucotrienes and thromboxanes (Pryanishnikova *et al.*, 1975; Bell *et al.*, 1986; Bolgova, 1993), and in regulating their transformations (Boukhchache and Lagarde, 1982; Aveldano and Sprecher, 1983; Fisher and Weber, 1984; Henderson *et al.*, 1985; Praag *et al.*, 1987; Kudryavtseva, 1991). Polyenoic acids underpin the adaptation of the fish to temperature, oxygen level and salinity, and for the high or low mobility displayed by the different types of fish. Evidence is provided from data on the iodine numbers of lipids (Shulman, 1974) from fish of the Sea of Azov and Black Sea (Table 7). It was realized some time ago that the level of total lipid unsaturation varies according to the content of polyenoic acids (Shulman, 1974). It then emerged that the unsaturation level as assessed from iodine numbers was the highest in such active migrants as anchovy and horse-mackerel. Next in line were kilka and sprat,

Table 7   Iodine values of lipids in fish of varying degrees of activity.

Azov anchovy	158
Black Sea anchovy	154
Horse-mackerel	150
Kilka	130
Sprat	126
Scorpion fish	121
Turbot	121
Red mullet	120
Whiting	114
Pickerel	107

which are also good swimmers but do not take part in lengthy migrations. The lowest iodine numbers were found in lipids from fish of moderate to low mobility: scorpion fish, turbot, red mullet, whiting and pickerel. The rating of fish by iodine numbers is here based on mobility, not temperature preference. The lipids of warm-water fish are more saturated than of those preferring cold water, but in almost all the cases we have studied, the natural mobility of the fish has the more important influence on the unsaturation of their lipids. This finding weakens the force of the well-known tenet of ecological biochemistry, that temperature adaptation is the key factor determining the level of lipid unsaturation. The principal factor is actually the natural mobility of the animal, a parameter strongly linked to its propensity to migrate.

As mentioned earlier, the polyenoic fatty acids from marine fish are mostly from the linolenic series ($\omega 3$). These are eicosopentaenoic ($C20:5\omega 3$) and docosohexaenoic ($C22:6\omega 3$) acids, the latter usually being the more abundant. Yakovleva (1969) reported that in Black Sea fish the ratio between fatty acids of the 3 and 6 series also displayed a close correlation with the natural mobility levels.

From published data (Shulman and Yakovleva, 1983) it is evident that the content of $C22:6\omega 3$ in both marine and freshwater fish is correlated closely with the level of their mobility. The highest values were found in tuna and flying fish, then in descending order sauries, mackerel and horse-mackerel: all highly mobile fish. Next in succession were fish of high-to-moderate mobility: wrasses, bluefish, fish of the herring family, anchovies, salmonidae and notothenidae. The final group included fish of moderate-to-low activity: congers, sea perches, sable fish and righteye flounders. A similar gradation in this fatty acid was found by Ueda (1967) in Thunnidae, Scombridae and Carangidae, 33 species in all. In freshwater fish, the descending series proceeds from live bearers to Salmonidae, Gadidae, Cyprinidae, the catfishes and finally freshwater eels.

Comparison between genera and between allied species of the same genus gives some interesting information. The content of $C22:6\omega 3$ in the liver lipids of four species of shark was highest in highly mobile predators, the blue shark and the smooth hammer-head, while lowest in the less active grey and thresher sharks (Akulin and Pervuninskaya, 1978). Rzhavskaya (1976) studied the docosohexaenoic acid content of the lipids from three species of notothenids from the Antarctic, and found that the values decreased from the most active species, Patagonian toothfish, through the marbled notothenia, of intermediate activity, to green notothenia, which was the least active. Reinhardt and van Vleet (1986) showed the same relationship for three other Antarctic species. Highly active horse-mackerel from the Black Sea was found to have 1.5 times as much $C22:6\omega3$ in the muscle tissue as in scorpion fish, which leads a more 'settled' life (Yakovleva, 1969).

As stated earlier, the iodine numbers of lipids are greater in the strongly migratory anchovy than in the sprat, which is more stationary. The finding agrees with that published by Yuneva (1990) on the $C22:6\omega3$ content of the phospholipids in the two species. In the summer-time, the values are signifi-cantly higher in anchovy than in sprat (37% and 30%, respectively). Zabelinsky et al. (1995) have shown that, in the gills of motile species like herring and horse-mackerel, the content of $C22:6\omega3$ in the phosphatidyl choline, phosphatidyl ethanolamine and phosphatidyl serine is 30–40% greater than in the more sluggish scorpion fish and plaice.

In five allied species of goby from the Sea of Azov (Table 2), the iodine numbers of liver lipids correlate closely with the living conditions of each species. Their natural mobilities lead them to select habitats with dissolved oxygen content appropriate to their activity, so the toad goby, the most active species, lives around rocks near the shore in warm, shallow water, while the most sluggish species, the syrman goby, resides on silty bottom grounds. The three other species are intermediate in their activities, and there is a significant correlation between the level of lipid unsaturation and the degree of mobility in these species of the same genus.

The content of $C22:6\omega3$ in the functionally more active red muscle of pink salmon is greater than that in white muscle, and the values are especially high in liver, the organ with the greatest metabolic activity (Yuneva et al., 1991). In Black Sea turbot, the level of this fatty acid is higher in the testis than in the ovaries (Yuneva and Bassova, personal communication to GES), possibly because the sperm are motile and eggs are not. In either case, gonad tissue contains more of it than is present in muscle (Henderson et al., 1984). Fish of the cod family, although sluggish, display high unsaturation in the deposits of oil in their livers (Shulman and Yakovleva, 1983).

The importance of docosohexaenoic acid in maintaining structural and functional integrity during high activity can also be seen when comparing individuals within a single species. Its concentration in the phospholipids and triacyl-glycerols of Black Sea sprat was found by Yuneva (1990) to increase in summer, when the elevated water temperature stimulated greater locomotory activity in this fish. In Atlantic and Pacific horse-mackerel the maximum levels of phospholipids coincide with the warmest season, when the fish spawn (Dobrusin, 1978; Eliseeva *et al.*, 1985).

These data reverse the supposed 'temperature paradigm' of polyenoic acid adaptation. Further work (Pomazanskaya *et al.*, 1979; Kreps, 1981) showed that the docosohexaenoic acid content of the phospholipids of the brain tissue in two species of flying fish (*Cheilopogon exsiliens* and *Hemirhamphus balao*), dwelling in tropical sub-surface waters, and also in their predator, the dolphin fish, is higher than in the longfin cod and lampfish from deep waters in the same region, and also greater than that in fish from the Arctic and the cold-water Baikal complexes. Kreps (1981) gave no explanation for these data and was puzzled at such an abnormality, but the phenomenon can easily be understood on the basis that activity, rather than environmental temperature, is the dominant effect here.

Rabinovich and Ripatti (1990) have shown that docosohexaenoic acid has conformational properties which keep its physico-chemical and, possibly, functional characteristics effective over a wide temperature range. This ensures the adaptation of cell membranes to changes of metabolic activity. Fluctuation in locomotory activity is one factor responsible for these changes. From their studies of the sea cucumber, *Cucumaria frondatrix*, Kostetsky *et al.* (1992) concluded that polyenoic acids of linolenic affinity did not exhibit a direct relatonship with temperature adaptation. In contrast to this, Zabelinsky *et al.* (1995) claim that C20:5ω3 (not C22:6ω3) and C18:1 are the fatty acids of key importance for temperature adaptation in marine fish.

The essential role performed by C22:6ω3 in the metabolism of fish is clearly seen when one compares its content in the red muscle of different males of Pacific salmon during spawning. These fish cover enormous distances from their oceanic feeding grounds upstream to the spawning grounds, where they start to fight aggressively for female fish. The 'spawning family' usually consists of a female and several males, the latter establishing an order of dominance (Chebanov *et al.*, 1983). Experiments conducted by Yuneva *et al.* (1987) on Kamchatka pink salmon showed that the docoso-hexaenoic acid in the phospholipids of the red muscle of the most dominant males in a group was more concentrated than that in 'recessive' individuals (Figure 21). In contrast, the content of the precursor – eicosopentaenoic acid (C20:5ω3) – in the triacyl-glycerols of the liver was lower in leaders than in satellite fish. This appears to indicate a more intensive synthesis of docoso-hexaenoic acid (by elongation and desaturation of eicosopentaenoic acid) in the dominant group. Curious results were obtained from comparison between the content of C22:6ω3 determined in the red muscle of leaders and competitors (dominant group), satellites and outsiders ('recessive' groups). By leaders, we mean male fish directly participating in spawning; competitors are males involved in fighting for females; satellites and outsiders are males

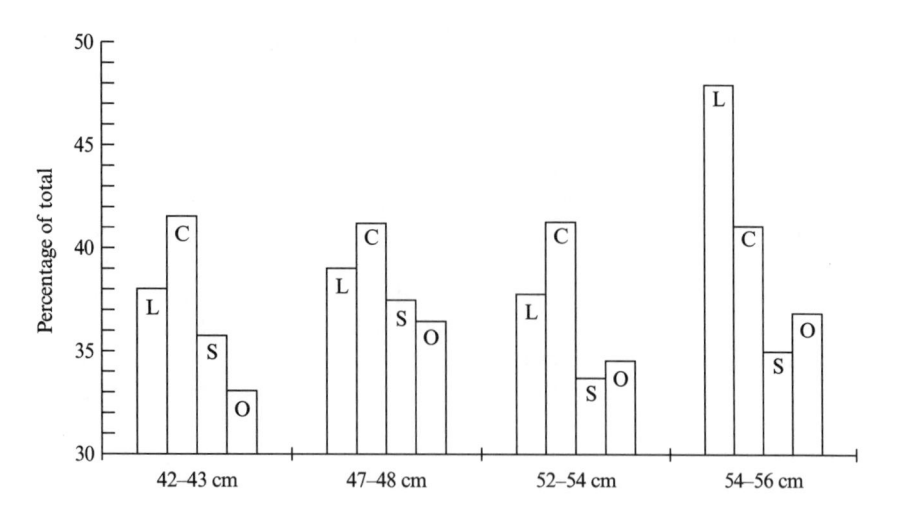

Figure 21   The content of C22:6ω3 in the phospholipids of red muscle of pink salmon males of increasing lengths at the spawning time. L, leaders; C, competitors; S, satellites; O, outsiders.

exhibiting passive behaviour during spawning, the latter being even less active than the former. Surprisingly, in three experiments out of four, the content of C22:6ω3 was greater in competitors than in leaders. Competitors, which do not take part in spawning, persistently follow or accompany leaders, and in so doing they dissipate much more energy as a result of stress. Semenchenko (1988) demonstrated experimentally that in Kamchatka sockeye salmon the mobility and oxygen uptake were greater in competitors than in leaders. This accords with their greater content of C22:6ω3 despite their defeat in the struggle for females. Higher concentrations of this fatty acid have also been found in the strongly swimming fry, compared with weaker ones, of Atlantic salmon and carp (Bogdan *et al.*, 1983; Bolgova *et al.*, 1985a, b; Sidorov *et al.*, 1985; Bolgova and Shurov, 1987).

In a flume, Black Sea horse-mackerel sorted themselves into two groups – good swimmers and bad swimmers (Yuneva *et al.*,1991). The first group maintained good swimming performance for many hours, while the second ceased to swim after 1 h. The content of C22:6ω3, total lipids, triacyl-glycerols, total phospholipids, phosphatidyl choline and phosphatidyl ethanolamine were found to be higher in all tissues examined in the first group compared with the second. A further observation made in these experiments was that, during long swimming, the levels of free fatty acids and of C22:6ω3 were maintained at steady levels. Such a situation would support intensive and efficient energy mobilization and locomotory activity. Only in physical fatigue did the concentrations of the two substances decline (Figures 22 and 23). Similar data have been obtained with horse-mackerel and red mullet (Shulman *et al.*, 1990).

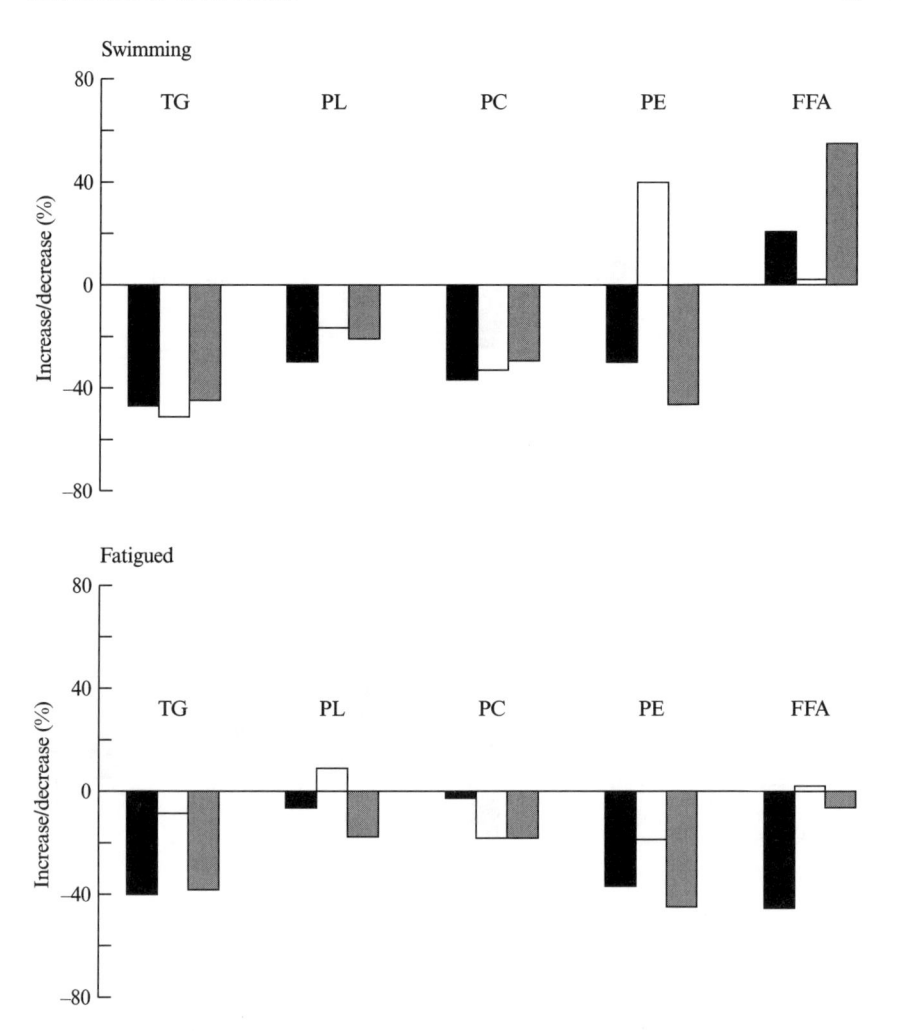

Figure 22 Dynamics of principal lipid constituents in horse-mackerel during cruising and fatigue. (After Yuneva *et al.*, 1991.) The free fatty-acid level increases in red muscle and liver when the fish swim, and decreases during fatigue. Other constituents decrease under both conditions. TG, triacyl-glycerols; PL, phospholipids; PC, phosphatidyl choline; PE, phosphatidyl ethanolamine; FFA, free (unesterified) fatty acids. Black columns, red muscle; empty columns, white muscle; shaded, liver.

The importance of maintaining the phospholipid 'frame' of membranes ($C22:6\omega3$ being a constituent) to ensure normal swimming in fish appears to be established by all these results. The stable level of this fatty acid in the phospholipids of muscle (Ando, 1986) and brain (Tyurin and Gorbunov, 1984) ensures the continuous swimming of Pacific salmon during spawning migration. The

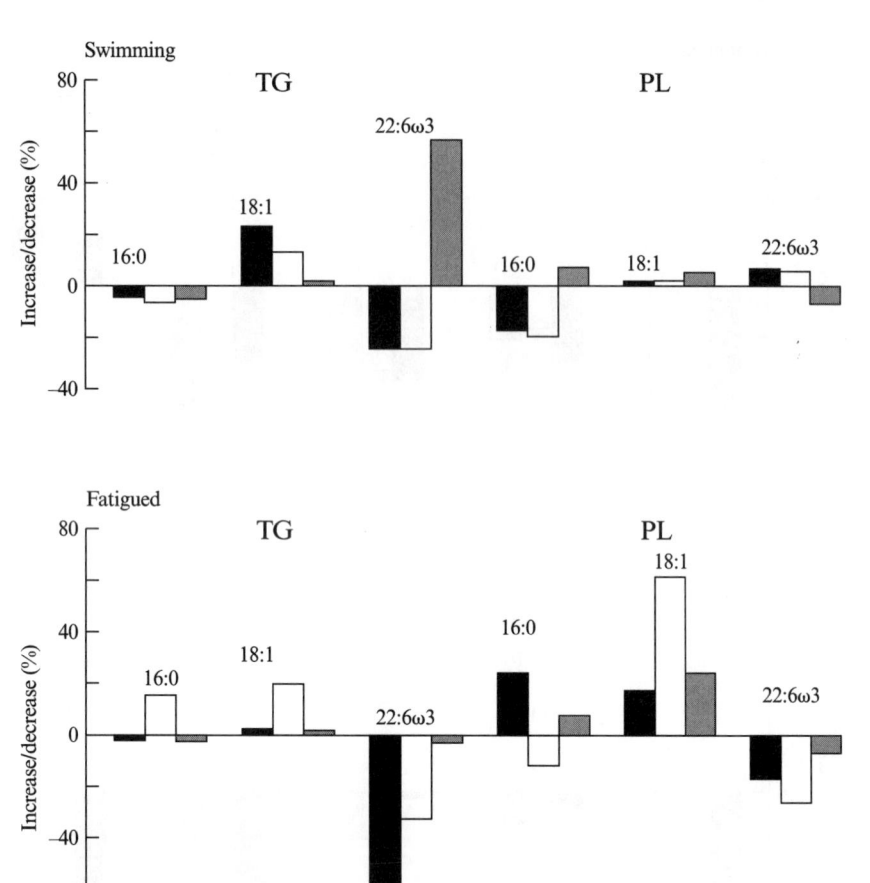

Figure 23   Dynamics of principal fatty acids in the triacyl-glycerols and phospholipids of horse-mackerel during cruising and fatigue. Symbols as in Figure 22. C22:6ω3
levels are more or less maintained during swimming but decrease in fatigue.

phospholipid content was found to keep stable during lengthy swimming in
horse-mackerel, a good swimmer, but only for a shorter time in red mullet, a
poor swimmer; in the latter, tissue homeostasis breaks down during swimming
much sooner than in the former. This confirms the conclusion of Belokopytin
(1978), that bad swimmers use more energy than good when swimming at the
same speed or over the same distance. Not only the intensity, but also the efficiency of locomotion during cruising is higher in the 'good' group.

During lengthy swimming, the balance of blood constituents is soon upset
in poor swimmers, while in good swimmers it persists for a long time

(Belokopytin and Rakitskaya, 1981). Moreover, poor swimmers tend to use more energy for the same effort, the unit for comparison being defined as energy consumed from swimming over a distance of 1 kilometre, per gram of body weight (Tucker, 1970). On reaching a state of fatigue, the content of cholesterol in the liver increases, indicating a disturbance of tissue metabolism (Shulman *et al.*, 1978).

The divergence of plastic metabolism between mobile and sluggish fish can also be found in other animals, including marine invertebrates, with similar structural–metabolic changes underlying it. A comparison between large taxa of molluscs, cephalopods and lamellibranchs, in particular, reveals that their content of docosohexaenoic acid is directly proportional to their degree of mobility.

As regards smaller taxa, in cephalopods, the content of docosohexaenoic acid is greater in squid than in octopods, while, in lamellibranchs, unattached scallops capable of leaping contain more than mussels or oysters which have a fixed mode of life. As regards squid, the more actively moving *Stenoteuthis pteropus* is richer in this fatty acid than the less mobile *Rhombus rhombus* (Yuneva *et al.*, 1994). Studies conducted on five pelagic species of euphausiids show that mobile predators which make lengthy vertical migrations contain much more than is found in the less active non-predatory forms (Yuneva *et al.*, 1992). Returning to squid for a moment, we should note that individual tissues also differ in the content of this substance according to their functional activity. In both the species mentioned above, the content was maximal in their swift catching-tentacles (Yuneva *et al.*, 1994).

In attached or 'fixed' marine organisms, e.g. sponges and corals, both docosohexaenoic and eicosopentaenoic acids are present, but in negligible amounts (Kostetsky, 1985).

In all probability, we are dealing with a universal phenomenon in which high functional activity of animals entails the most unsaturated fatty acids being incorporated into the phospholipid framework of cell membranes. This structural framework can have 'coverings' which vary in their function, generating a wide variety of types of metabolism. For energy metabolism, as stated earlier, this refers to the relative proportions of lipid, protein or carbohydrate substrates used in ATP resynthesis. However, the plastic (structural) base and the superstructure of energy supply must fit each other. This is a manifestation of the integrity of the two parts of metabolism that determine the forms of energy transformation in an organism.

In higher terrestrial animals and, to some extent, in freshwater fish, it is arachidonic acid ($C20:4\omega6$) which serves as the functional analogue of docosohexaenoic acid, with which it shares many characteristics (Parnova and Svetashev, 1985; Parnova, 1986).

Despite these findings, a large amount of $C22:6\omega3$ is also required by higher terrestrial animals and freshwater fish to ensure sufficient functional activity. Studies made by Sidorov (1983) and Ripatti *et al.* (1993) showed that juvenile freshwater salmon in the wild contained more of this acid than those which had been artificially propagated. In order to grow normally, young freshwater fish must acquire a certain amount of docosohexaenoic acid in their feed (Lee *et al.*, 1967; Sinnhuber *et al.*, 1972; Atchison, 1975; Takeuchi and Watanabe, 1976; Watanabe and Takeuchi, 1976; Sargent and Henderson, 1980; Golovachev, 1985; Bolgova *et al.*, 1985a; Henderson *et al.*, 1985; Sergeeva, 1985; Bell *et al.*, 1986; Bolgova and Shurov, 1987; Sergeeva *et al.*, 1987; Gershanovich *et al.*, 1991). Data obtained by Burlachenko (1987) on freshwater fish during early ontogenesis when the triacyl-glycerol and phospholipid contents of the yolk sac had diminished showed unexpectedly that the absolute amount of docosohexaenoic acid was increasing, apparently by transformation of other fatty acids of linolenic affinity. It is also known that the activity of the spermatozoa of higher animals is directly related to their content of this fatty acid (Nikitin and Babenko, 1987).

Is it possible that the essential role of ω3 acids, and docosohexaenoic acid in particular, is evidence for a genetic similarity between terrestrial and freshwater animals, implying a common marine origin? The composition of the blood of higher terrestrial animals seems to furnish evidence for this suggestion. It may not be coincidence that, in modern medicine, there are remedies based on polyenoic fatty acids which are very effective in curing a number of illnesses. In fish, the same fatty acid increases immunity from infection and parasitic diseases (Shchepkina, 1980a; Bolgova, 1993; Ripatti *et al.*, 1993).

Up to now, we have linked functional activity with mobility. However, as a concept, 'functional activity' is much broader and more complex, embracing all the vital reactions of the organism. We have already cited examples of the developmental stages of freshwater fish. It is relevant here to mention data obtained by Yuneva *et al.* (1990) on the relationship between the docosohexaenoic acid of the triacyl-glycerols present in white muscle of female salmon and the survival of their eggs and larvae (Figure 24). Fish that are deficient in docosohexaenoic acid have disorders of the reproductive system, low fecundity and sperm count, difficulties with fertilization and hatching (Watanabe, 1982; Gershanovich *et al.*, 1991) and the number of abnormal larvae developing (Walton and Cowey, 1982). It has to be borne in mind that triacyl-glycerols are the main source of lipid 'plastic materials' used in the reproductive process. The relationship between the content of the same acid and the intensity of the photoreaction in the eye has recently been established in studies made on the Pacific saury from the Far East (Yuneva *et al.*, unpublished data; Table 8). Yet another finding has been the high level of C22:6ω3 in the membranes of the photoreceptive cells (Berman *et al.*, 1979), to a value of up to 50% of the total fatty acids of their phospholipids (Nicol *et al.*, 1972; Meyer-Rochow and Pyle, 1980; Tyurin *et al.*, 1982; Bell *et al.*, 1986). In the adaptation of steelhead salmon

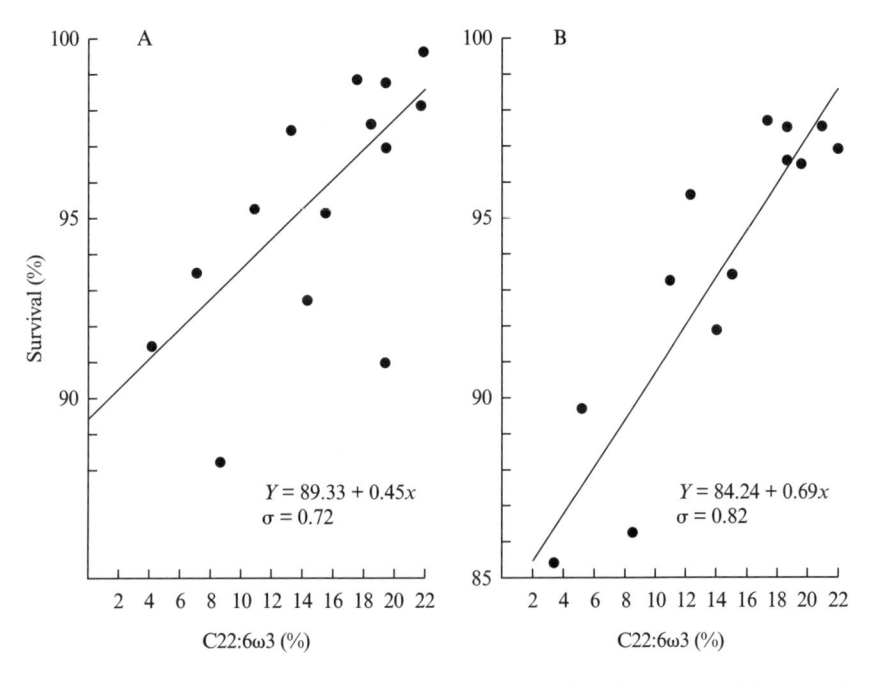

Figure 24　Relationship between the docosohexaenoic acid content of the triacyl-glycerols of the white muscle of female pink salmon and the subsequent survival of (A) eggs and (B) larvae.

Table 8　Docosohexaenoic acid (DHA) content of the eye lipids of saury-pike.

	Rate of photoreaction			
	I	II	III	IV
DHA in phospholipids (%)	38.5	37.6	43.5	45.9
DHA in triacyl-glycerols (%)	6.3	6.2	15.1	15.1

to a hyperosmotic environment, the docosohexaenoic acid content in polar lipids increases (Gershanovich *et al.*, 1991). It decreases when pink salmon are introduced into fresh water (Tyurin and Gorbunov, 1984). The copepod *Calanus euxinus* was found to produce a greater content of docoso-hexaenoic acid when entering the hypoxic zone of the Black Sea (Yuneva and Svetlichny, 1996).

All these data provide evidence that, whenever an organism has to enhance its functional activity by compensatory means, the membranes alter their fatty acid composition. This points to the multifunctional role played by docoso-hexaenoic acid in a wide variety of adaptation processes; it is an 'adaptogen', assisting the complicated processes which occur in cell membranes.

We now consider the composition and content of protein in the present context of compensation. Kondratyeva and Astakhova (1994) found that, as in other fish species (reviewed by Love, 1970), the content of total proteins in the white muscle of Black Sea fish was always greater than in red muscle, although the proteins in the latter were the more heterogeneous. The concentration of proteins in the serum of the sluggish round-goby was considerably higher than in that of horse-mackerel: 4.0–6.0% compared with 2.6–3.6%, respectively (Kulikova, 1967). The data apparently indicate a special function of proteins in the plastic metabolism of sluggish fish. Indirect confirmation comes from the fact that the number of leucocytes, and particularly of lymphocytes, are inversely proportional to the rate of natural mobility (Tochilina, 1994). Another inverse relationship is that between leucocyte and erythrocyte numbers. Enhanced numbers of leucocytes probably promote intensified proteolysis, and in general more efficient assimilation by the tissues of organic substances from the diet.

Proteolytic digestive enzyme activity in sluggish Black Sea fish (e.g. toad goby, scorpion fish) has been found to be equal to and sometimes greater than that of the active fish (e.g. horse-mackerel, annular bream, pickerel). The aminolytic activity of fish of low mobility is also greater (Ugolev and Kuzmina, 1993). It will further be shown in Chapter 5 that the consumption and assimilation of food occur more efficiently in sluggish than in active fish.

The relationship between the quantity of serum albumin and activity is, however, not clear. Although it forms up to 50% of the serum proteins in tunas, 40% in horse-mackerel, 14% in scorpion fish and 5% in gobies (Shulman and Kulikova, 1966), such a relationship is not distinct in other orders of fish. In the blood it performs transport functions and is an anticoagulating agent.

The intensified carbohydrate metabolism observed in fish of low mobility is stimulated by enhanced work by the endocrine glands responsible for carbohydrate mobilization. As Plisetskaya (1975) observed, scorpion fish display a greater activity of adrenocorticotropic hormone (which is produced in the hypophysis and regulates the secretion of catecholamines) than do the more active horse-mackerel, and this is directly linked to carbohydrate mobilization in tissues. It is perhaps surprising that the activity of the thyroid glands in scorpion fish was the greater of the two species (E.V. Ivleva, 1989a). Possibly this attests to compensatory metabolism, which sustains vitality in sluggish species. Some investigators, for example, Epple (1982), suppose that the thyroid gland of lower vertebrates fulfils precisely this function – supporting a high vital tone. In fish, the activity of the thyroid gland and the intensity of oxygen consumption may not be directly related (E.V. Ivleva, 1989b).

The two alternative strategies of metabolism explored in active and sluggish fish therefore apply both to the divergence of energy metabolism and to the closely associated plastic metabolism, which provides normal functioning of energy transformation mechanisms. Active fish are characterized by more intensive consumption and transformation of food, greater content of total phospholipids in red muscle, and of polyenoic fatty acids, especially docoso-hexaenoic acid, which creates the essential structural–functional basis for energy metabolism at a high level. Sluggish fish are characterized by strong compensatory mechanisms characteristic of plastic metabolism. These readily allow the emergency mobilization of energy reserves. Among the compensation strategies are high contents of phospholipids in white muscle, high protein concentrations in muscle and blood serum, and the elevated activity of endocrine glands, which promotes efficient carbohydrate metabolism and a proper 'vital tone' in the organism. A further item to add to this list is a very efficient utilization of food (Chapter 5).

4. Molecular and Metabolic Aspects of Life Cycles

The relative contributions of endogenous and environmental factors to life cycles are not completely understood as yet, but physiological and biochemical phenomena that have been observed help to clarify matters. Many observations, reported in Russian publications, and little known to the rest of the world, constitute the bulk of accumulated knowledge. The basic studies from which this work developed were on ecological physiology (Kalabukhov, 1950; Slonim, 1971; and others) and developmental physiology (Nagorny, 1947; Orbeli, 1958; Kreps, 1977; and others). Studies on the early stages of ontogenesis have yielded many important results (Milman and Yurovitsky, 1973; Vladimirov, 1974; Ryzkov, 1976; Neifach and Timofeeva, 1977; Reznichenko, 1980; Shatunovsky, 1980; Sidorov, 1983; Konovalov, 1984; Gosh, 1985; Ozernyuk, 1985; Zhukinsky, 1986; Gershanovich *et al.*, 1987; Novikov, 1993). The changes in fish metabolism over the stages from embryogenesis to old age and death (Shatunovsky, 1980) and during the annual cycles (Gerbilsky, 1956; Shulman, 1960a, b, 1972a; Krivobok, 1964; Barannikova, 1975; Shatunovsky, 1980; Shulman and Urdenko, 1989; Minyuk *et al.*, 1997) have been examined in detail. The daily and interannual rhythms, however, have failed to attract the proper attention of research workers.

4.1. ONTOGENESIS

The following stages* are traditionally demarcated in ontogenesis: (1) embryonic; (2) larval, when the larvae grow and develop using exogenous

---

*The words 'stage', 'period' and 'phase' are often used interchangeably in ontogenesis. Here we should like to differentiate them, thus: 'stage' refers to ontogenic developments of long duration; and 'phase' indicates minor qualitative changes taking place within a stage and for developing reproductive products; 'period' refers to specific qualitative states occurring over the annual cycle.

supplies; (3) juvenile, until fry are sexually mature; (4) the period of becoming mature; (5) sexual maturation in fully grown fish; and (6) senescence (Severtsev, 1934; Rass, 1948; Kryzhanovsky, 1949; Dryagin, 1961; Shatunovsky, 1980).

Ontogenesis provides: (1) optimum quality of embryos and eggs, making them capable of survival *en masse*; (2) optimum development to the beginning of exogenous feeding; (3) maximum growth rate of fry, which speeds their rescue from predatory stress; (4) genetically determined growth and development up to maturation; and (5) reproduction, the most prolonged stage in most species. Together with somatic increase, reproduction accounts for all the 'production' of a population. As regards ecology, even the final stage (including growing old and death) is important for a species, as it eliminates genetically depleted individuals from a population and supplies the ecosystem with organic substances that arise from their decay. The chemical composition and dynamics during early ontogenesis have been reviewed by Needham (1963) and Cetta and Capuzzo (1982).

Fertilized eggs contain the complete set of substances required for the growth of the embryo: nucleic and free amino acids, vitamins and inorganic substances, in addition to proteins, lipids and carbohydrates. In the lipids, the triacylglycerols provide the reserve of energy, which explains the high lipid content of the eggs of sturgeons and salmonids, which have a long embryological development, but the pelagic eggs of cod, flounders and clupeoids are less rich in neutral lipids (Henderson and Tocher, 1987). Phospholipids, along with proteins and cholesterol, form the stable structures of tissues and organs of the embryo (Gershanovich *et al.*, 1991). Reserve components such as triacyl-glycerols, cholesterol ethers and glycogen are found mostly in the yolk sac, and generate energy for growth and differentiation. Wax esters are less dense than triacyl-glycerols and so are better for regulating buoyancy, and at low temperatures they afford a better storage of reserve energy in the eggs of mullet and some other species. The eggs of some species contain carotenoids, mostly astaxanthin and canthaxanthin, sometimes in considerable quantity. These compounds can act as respiratory pigments, supporting metabolism under conditions of oxygen deficiency, but Deufel (1975) has presented evidence to show that carotenoids also fulfil important functions in reproduction, being sperm activators and acting as hormones influencing growth, fertilization, maturation and embryonic development. Carotenoids seem to be particularly important in the eggs of salmonids, which are laid in shallow water and may be protected by them from sunlight (Hata, 1977, personal communication to RML). Energy substrates utilized vary according to the phase of embryogenesis. The initial phases – morula, blastula, gastrula and early organogenesis – display a prevalence of glycolysis and glycogenolysis (Milman and Yurovitsky, 1973; Boulekbache, 1981; Timeyko, 1992; Novikov, 1993). Subsequently, right up to hatching and mixed feeding, it is triacyl-glycerols which are utilized, these being rich in energy. Like polar lipids, these neutral lipids are partly involved in plastic metabolism by supplying energy for biosynthesis. Figures 25 and 26, and Table 9 (Shatunovsky, 1980; Chepurnov and Tkachenko, 1983; Sidorov, 1983; Novikov, 1993) illustrate the dynamics of essential organic components in the developing embryo and larva.

During the embryonic and larval stages, there is a decrease in the amount of lipids available from the yolk sac and oil droplet, and a probable increase in material available from reserve depots in the embryo body. During this process, the relative proportions of lipid fractions and fatty acids change. Both neutral and polar lipids decrease in absolute terms during the period preceding exogenous feeding; and the contents of saturated (except for 16:0) and monoenoic fatty acids are much reduced, having supplied energy to the growing embryos and larvae (Ando, 1968; Nakagawa and Tsuchiya, 1971; Hayes *et al.*, 1973; Cetta and Capuzzo, 1982; Gershanovich *et al.*, 1987, 1991). The total content of calories also decreases (Wang *et al.*, 1987).

In contrast, the relative proportion of polyenoic acids increases steadily. It was found that ω3 fatty acids are important to normal growth during early development (Golovachev, 1985). Surprisingly, this applies to both marine and freshwater fish, the latter as adults being richest in acids of the ω6 series. During the early stages of life, the content of C22:6ω3 and C16:0 steadily

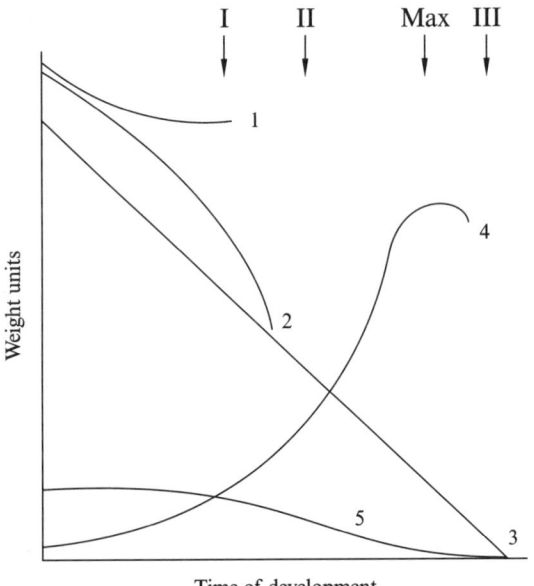

Figure 25   Generalized scheme of protein and lipid changes during the development of embryos and larvae (endogenous feeding). Protein in: 1, whole egg; 2, egg capsule; 3, yolk; and 4, embryo. 5, Total lipids in embryo. I, hatching from egg; II, transfer to mixed feeding; III, total yolk resorption: Max, maximum body mass.

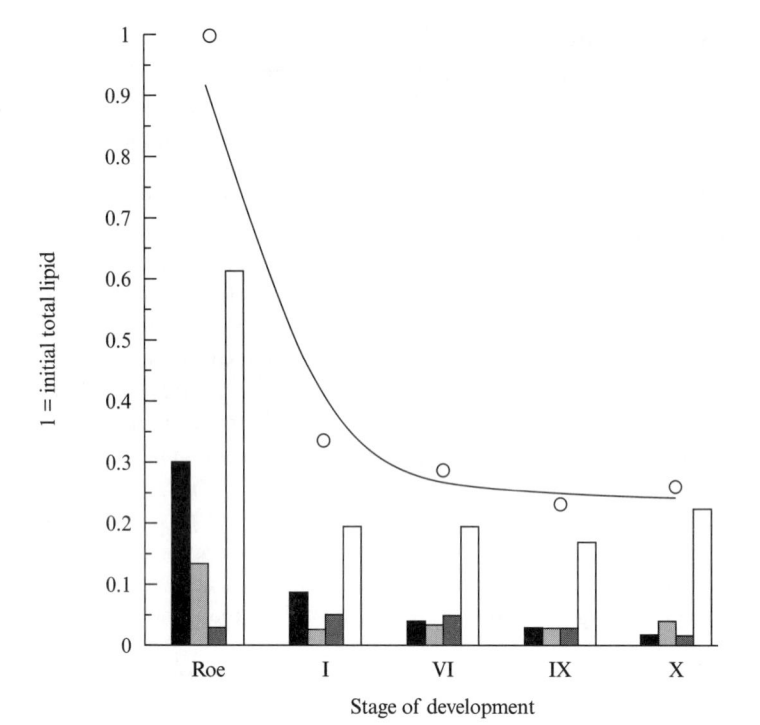

Figure 26   Dynamics of lipid fractions in round goby, from the Sea of Azov, during embryonic development. (After Chepurnov and Tkachenko, 1983.) Columns at each stage, left to right: Phospholipids, cholesterols, non-esterified fatty acids, triacyl-glycerols; line curve, total lipids.

Table 9   Variation of lipid fractions in early stages of striped bass (% of total lipids). (After Dergaleva and Shatunovsky, 1977.)

	Days after hatching		
	7	13	17
Phospholipids	18	19	32
Cholesterol	10	18	29
Non-esterified fatty acids	6	10	19
Triacyl-glycerols	20	13	3
Sterol esters	46	40	17

increases (Hayes *et al.*, 1973; Atchison, 1975; Bolgova *et al.*, 1985a; Pionetti *et al.*, 1986; Isuev and Musaev, 1989), indicating synthesis of phospholipids which consist for the most part of these fatty acids. In the process, C22:6ω3 may be synthesized from C:18ω3. As regards phospholipids, the highest rates of quantitative increase are shown by the phosphatidyl choline and phosphatidyl ethanolamine (Gershanovich *et al.*, 1991). During gastrulation, the content of docosohexaenoic acid and the cholesterol/phospholipid ratio increase in the egg membrane (Nefedova and Ripatti, 1990).

The main organic constituent of the bodies of embryos and larvae is protein, of which the relative content steadily increases on a dry weight basis and also in absolute terms, as a result of input of protein and other nitrogenous components from the yolk sac (Ando, 1968; Cetta and Capuzzo, 1982; Hinterleitner *et al.*, 1987; Novikov, 1993). In this organ they combine to form lipoproteins, glycoproteins and phosphoproteins. It is only when the yolk sac is completely exhausted, immediately before the switch to exogenous feeding, that the total protein mass stored in the larva shows a tendency to decrease slightly (Novikov, 1993). As reported elsewhere (Krivobok and Storozhuk, 1970), as much as 60% of the lipid and 40% of the protein reserves are used up in the period from fertilization until the onset of exogenous feeding. At the same time, the content of free fatty acids and amino acids increases, corresponding with intensive protein breakdown (Ando, 1968; Chepurnov and Tkachenko, 1983; Shcherbina *et al.*, 1988). Glycogen diminishes during phases III–VII (Figure 27), then increases in phase VIII through gluconeogenesis involving lipids and proteins, finally receding again at phase IX which is followed by hatching (Timeyko and Novikov, 1991;

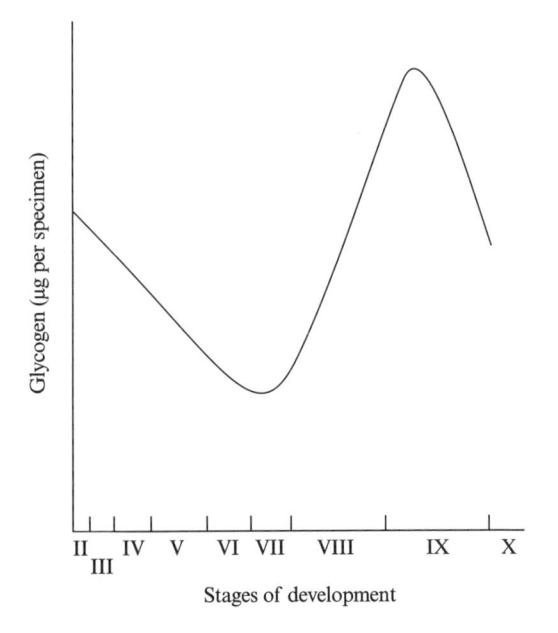

Figure 27 Dynamics of glycogen content in eggs of Atlantic salmon. (After Novikov, 1993.) II, cell division; III, blastula; IV, gastrula; V, organogenesis; VI, start of mobility, heart pulsation; VII, appearance of blood; VIII, liver-yolk blood circulation; IX, start of continuous moving of burst fins, preparing for hatching; X, hatching.

Timeyko, 1992; Novikov, 1993). It is during this last phase that the vigorous 'jerks' of the pectoral fins appear, requiring rapid availability of energy.

There are also changes in the rates of metabolism as red blood cells appear and aerobic processes intensify (Lasker and Theilacker, 1962; Laurence, 1975; Timeyko and Novikov, 1991) during the early phases of ontogenesis. Oxygen consumption increases, as do the number of mitochondria and their protein contents (Abramova and Vasilyeva, 1973; Ozernyuk, 1993). The adenyl nucleotide pool (ATP and ADP) decreases (Milman and Yurovitsky, 1973; Boulekbache, 1981), while the activity of cytochrome oxidase increases (Ozernyuk, 1993). The increased energy metabolism corresponds to a considerable extent with motor activity (Reznichenko, 1980). In the yolk sac, the activity of proteinase, which supplies nitrogenous materials to the embryo, increases, as does the rate of amino acid incorporation into the body proteins.

In proteins of the growing embryo, SH- groups become more abundant (Konovalov, 1984). The antioxidative system is established in the early phases of ontogeny (Rudneva-Titova, 1994). This can be deduced from increased activity in such enzymes as superoxide dismutase, ketolase, peroxidase and glutathione reductase, which neutralize peroxides generated during lipid oxidation. The rise of free radical activity becomes evident also from the increasing content of the 'pro-oxidant' lipoxygenase, which lowers the natural content of antioxidants: the glutathione, carotenoids, and vitamins A, E and K contents are lower in larvae ready to switch to exogenous feeding than they were in the embryos; these substances are widely used for eliminating the harmful effects of peroxides. Interestingly, the interaction between pro-oxidants and anti-oxidants is more intensive in pelagic than in demersal eggs.

In this context, there is a bottleneck in early ontogenesis at the time of switching to exogenous feeding, when energy and structural reserves from the yolk come to an end. This is preceded by a period of mixed feeding, when decreasing endogenous nourishment is more or less supplemented by exogenous material; high mortality is usual at this time (Blaxter, 1969; Eldridge et al., 1981).

When the organism starts to feed exogenously, there is a burst of diverse metabolic activity (Ozernyuk, 1985; Semenchenko, 1992): a good example is oxygen uptake (Figure 28). At the same time there is a sudden increase in the activity of the enzymes that are related to energy metabolism. The balance between aerobic and anaerobic processes shifts towards aerobic and, surprisingly, the rate of energy metabolism rises with increase in body weight (Ozernyuk, 1985; Ryzkov, 1976). As the fry continue to grow, however, then mature and age, the metabolic rate becomes inversely related to the weight of the body. At the changeover from endogenous to exogenous feeding, a sharp increase in both anabolism and catabolism of proteins, lipids, carbohydrates and compounds of phosphorus has been observed (Hinterleitner et al., 1987). The highest specific growth rate coincides with the lowest lipid content

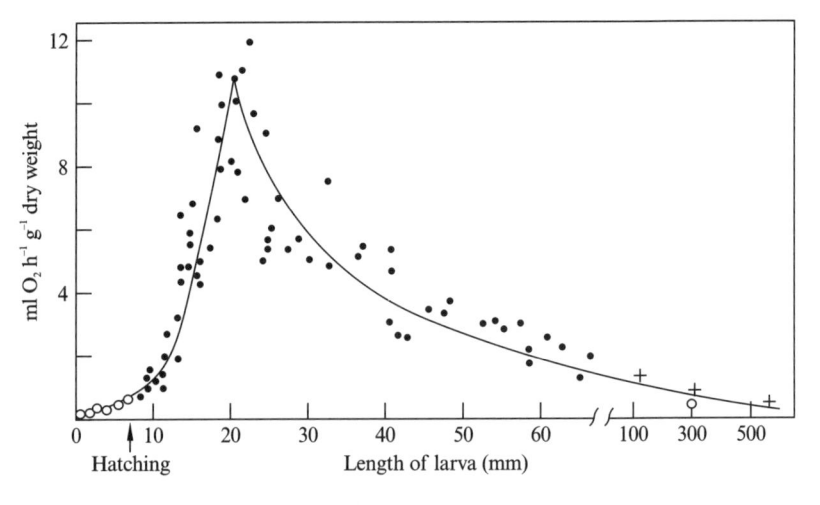

Figure 28   Oxygen uptake in pike during embryonic and larval development. ○, embryos; ●, larvae. (After Shamardina, 1954.)

(Gershanovich *et al.*, 1987; Semenchenko, 1992). In the brain tissues of the fry, the levels of L-dopa and dopamine rise, fostering their 'searching' activity (Nechaev *et al.*, 1991). The white muscle develops particularly strongly in fish larvae, because the manner of swimming at this stage is in sharp bursts alternating with rests (Nag and Nursall, 1972). The high rates of energetic and synthetic metabolism make nutritional demands that are often too much for the larvae, hence the high mortality rate at this time.

The next stage is the juvenile, which precedes sexual maturation and is characterized primarily by intensive protein production, and increases in the weight and length. It is accompanied in fry by enhanced incorporation of amino acids into proteins, activity of protein synthetases and aminotransferases, and high RNA/DNA ratios (Shulman, 1974; Haines, 1973; Shatunovsky 1980; Kurant *et al.*, 1983; Weatherley and Gill, 1987). Enhanced activity of proteinases in the alimentary canal provides substrates to support protein synthesis, while at the same time mineral metabolism intensifies to form and develop the skeleton. The mode of swimming of the fry then becomes more continuous, and the proportion of red muscle increases (Mosse and Hudson, 1977).

This period is also marked by the deposition of energy substrates – mostly triacyl-glycerols in fatty fish and liver glycogen in lean fish. The rate of accumulation is much lower than that of protein production (O'Boyle and Beamish, 1977), a fact explicable from the life strategy of the fish: they must attain maximum size and weight as soon as possible before becoming sexually mature, so as to be able to escape predators, to search effectively for food and to evolve optimum reproductive power. Fecundity is directly proportional to their weight and size. Hormones, such as thyroxin, somatotrophin and prolactin, regulate the direction and intensity of metabolic processes from early ontogenesis onwards (Boeuf *et al.*, 1989).

The stage of sexual maturity is the longest in the life of most species of fish, apart from freshwater eels and Pacific fish of the salmon family. A fundamental change takes place through a co-ordinated increase in the hormones of the hypothalamus, hypophysis, adrenal cortex and gonads. Neuroendocrine regulation switches the protein synthesis from solely somatic to part-somatic and part-generative tissue production. In Black Sea fish, the annual yields of these two types of tissue are roughly equal (Shulman and Urdenko, 1989). With increasing age, the generative production comes to dominate the somatic, the ratio of 1:2 in first spawners becoming 2:1 in later years.

It is widely recognized that sexual maturity is accompanied by a sharp deceleration in linear growth, but the body weight continues to increase, usually at a higher rate (Shulman, 1974). Such a high production of somatic tissue and sexual products is possible only through intensive food consumption. The retardation in linear growth is simply the result of a change in the geometric proportions of growing fish (Shulman, 1974), noted earlier by Berdichevsky (1964) and Smirnov (1967). The idea is easier to understand if one regards the growth of the fish as three-dimensional, like a cylinder or cone in geometrical terms, rather than in one dimension. Thus the longer a fish becomes, the greater is the increase in weight and protein per unit increment in length – a relationship similar to that seen in the growth of trees.

The effect of hormonal changes in sexual maturation can therefore be regarded as the acceleration of protein biosynthesis, directed to both somatic and generative tissues. There is also an accelerated production of reserve energy, mostly triacyl-glycerols, stored in special depots in the muscular and connective tissues, peri-intestinal spaces, liver and sub-dermal spaces (Shulman, 1974; Shatunovsky, 1980; Lapin and Shatunovsky, 1981; Sidorov, 1983). The purpose of the lipid reserves is to supply energy for the development and spawning of the generative tissue.

In Black Sea fish with a short life span, the period of sexual maturity compared with the juvenile period is 3–5 to 1, whereas in longer-lived species it is 10–15 or more to 1. These fish are capable of spawning until the very end of their lives.

4.1.1. Sex Differences

In many species of fish, males and females differ in their rates of growth and so in their rate of protein synthesis (Shulman, 1974). Energy reserves are consumed more rapidly in female fish than in males during maturation, because female gonads are larger. The reserve substances include triacyl-glycerols, glycogen, serum proteins (albumins, alpha- and beta-globulins), which are used in the synthesis of structural components and as sources of energy (Sand *et al.*, 1980; Henderson *et al.*, 1984; Ando, 1986; White *et al.*, 1986). The presence of greater quantities of non-esterified fatty acids, glucose and free amino acids in females indicates that enzymic activity, both anabolic and catabolic, is also higher in females (Liu *et al.*, 1985).

At the climax of spawning, male fish incur greater energy losses than females because of greater motor activity – they fight each other. The relative compositions of male and female gonads vary according to a number of factors, the most important being that the female has to support normal development of the embryo which requires a high concentration of plastic substances (precursors for tissue synthesis) and energy. Pelagic eggs require a relatively large quantity of lipids in the oil globule in order to provide buoyancy, while demersal eggs need less (Chapter 2, section 2.4). Lizenko *et al.* (1983) and Sidorov (1983) demonstrated substantial differences between the phospholipid fractions (Figure 29) in mature oocytes and spermatocytes. Sidorov (1983) also showed that there were more triacyl-glycerols and cholesterol esters in the former than the latter.

Along with processes going on in the reproductive tissues themselves, changes in the quantity and composition of organic substances in the liver are

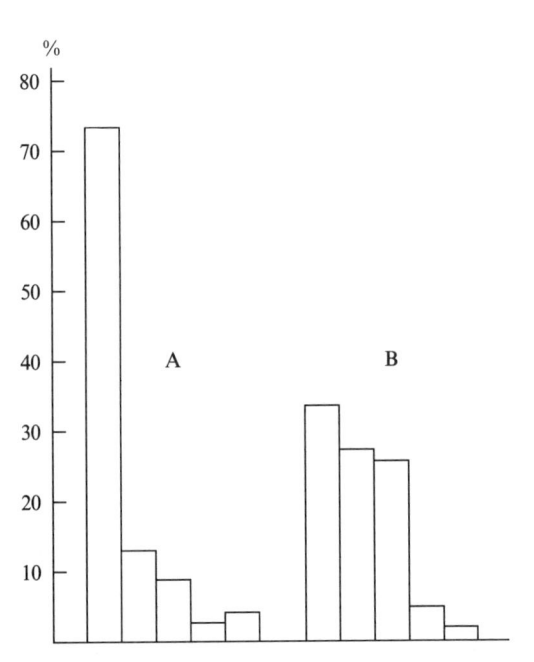

Figure 29   Phospholipid fractions in: A, mature oocytes; and B, testes in freshwater fish, % of total phospholipids. (After Sidorov, 1983.) 1, Phosphatidyl choline; 2, phosphatidyl ethanolamine; 3, sphingomyelin; 4, cardiolipin; 5, lyso-phosphatidyl choline.

important in maturation. Krivobok (1964) and Shatunovsky (1967) recognized this as a measure of the increasing metabolism induced by the synthesis of reproductive products. Increases in the content of lipids and proteins in the gonads are accompanied by increases in the quantities of glycogen, vitamins and other biochemically active substances (Plack *et al.*, 1961; Braekkan and Boge, 1962; Timeyko and Novikov, 1991).

During vitellogenesis and yolk formation, maturation stages II and III, the main activity is the retention of protein and the formation of structural lipids. At stage IV (extensive growth), triacyl-glycerol accumulation prevails. Yolk lipids are mostly lipoproteins and vitellogenin (De Vlaming *et al.*, 1984; Gershanovich *et al.*, 1991). A proportion of the proteinases and other hydro-lases of female fish are transferred into the yolk structures of the oocytes, where they begin working as soon as fertilization has taken place (Nemova, 1991). Lysosomal cathepsin D has been studied by Nemova and Sidorov (1990). In the early stages of spermatogenesis, the quantities of neutral lipids, phospholipids and sterol ethers increase, the phospholipids rising as high as 90% of the total lipids. Aerobic metabolism is greater than glycolysis (Gosh, 1989). Most of the proteins and lipids entering the gonads in Stages II and III originate in the feed, but at Stage IV they come mostly from the reserves in the liver and muscle (Wang *et al.*, 1964). The greater quantity of water present at Stage V assists oxidation of the fatty acids. The main form in which material is transported to the gonads is via lipoalbumins and lipoglobulins (Kulikova, 1967; Ipatov and Lukyanenko, 1979; Kychanov, 1981; McKay *et al.*, 1985). The accumulation of materials in the gonads is under the control of hormones such as gonadotrophins, insulin, corticosteroids and prolactin (Fontaine, 1969; Plisetskaya *et al.*, 1977; Donaldson *et al.*, 1979; De Vlaming *et al.*, 1984; Sautin, 1985).

In fish that spawn more than once a year, the dynamics of chemical compo-sition differ between groups. The early spawners are both absolutely and rela-tively richer in their content of protein and lipid. In its post-spawning period, fatty degeneration of the gonad takes place. Figure 30 shows the effect of gametogenesis on the lipid content of the female gonads; the main change in lipid deposition and withdrawal is in the triacyl-glycerols, although small changes in phospholipids follow the same pattern, indicating alterations in the amounts of cellular material. In most fish species, the formation of sexual products in the gonads uses up the internal stores of energy and plastic reserves no matter how great their food intake. The situation is even more fraught in species that spawn at the end of winter and really are starving. A classic case is that of Atlantic cod, in which the internal reserves are progres-sively depleted with each additional year of spawning (Love, 1960) until the fish is unable to recover from spawning and dies (Love, 1970).

Over a long life span, the ratio of protein to lipid in the body shifts towards lipid. The phenomenon is controlled by a change of activity of somatotrophin

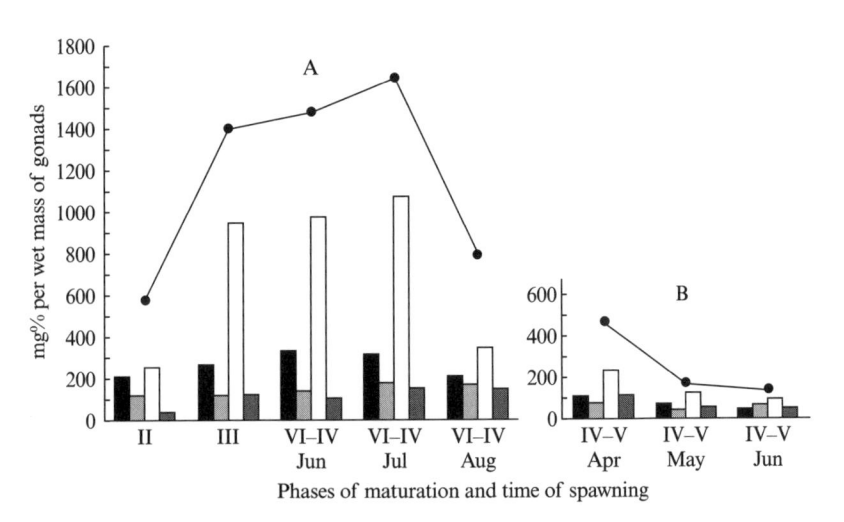

Figure 30   Changes in the lipid fractions of female gonads during gametogenesis. (After Chepurnov and Tkachenko, 1983.) Black columns, phospholipids; light stipple hatching, cholesterol; open columns, triacyl-glycerols; dark stipple, cholesterol esters. Line and closed circles, total lipids. (A) Horse-mackerel; (B) turbot.

(a growth hormone) and prolactin, which controls lipid accumulation (Donaldson *et al.*, 1979; Sautin, 1985). At the same time, the relative contents of triacyl-glycerols and of saturated and mono-unsaturated fatty acids increase. The activity of nucleases declines (Berdyshev, 1968), and the content of serum protein and its ratio of albumin to globulin decreases (Ipatov, 1970; Figure 31). On the other hand, the share of lipoproteins and non-esterified fatty acids increases (Shatunovsky, 1980). Figure 31 convincingly demonstrates the steady decline in efficiency of conversion of food for constructive processes with the passage of time (Shatunovsky, 1980; Shulman and Urdenko, 1989).

The changes in fish metabolism with age mirror many of those described in higher animals (Nagorny, 1940; Comfort, 1964; Nikitin, 1966; Parina, 1967; Frolkis, 1975). An obvious difference, though, is the fact that, as in many other water organisms, the generative synthesis in fish does not die out or slow down with age, but is enhanced. The quantity of reproductive material thus produced is of key importance for the replenishment of the stock as a whole. However, a frustrating feature of trying to describe phenomena peculiar to fish is that they often do not apply to every single species. In this case, Siamese fighting fish can live for a considerable period after the cessation of their reproductive lives. Both testes and ovaries become shrunken and infiltrated with connective tissue, any remaining germ cells being changed in appearance (Woodhead, 1974a,b). The latter author concluded that there had been a decline in gonadotrophin secreted

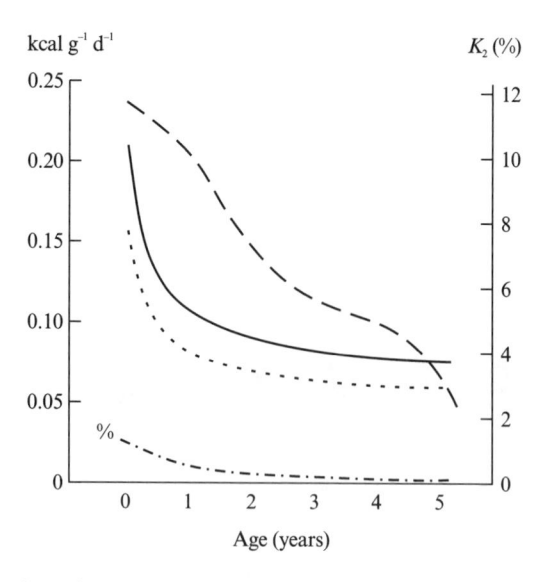

Figure 31   Age changes in horse-mackerel. (After Shulman and Urdenko, 1989.)
– – –, Efficiency of using food for growth; ——, intensity of food consumption;
······, energy metabolism; –·–·, production.

by the pituitary. A few other species that can show signs of reproductive
degeneration with age are reviewed by Love (1980, pp.60–61).

The next question is the age at which fish shed reproductive material of the
highest viability. This topic has been studied at Kiev (Kim, 1974; Vladimirov,
1974; Konovalov, 1984; Gosh, 1985; Zhukinsky, 1986). These workers found
that the highest quality of eggs and progeny came from middle-aged fish.
Their eggs and sperm were the best from a biochemical viewpoint, being
richer in protein, polar and neutral lipids and polyenoic fatty acids. The rate of
tissue respiration, the conformity between respiration and phosphorylation,
the activity of cytochrome oxidase and the dominance of respiration over
glycolysis were also higher.

Using the quality of the reproductive products as the criterion, it is possible
to rank fish in decreasing order of reproductive excellence, as follows: middle-
aged, young, aged. The progeny from eggs and sperm from spawning fish of
different ages descend in the following order of viability:

- middle-aged mating with middle-aged
- middle-aged with young
- young with young
- middle-aged with aged
- young with aged
- aged with aged.

This pattern of matching applies not only in fish-farming but in nature, and, in their search for a mate, zygotes observe the same principle; the reproduction of a natural population proceeds in this way. Experiments using radioactive phosphorus have shown that individual sex cells appear to compete for 'partners' spawned by the most compatible age-group that is likely to give the best progeny (Zhukinsky, 1986). Cases are known where reproduction ceased or was impeded through the presence of incompatible age groups. With optimal matching, the embryogenesis would display much less teratology and better tolerance of unfavourable environmental conditions.

4.1.2. Aging

The final stage of ontogeny of fish, aging, has not been given due attention by researchers, particularly as applied to marine fish. This may be because few fish reach their maximum age on account of predators (including man) and death following spawning, winter starvation or migration. One reliable fact is that protein, and so body weight, still increases in the oldest group of fish, although by smaller increments (Figure 32). What in higher animals is termed 'wild' synthesis of protein (proceeding at a very high rate) occasionally occurs in fish at this ontogenetic stage (Nikitin, 1966). Lipid accumulation is enhanced at the same time and overtakes that of protein in fatty fish such as herring, anchovy, mackerel, saury-pike, *Notothenia*, whitefish, smelt, halibut and eel families. The rates of consumption and assimilation of food, metabolic expenditure and efficiency of conversion of food into growth then drop dramatically (Figure 31) through a destructive change in the ratio of anabolism and catabolism of protein. In gerontology this shift is known as 'fading self-regeneration' (Nagorny, 1947; Nikitin, 1966; Parina, 1967; Parina and Kaliman, 1978). The immune system of old fish becomes very weak and can be the cause of death through disease (Mikryakov, 1978).

It is not easy to trace the final stage of ontogeny in fish of the Black Sea, since they have short life spans. The task is easier in the Baltic, White and Barents Seas, where the fish live much longer. According to Shatunovsky (1980), the fat content of fish in their final ontogenic stage drops markedly. This is surprising since, as we stated above, lipid accumulation overtakes that of protein in older fish. A reduced lipid content in the oldest fish indicates a reduced ability to assimilate or synthesize it. Under unfavourable foraging conditions, the anchovy and sprat of the Black Sea tend to develop an inverse relationship between fatness and age (Shulman, 1962) instead of the normal positive relationship. All evidence points to an unbalanced metabolism rather than a dramatic reduction of biosynthetic capacity in aged fish. The synthesis of DNA and RNA, although not as intensive as it was formerly, does not cease completely in old fish, which are similar to other animals which exhibit

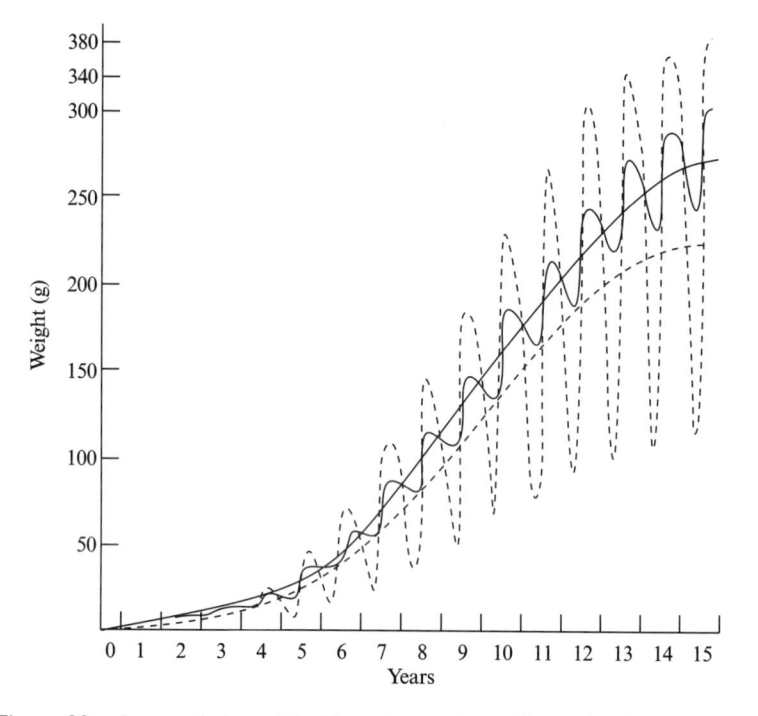

Figure 32   Accumulation of lipids and proteins with age in the 'large' form of horse-mackerel showing overall trends (smooth curves) and annual cycles. (After Shulman, 1972a.) Solid lines: proteins, broken lines: lipids. Note the greater annual fluctuations of lipids.

continuous growth. This feature provides proof at the molecular level (Berdyshev, 1968; Pirsky *et al.*, 1969). Confirmation that the heart of the problem is disturbed homeostasis is provided by the fact that the activity of endocrine glands is reduced in old age (thyroid: Woodhead, 1979; hypophysis: Khristophorov, 1975).

A classic example of this unbalanced homeostasis is found in the spawning mortality of Pacific salmon. Death comes as a result of depleted energy supply and decay of the protein–phospholipid structures, but the immediate cause is the enormous quantities of corticosteroids discharged from the adrenal glands into the blood (Idler and Clemens, 1953; Ardashev *et al.*, 1975; Maksimovich, 1988). Tyurin and Gorbunov (1984) and Kreps *et al.* (1986) have deduced mechanisms connecting this effect with a stress reaction and dramatic rise of free radicals in tissues. Dying mature Atlantic cod are also found to have an abnormally high concentration of cortisol in their blood (Idler and Freeman, 1965). The cortisol itself appears to be the lethal agent rather than a coincidental presence. Robertson *et al.* (1963) implanted pellets of cortical hormone into immature rainbow trout so that cortisol was continuously released into the blood stream. The fish lost weight, developed the characteristic infections of the skin and died within 9 weeks. Deaths occurred much more rapidly and with smaller doses of cortisol at higher water temperatures.

Post-spawning mass mortality probably occurs in fish other than Pacific salmon, but it is difficult to be certain. The round goby from the Sea of Azov may furnish an example of energy depletion. After spawning, the male fish guard the spawned egg clutch over the entire period of their development. They do not feed, so the fat stored in the liver is used up (Shulman, 1974). The males turn black (an alternative name of this species is dark goby), their endocrine system degenerates (Moiseeva, 1969) and death ensues. Cod furnish another case, whether from the North Sea (Love, 1960) or the Barents Sea (Borisov and Shatunovsky, 1973). Here the water content of the musculature after the fish have spawned is greater with each successive year, indicating progressive exhaustion until the fish are unable to recover from spawning and die (Love, 1960, 1970). Salmon of the genus *Oncorhynchus* are the most impressive example of self-programmed mortality (genome-conditioned death). But is it actually true that in all the cases death is a rigidly programmed phenomenon, or does it result from errors that accumulate in the genome? The question is still open. It is noteworthy that other salmonids, for instance *Salmo salar*, can spawn more than once in their lifetime. Other species, such as sturgeon, pike and halibut, would live for many years if predators did not kill them.

It should again be emphasized that fish differ radically from most of the higher animals in that growth does not stop during the final stages of ontogeny. Perhaps terrestrial animals have to limit their size through having to support their full body weight. Fish and other aquatic animals living in the 'hypogravitational environment' (Korzhuev, 1964) have only slight problems of this sort (Brawn, 1969), so can grow to tremendous size (e.g. whales, sharks and giant squid). It is the failure of synthesis, rather than a slowing down, which characterizes ageing in fish. Particular species of fish cannot be characterized by life span, since there is so much variation between individuals (Zotina and Zotin, 1967; Nikolsky, 1974) and between populations, ecological influences affecting the length of life by whole orders of magnitude. One factor is the ambient temperature. Atlantic cod in the North Sea mature in 3 years and die at 8 years of age (Love, 1970), while in the Barents Sea they mature at 11 years and die at 25 years (Trout, 1954) (all ages approximate). The annual fish (*Cynolebias adloffi*) lives for about a year at its usual temperature of 22°C, but lives much longer in water at 16°C (Liu and Walford, 1966).

Food supply appears to be the most important ecological factor governing individual variation in life span. Evidence of this arises from studies of diadromous and freshwater forms of the smelt (Ivanova, 1980), normal and dwarf salmon (Krogius, 1978) and whitefish (Reshetnikov, 1980). The life spans of the different forms of these species may differ by several years. Poor food supply would entail reduced food consumption, lower metabolic and growth rates that finally result in premature aging and death (Shatunovsky, 1980). The reverse is seen when food is sufficient. Large forms of Baltic cod, herring and flounder mature later and live longer than their smaller counterparts. Variation in life span is most obvious in large Black Sea horse-mackerel. From the suggestions of Aleev (1957) and Shaverdov (1964) that the small and large varieties of this species are two conspecific races, it can be concluded that the large form arose as a result of the very favourable supply of nutrients that developed by the late 1940s–1950s, which must have governed the amount of food consumed and given rise to an intensified metabolism and a 2–3-fold lengthening of life. The concept could be extrapolated to engender hope of increasing the life expectancy of humans also. (Perhaps we should exercise caution in equating long life with bountiful nutrition in animals other than fish. Experiments have been carried out which showed that rats on reduced rations for prolonged periods lived longer, and richly nourished humans are not always noted for longevity: RML.)

4.2. ANNUAL CYCLES

Annual cycles occur in many characteristics of fish, regardless of any endogenous rhythms that may proceed at the same time. The literature abounds with examples of physiological and biochemical rhythms that are collectively termed 'seasonal variations'. However, the phrase does not help our understanding and is best avoided, particularly because some variations do not conform with season. For example, different groups of North Sea herring spawn in spring or autumn (Graham, 1956; Parrish and Saville, 1965; Burd, 1974; Shatunovsky, 1980) and the spawning periods of different species also come at different times. As regards fish from southern seas, autumn is the post-spawning period for fish that prefer warm waters and the pre-spawning period for those from cooler environments. Finally, the spawning season of fish from the Mediterranean is very extended and may encompass spring, summer and autumn.

A number of basic studies have suggested that the annual cycles of fish evolved as programmes stored in the genome, implemented so as to relate to environmental conditions (Fontaine, 1948; Hoar, 1953; Gerbilsky, 1958; Idler and Bitners, 1958; Shulman, 1960b, 1972a; Love 1970, 1980; Beamish et al., 1979; Shatunovsky, 1980). These and further studies on the rhythms of oxygen uptake, nitrogen excretion, activity of enzymes, dynamics of protein increase and lipid accumulation, variations in composition of tissue lipids, constituents of blood and endocrine activity have yielded a mass of results awaiting interpretation.

The first step in understanding the principles underlying the annual cycles is to collect information on seasonal rhythms. There are many publications that merely describe metabolic activity in fish during the course of a year, without attempting to demonstrate underlying principles. Even so, they record very large differences in the metabolic patterns between species. For example, while the accumulation or utilization of triacyl-glycerols occur respectively during the pre-winter feeding and spring spawning in fish from warm waters of the Black Sea, the opposite occurs in Black Sea scorpion fish (Khotkevich, 1975), where ^{14}C-acetate incorporation into the liver proteins is higher just before spawning. It is noteworthy that the temperature of the waters from which the scorpion fish were taken was the same in the spring as in the autumn. This example shows that one could never succeed in understanding the pattern of metabolism in a fish species from data obtained in one season: preferably it should be from all four, and best of all over several years. A second task is to identify the point where metabolism changes or reverses its orientation.

Somatic growth, energy accumulation and reproduction can all develop at any time in the annual cycle, given the right conditions. In fish from many seas it is the spawning season that appears to be the crucial point around which most of the metabolic processes revolve. Meyen (1939) had every reason to claim that the biology of fish in general pivots around spawning. This concept is supported by data obtained in the Institute of Biology of the Southern Seas (IBSS) in Sevastopol, and in the Azov–Black Sea Institute of Marine Fishery (ABIMF) in Kerch, using warm water species including anchovy, horse-mackerel, red mullet, pickerel and scorpion-fish (Shulman, 1960a,b, 1964a,b, 1967b, 1972a, 1978a; Kulikova, 1967; Shchepkin, 1972, 1979; Belokopytin, 1978, 1990, 1993; Trusevich, 1978; Shchepkina, 1980a; Rakitskaya, 1982; Emeretli, 1990, 1994a,b; Stolbov, 1990). The spawning period for the majority of Black Sea fish is June–July and it was found that the triacyl-glycerol reserves stored in the body are greatly diminished in order to supply energy for intensive generative synthesis. This also entails the mobilization of serum proteins (albumin, α- and β-globulins). The high temperature stimulates oxygen uptake and nitrogen excretion (energy production and catabolism of protein) and also high activity of the enzymes (ATPase, LDH and succinic dehydrogenase) of red and white muscle and liver. The numbers of red blood cells and their haemoglobin content increase, improving the oxygen transport into the tissues.

The post-spawning feeding (August–September in this case) results in somatic growth and the accumulation of energy. The first half of this season is characterized by steadily increasing protein biosynthesis and lipogenesis, which attain their highest point in early autumn and then gradually decline (Shulman, 1974; Shchepkin, 1979; Shchepkina, 1980a). A similar pattern has been observed in the activity of scale phosphatase, accounting for the formation of sclerites and indicating the periods of skeletal development (Senkevich, 1967; Shulman, 1972a, 1974).

During the winter, there is a switch from deposition to mobilization of proteins (Shulman, 1974). As with the utilization of triacyl-glycerol, this

sustains the metabolism of the fish ('endogenous feeding') at the required
level, if the external food supply suddenly decreases or disappears. Because
the winter season marks the utilization of reserves of lipid (Figure 33), the
degree of unsaturation of the remaining lipids increases because the
polyenoic fatty acids are utilized less, or not at all, at this time (Yakovleva,
1969; Shulman, 1974; Yuneva, 1990). The metabolic activity of the fish is
thus kept at normal levels despite the low ambient temperature. It is quite
possible that in fish dwelling in other water bodies and other climates, the
biological pivots or factors inducing metabolic change may be different from
the above, and it would be of great interest to identify both them and the
season at which they operated.

In addition to endogenous factors, the environment also influences rhythmic
processes, and a third research task would be to identify and measure the
impact of this. We share the view of Nikolsky (1974) that the chief of such
environmental factors is food supply, to be interpreted as the ratio between
food consumed and food essentially required. It is an integrated characteristic,
comprising availability of food and the rate and efficiency of food uptake.

The food supply is strongly influenced by the temperature of the water
where the fish live. For this reason, the amplitude of metabolic variations
during the annual cycle of a given fish is specific for each climatic zone,
whether tropical, temperate-warm, temperate-boreal or polar. The seasonal
variations in metabolism (e.g. lipid dynamics) display the greatest range in the

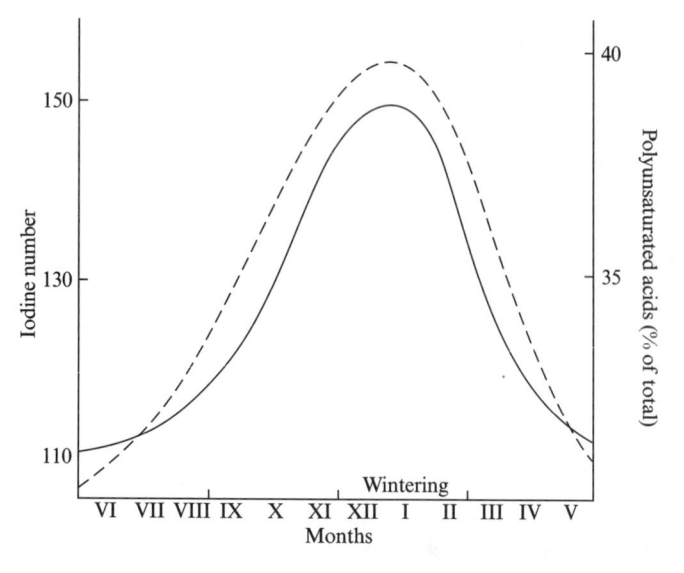

Figure 33   Seasonal dynamics of lipid unsaturation in anchovy. (After Shulman,
1978a; Yuneva, 1990.) Continuous line, iodine value; broken line, total polyenoic acids.

temperate-boreal zone – the North Pacific and North Atlantic Oceans, including the North, Norwegian and Baltic Seas. This is caused by large swings in temperature and resulting unevenness of the food supply typical of this climatic zone. Again with fatness as the criterion, the second-greatest range is found in temperate-warm zones, encompassing the Mediterranean, Black, Azov and Caspian Seas, in which the temperature regime and nutritive base are more stable. Further stability is shown in the tropical zone, where the fish need not store much lipid, the more so because their spawning season is more extended than in temperate seas. The steadiest metabolic course has been found in fish from Arctic waters, for instance in Arctic char (Nikolsky, 1974). It seems surprising, but the explanation is simple: the ambient conditions are so severe that fish have to feed continuously in order to maintain their lipid reserve at the required level throughout the annual cycle.

However, some tropical pelagic fish, such as horse-mackerel and sardine, have clearly marked periods of intensified lipid accumulation and expenditure (Shulman, 1974). They are caused by alternating dry and rainy conditions in the tropical belt of the world ocean (Lowe-McConnell, 1979). Snappers, horse-mackerel and herrings from seas off Cuba show distinct variation in their lipid contents during the annual cycle (Klaro and Lapin, 1971; Reshetnikov and Klaro, 1976; Bustomante and Shatunovsky 1981; Reshetnikov *et al.*, 1984). This variation is even more pronounced during the annual cycles of Antarctic notothenids, primarily *Notothenia marmorata* (Kozlov, 1972). There is thus a general trend, rather than a strict relationship between rhythm and geographical zone. Illumination could be the key to some of the variation between the tropics and the polar regions. The dry and rainy conditions in the tropics represent alterations between brilliant light and subdued light, while in polar regions the prolonged period of complete darkness is bound to reduce the food supply of the fish.

Fish from temperate zones have been most studied, and from these we learn that temperature may influence the character of metabolic rhythm both indirectly, through food supply, and directly, through the limited temperature ranges within which one or other vital activity can or cannot occur. It is the response or the preference that fish show to temperature that shapes the rhythmic pattern. This is well shown in two diametrically opposite fish from the Black Sea, the warm-water anchovy and the cold-water sprat (Figure 34). The curves that describe the course of triacyl-glycerol accumulation in these species mirror each other, as the fish oppose one another in their time of spawning.

While the character of metabolism is greatly influenced by temperature, its rate depends on foraging conditions. In Chapter 2, distinctions were made between the lipid contents of three races of anchovy. Here, the curves in Figure 35 show the rates of lipid accumulation in *Engraulis encrasicholus maeoticus*, *E. e. ponticus* and *E. e. mediterraneus* from the Azov, Black and Mediterranean Seas, respectively. The abundance of zooplankton in the three

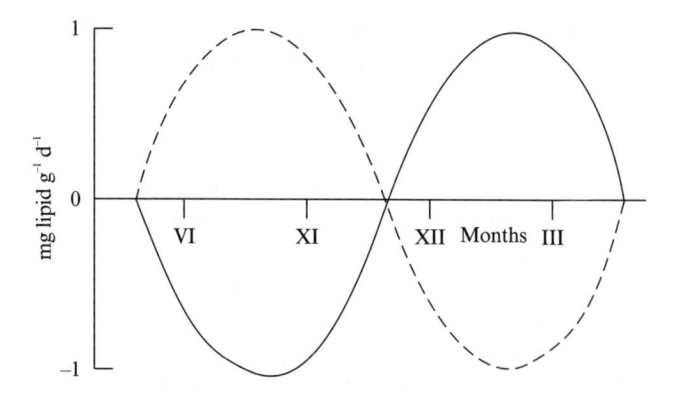

Figure 34   Lipid accumulation and withdrawal (schematic) in different months
by the anchovy and sprat. (After Shulman, 1978a.) Continuous line, sprat; broken
line, anchovy.

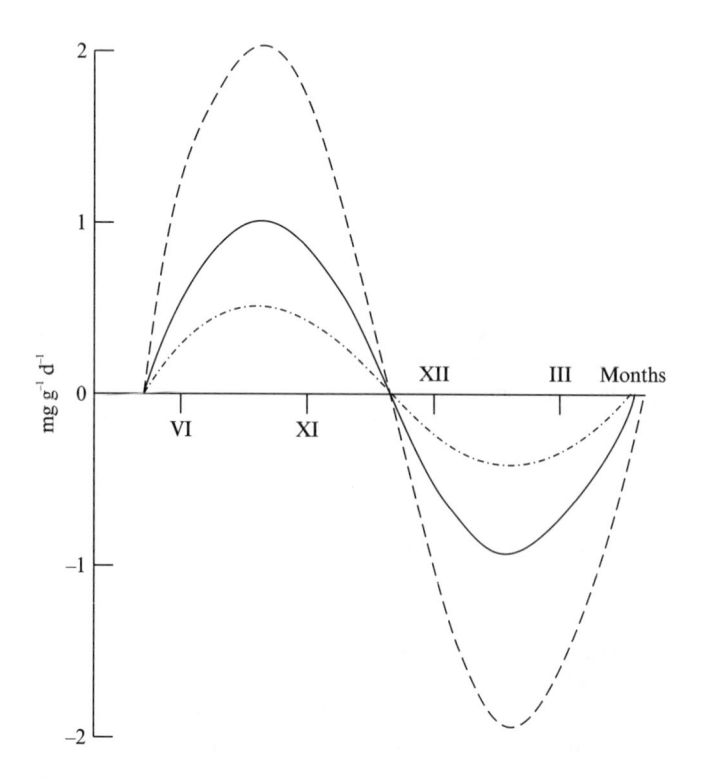

Figure 35   Lipid dynamics of three races of anchovy. (After Shulman, 1978a.)
Broken line, Azov Sea anchovy; solid line, Black Sea anchovy; dotted and broken line,
Mediterranean anchovy.

seas reflects the rate of lipid accumulation, Azov yielding the most and Mediterranean least.

Temperature affects food supply in another way also. The Mediterranean sprat, which prefers cold waters, exploits a wider feeding area than fish that inhabit only warm waters, because it takes advantage of a greater water depth and can feed all the year round. It possesses a much greater lipid reserve than the warm-water anchovy, and its range of fatness over the annual cycle is wider (Figure 36). The feeding conditions in the warm Mediterranean waters are therefore more favourable to fish that prefer cooler waters rather than warm.

The curves describing metabolic variations during the annual cycle are so diverse as to be confusing. In fact, metabolic rhythms are environment specific and can be classified by groups. Each of these environmental groups displays a specific rhythmical pattern. Based on this principle, fish could be grouped as warm- or cold-preferring, pelagic or benthic (or else benthopelagic), migratory or non-migratory, wintering or non-wintering, plankton- or benthos-eaters and predators, those spawning intermittently or those spawning all at once, short-lived/long-lived and so on. Many examples of specific metabolic curves referring to particular environmental groups

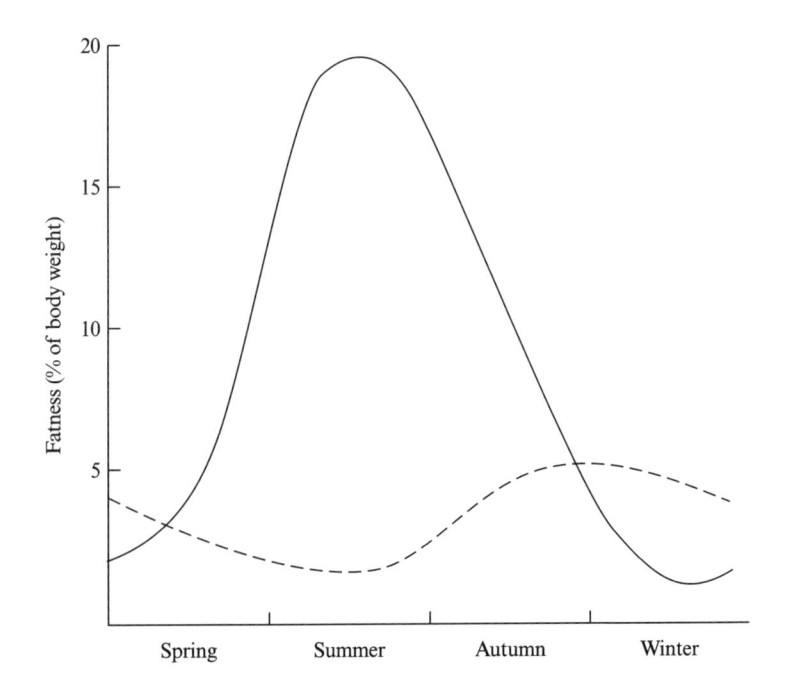

Figure 36   Annual lipid cycle in anchovy (broken line) and sprat (continuous line) in the Mediterranean. (After Shulman, 1978a.)

have been published. Adding to these, Figure 37 indicates that very active fish have higher metabolic rates and a greater range of metabolic variations during the annual cycles when compared with more sluggish fish.

Analyses of rhythmic behaviour give indications of their quantitative significance and make comparisons possible. When taken with other metabolic elements, study should include harmonic, parabolic and exponential functions, which must be examined separately in order to assess the actual and specific rates of the various processes.

Harmonic analysis is sufficient for dealing with simple cases. The dynamics of lipid stored in the body of a fish over the annual period would then be described by:

$$F = a + b \sin\left(\frac{\Pi}{6}T\right) + c \cos\left(\frac{\Pi}{6}T\right) \tag{5}$$

where F is the percentage fat (lipid) content of the fish, T is time in months and a, b and c are coefficients.

A more complex aproach is required in analysing the protein or weight increase. This would be represented graphically by a curve resulting from two processes: progressive, the quantitative increment of protein content arising from ontogenesis; and oscillating, which depends on the expenditure of reserve and structural proteins on wintering. Only that part of the curve which features the variation within the year is considered when studying a single annual cycle.

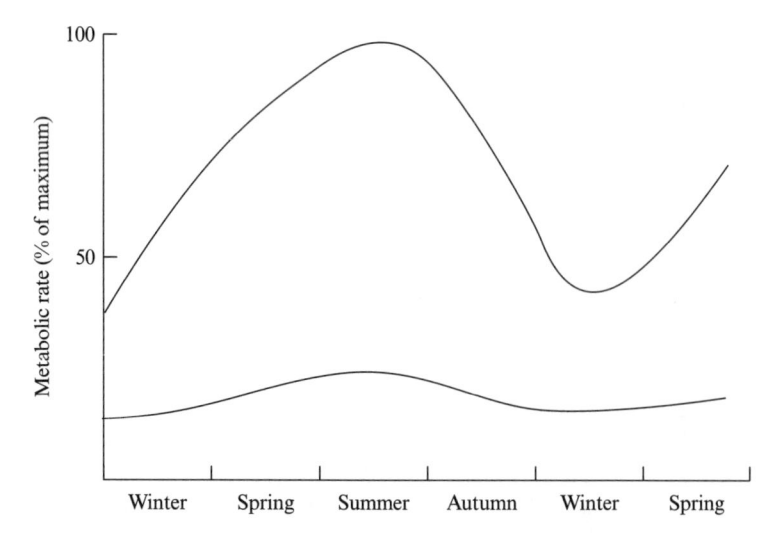

Figure 37   Annual cycle of averaged metabolic rates of active (upper curve) and sluggish (lower curve) fish. (After Shulman, 1978a.)

The complex curve (P_e) may be differentiated into two constituent parts (P_1 and P_2). One of them characterizes the average trend of protein increase:

$$P_1 = a_1 T^n \tag{6}$$

where P is the absolute amount of protein (in grams), T is time in months and a and n are coefficients. The other describes the amplitude of a particular rhythmic process (Figure 38):

$$P_2 = a_2 + b_2 \sin\left(\frac{\Pi}{6} T\right) + c_2 \cos\left(\frac{\Pi}{6} T\right) \tag{7}$$

A third constituent (P_3) can be distinguished, representing the key harmonic trend. These calculations describe the variations in the wintering conditions which influence the scope of protein expenditure. A similar approach has been employed by Smetanin (1978). The method for analysing weight increase curves was originally designed by Kokoz (Shulman and Kokoz, 1968).

The two processes proceed together as the seasonal and age-induced changes in protein add up. The ratio between oscillating and progressive changes in protein is markedly higher in warm-water fish than in cold-water fish, so that the latter expend less protein in the winter compared with the former.

Thus far we have examined the rhythmic processes as though they were proceeding independently. Now, as the sixth stage, we should look upon them as interacting elements. The protein, phospholipid and ATP in tissues and the γ-globulin in blood serum have the most stability. Seasonal variability, although present (Kulikova, 1967; Shulman, 1974; Sidorov, 1983; Malyarevskaya et al., 1985; Bilyk, 1991), is not significant, which shows the importance of these substances to the organism. The proteins and phospholipids make up the structural

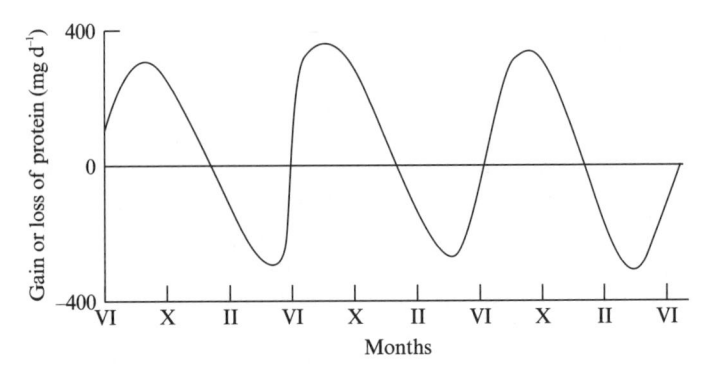

Figure 38 Deposition and removal of protein from anchovy over a 3-year period. (After Shulman, 1972a.)

base of tissues and membranes, ATP is constantly being resynthesized and broken down and the γ-globulin provides immune defence.

The second group comprises triacyl-glycerols in muscle, liver and other tissues, and the third the albumins and α- and β-globulins in serum. Both groups exhibit a pronounced, monocyclic variation. The triacyl-glycerols are the strategic energy reserve used especially for ATP resynthesis. The serum proteins apart from γ-globulin are also a reserve, but are used as sources of plastic materials for the synthesis of reproductive tissue.

The fourth group consists of glycogen in muscle and liver, and creatine phosphate in muscle. Their variation has a smaller amplitude than that of the previous two groups but is multicyclic or occasionally bicyclic, these variations being inversely related to changes of lactate in muscle and liver, and to changes of lactate and glucose in the blood. In strong swimmers, glycogen is inferior to triacyl-glycerols as a source of energy for long-term swimming (see Chapter 3). Similarly, creatine phosphate is a secondary energy source compared with ATP. Glycogen and creatine phosphate function at the start of swimming. Their multicyclic behaviour may shed light on their tactical importance compared with the strategic importance of triacyl-glycerols and ATP.

These degrees of variability represent a hierarchy of homeostases in the organism. Variation in the glycogen level smoothes out the accumulation and mobilization of triacyl-glycerols. In its turn, the wide range of variations in the latter ensures stable levels of proteins and phospholipids in the body. Glycogen and creatine phosphate can be compared to the small change in one's pocket, rapidly changing and readily spent. Triacyl-glycerols are more like bank notes, which are spent more carefully and gradually decrease between pay-days. Finally, proteins and phospholipids are like gold reserves, used only in emergency. ATP levels also appear to be stable for much of the time, but this is illusory because it is constantly being destroyed and resynthesized. Proteins and phospholipids turn over continuously, but at a much slower rate: it is the apparent invariability of these last three compounds which distinguishes them from the others (Shulman, 1978a, b).

In some species, like the cold-water fish of the Black Sea, protein and lipid accumulation may occur together, while in fish which prefer warmer water in the same sea, such as anchovy and horse-mackerel, they do not. Both the degree and duration of such concordance are important in the adaptation of fish.

It is necessary to understand the mechanisms governing the integration of individual metabolic rhythms into the biological cycles of a population. The seventh stage of investigation is therefore to extrapolate data gained at subcellular, cellular, tissue and organismic ('whole animal') levels to the level of a population. The task is difficult. To take one example, the lipid reserves of migratory fish must reach a certain level before they can switch to the migratory state, but only a part of the population develops the critical level. However, the altered pre-migratory behaviour of this part governs the behaviour of the whole population. This has been observed in the Azov anchovy (Shulman, 1974) and will be discussed further in the following section. Dolnik (1965) reached the same conclusion when studying the migration of birds.

As stated earlier, the annual cycles of species and populations divide into periods. The overall description would be the 'syndrome' of the population, while individual metabolic characteristics would be 'symptoms', borrowing terms from the medical profession. Each period manifests a specific set of metabolic features which support the normal course of vital processes in the population, and the population would enter the next period of the cycle only if the preceding step had been completed. If it had not, the situation would lead to that mass mortality which is so common in nature. The change in metabolism is triggered with the help of the endocrine system, as will be described in the next section.

The following summarizes the annual cycle as it appears in adult warm-water migratory fish of the Black Sea.

Pre-spawning period (feeding). This coincides with pre-spawning migration, and is characterized by intensive protein synthesis coupled with differentiation and growth of reproductive tissue. Energy supporting the generative synthesis is supplied by mobilization of reserve lipids, primarily triacyl-glycerols.

The *spawning period* is characterized by a high rate and high efficiency of anabolism and catabolism. Oxygen uptake reaches its maximum, both at tissue and organismic level. Fish develop higher ATPase and other enzymic activities, and the haemoglobin and red cell contents of the blood also increase. A switch from mobilization to accumulation of energy reserve (mostly lipid) then occurs, and the accumulation of lipids coincides with an increase in the total protein of the fish.

Post-spawning (pre-wintering feeding). This period is marked by intensive lipid accumulation that will allow normal living of the population later, when food consumption has ceased or been much curtailed. Fish accumulate substantial reserves of triacyl-glycerols, and the content of creatine phosphate in the muscle and glycogen in the muscle and liver increase. A similar increase is found in the content of serum proteins, albumin in particular, which provides for future gonad development. The increase in protein continues, but is less than the accumulation of lipids.

Pre-wintering migration. Metabolism is adjusted to a stable state; lipid accumulation and increase in protein are at a standstill; the activity of alkaline phosphatase in the scales, which regulates the development of sclerites, decreases markedly; the concentration of protein in the blood serum remains stable. During the wintering period, the metabolic activity of the population declines considerably. Lipid reserves decline steadily and tend to exhibit an increased degree of unsaturation in these warm-water fish because of the lower temperatures. A further factor leading to greater overall unsaturation is the preferential utilization of triacyl-glycerols, which are less unsaturated than phospholipids. The content of glycogen and creatine phosphatase decreases, and eventually a large part of the protein reserves is used to supply both plastic and energy metabolism.

This summary of the annual cycle was deduced from data on the Azov anchovy (Shulman, 1960b) and other fish of the Black Sea (Shulman, 1972a). The same principles underlie the periods of the annual cycle of fish from the northern seas described by Shatunovsky (1980), the Okhotsk and Japan Seas by Shvydky and Vdovin (1991), and Caspian Sea by Rychagova (1989). The corresponding cycle in mussels of the Black Sea has been described by Goromosova and Shapiro (1984) and Shcherban and Abolmasova (1991). The scheme is still incomplete and needs further development. Fish from other water bodies exhibit metabolic patterns which may be widely different from those described and will also need to be characterized. A case in point is the resource of protein in the musculature of North Sea cod, which continues to be depleted for a few weeks after the completion of spawning in the first week in March (unpublished observations, RML), presumably because food is still scarce during such a cold period. Atlantic salmon show a sex difference: males stop feeding some 8 weeks before spawning and several weeks after, while females continue to feed throughout the spawning period. Consequently, at least among farmed salmon, the plumpest fish are nearly all females (I. MacFarlane, personal communication).The ninth and last stage of the study of metabolic rhythms is identification and characterization of the 'degree of well-being' of organisms and populations of different species of fish during the course of their annual cycles. This stage is discussed in Chapter 6.

4.3. DAILY (CIRCADIAN) RHYTHMS

Rhythms of the annual cycle depend to a considerable degree upon climatic factors, which in turn are largely governed by the amount of solar radiation. Light intensity and day length influence metabolic rhythms indirectly through ambient temperature and food supply, and directly, as discussed in section 4.5.

The photoperiod affects the daily metabolic rhythm by setting and adjusting the 'biological clock'. Fish and other aquatic animals have been studied less in this connection than have terrestrial animals, probably because of the technical difficulties of observation in the wild. Many systematic observations must be made to establish the character of daily rhythms. Small numbers of irregular samplings are insufficient: 12 evenly spaced samples per day would not be excessive. Such sampling must be repeated over many days, and this explains why daily metabolic rhythms have been less studied or understood in comparison with seasonal rhythms. Variations in oxygen uptake within a day were, however, reported more than 40 years ago (Vinberg, 1956).

A number of diurnal effects on metabolism are of considerable magnitude, as during vertical migrations for food during the day or the need for rest at night. Black Sea anchovy prefer to spawn at night time and at certain hours only (Safyanova and Demidov, 1955; Oven, 1976). Cod from the Faroe Bank stay on the sea-bottom at night, so can be caught by bottom trawling only then (unpublished observation, RML). Daily changes in metabolism are reversible and usually return to the same value at the same time the following day, in contrast to characteristics which change more fundamentally and irreversibly between one season and the next.

Unlike diurnal changes in metabolism, changes in locomotion of fish have been widely investigated (reviewed by Woodhead and Woodhead, 1965a,b; Pavlov, 1970, 1979; Poddubny, 1971; Zusser, 1971; Radakov, 1972; Kelso, 1973; Vyskrebentsev, 1975; Karmanova et al., 1976; Schwassmann, 1979; Manteifel, 1980, 1984, 1987; Mochek, 1987; Quinn, 1988). Locomotor activity correlates with light intensity, vertical migrations, nutritional preferences, spawning behaviour and avoidence of predators.

The daily dynamics of locomotor activity in Black Sea fish under natural conditions have been described by Belokopytin (1990, 1993), as shown in Figure 39. The daytime predators horse-mackerel and pickerel, as well as the planktonivorous anchovy and Black Sea silverside, are most active in daylight and rest at night. In contrast, the scorpion fish, a night-time predator, and whiting, are most active in the dark. The normal courses of daytime activity in predatory and planktonivorous fish were found to change under laboratory

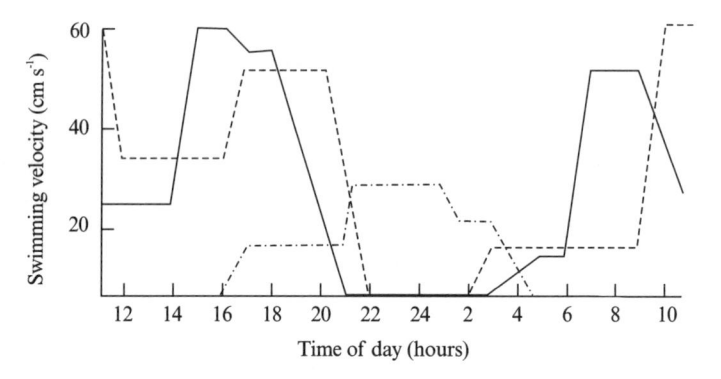

Figure 39   Average swimming velocity of fish at different times of the day. (After Belokopytin, 1993.) Solid line, horse-mackerel; broken line, pickerel; dotted and broken line, scorpion fish.

conditions, since the fish continued to swim at night. Belokopytin (1993) suggested that the fish were making up for exercise deficiency caused by the conditions of captivity.

As pointed out earlier, locomotor activity is linked to oxygen uptake, and so to the level of energy metabolism. Belokopytin (1978) related swimming velocity (V) to the rate of energy metabolism (Q) (see also Chapter 2, section 2.2) by the formula $Q = qb^v$. From this, the data shown in Figure 39 describe quantitatively the daily dynamics of total energy metabolism (consumption of oxygen) in a fish. At night when the fish are at rest, it is necessary for the consumption of oxygen to be adequate for basal metabolism (Belokopytin 1968; Table 3, Chapter 3, section 3.2).

A remarkable fact, which has not yet received much coverage in the literature, is that fish of high innate activity make a switch each day from high-energy metabolism to basal metabolism, in order to maintain the energy balance and save resources. The resting condition may last for as long as a quarter of a day. Such fish are those which eat plankton and those which are daytime predators. That herrings become torpid in the dark has been demonstrated by observation from a submarine (Radakov and Solovyev, 1959). This 'primary sleep' is inherent in a wide variety of fish (Marshall, 1972; Karmanova et al., 1976) and differs significantly from 'real' sleep in higher animals. However, both lower and higher animals need rest to reduce energy expense and to restore the energy potential.

According to Karmanova et al. (1976), fish have distinct daily rhythms of opercular (respiratory) movement and of cardiac contractions that have been revealed by electrocardiograms and electromyograms. These studies were conducted on several Black Sea fish: spiny dogfish, sting ray, grey mullet, shore rockling, scorpion fish and ombre (Sciaena umbra). It is usually assumed that the daily variations of total energy metabolism relate to changes in the supporting metabolic processes occurring at lower levels (shown in detail in Chapter 3, section 3.2), for which there is evidence from birds and mammals (Dolnik, 1965; Slonim, 1971; Shilov, 1985). Since the daily rhythms are associated with food consumption as well as energy aspects, it is reasonable to suppose that regular changes must also take place in the transformation of proteins, lipids, carbohydrates and mineral substances. Some indication of this can be found in the work of Ugolev and Kuzmina (1993), who studied the daily dynamics of digestive and other enzymes of fish. It seems significant that the peak of physical activity of daytime predators, such as pike-perch and perch, occurs in the morning and twilight, when they feed most energetically.

In the present context, the data referring to daily variations in the ratio between energy and plastic metabolism are of special interest. It has been found that the processes of energy catabolism predominate in the daytime, while duplication and replication of DNA and protein synthesis dominate the dark hours (Shilov, 1985; Kondrashova and Mayersky, 1978). One can assess the highly efficient protein metabolism from the excretion of nitrogen. According to Revina (1964), the peak of nitrogen excretion in Black Sea horse-mackerel takes place between 18.00 and 0.00h, after which it declines to a minimum value at 06.00h. At other times it keeps to an intermediate level. Arkhipchuk and Makarova (1992) have shown that the nucleolar activity associated in carp with protein biosynthesis is greater in the morning than the evening, but it is not yet known whether such activity is in fact linked to the alternation between light and darkness. Another factor which regulates the 'biological clock' is the ratio between the catabolism and anabolism of lipids and carbohydrates (Selkov, 1978). An important controller of the daily rhythms of metabolic processes is the neuroendocrine system hypothalamus, epiphysis and hypophysis and the hormones secreted under their control, including among others somatotrophin and prolactin (Lee and Meier, 1967; Polenov, 1968, 1983; Leatherland et al., 1974; Semenkova, 1984). In the brains of fish there is a clear rhythm of changes in the amount of serotonin and noradrenaline present (Nechaev, 1989). The maximum levels of somatotropin and prolactin are found in kokanee salmon in the second half of the dark period, and maxima of free fatty acids occur in the morning (Leatherland et al., 1974). Circadian variations in carbohydrate parameters were observed by Narasimhan and Sundararaj (1971) in featherback and giant gourami.

Data also exist concerning the daily rhythms of energy and plastic metabolism and their ratios in aquatic invertebrates. Calanus, a copepod that makes daily vertical migrations in the Black (Petipa, 1981) and Barents (Sushkina, 1962) Seas uses up stored lipids (triacyl-glycerols) in a most intensive way in the twilight, so as to be able to ascend to the surface of the sea. Feeding on phytoplankton is a nocturnal activity that replenishes the lipid resources. In the morning, when the copepod descends into deep water, lipids are again consumed, but not as extensively as during ascent. During the daytime, Calanus rest, and slowly but steadily consume lipids. There may be a four-step cycle in some nocturnal predatory and planktivorous fish that ascend rapidly to the sea surface for their prey. This type of vertical migration made by copepods and fish has been thoroughly explored by Manteifel (1984, 1987).

The ratio of oxygen uptake to ammonia excretion in Black Sea mussels has been found to exhibit a distinct daily rhythm (Slatina, 1986). As a rule, the ammonia coefficient (O/N) tends to increase greatly at night. As this rise in energy metabolism does not result from an enhanced locomotor activity, it is difficult to explain or to find any analogue in fish. Farbridge and Leatherland (1987) demonstrated a strong effect of the lunar cycles on amino acid uptake by the scales, also on nucleic acids, metabolic reserves and plasma thyroid hormones in coho salmon.

4.4. INTERANNUAL FLUCTUATIONS

Variations in the biochemistry and physiology of fish from one year to another must be accepted as real. Such a cycle would be linked to long-term changes in the climate resulting from solar activity (Chizhevsky, 1976). The trouble is that observations are insufficiently representative to yield clear patterns. As with diurnal variation, much of the published work relates to terrestrial organisms rather than fish, and much of the study has centred on fluctuations in the abundance of species which tend to develop 'outbreaks' – sudden marked increases in numbers. However, fish such as salmon, cod, herring, sardines and other species have also been shown to exhibit long-term fluctuations in their numbers. Klyashtorin (1996) has found a close correlation between the velocity of rotation of the earth, which affects the intensity of circulation of the water in the oceans, and the abundance of stocks of many species of fish.

Since methodical studies on the metabolism of fish are relatively recent, the data are rarely sufficient to establish complete interannual cycles, so we are limited here to describing independent interannual fluctuations of various kinds.

4.4.1. Lipids

Comparative surveys covering only 3–4 years do not yield firm conclusions, although Shatunovsky (1980) and Siderov (1983) have analysed data from such short-term studies.There are more extensive studies by Lasker (1962, 1970) on the Pacific sardine off California over the period 1932–1956, and comparable long-term work by Smith and Eppley (1982) on the anchovy from the same region, *Engraulis ringens*.

Long-term fluctuations in lipid accumulation by Black Sea sprat during the summer post-feeding period, when their fat reserve is at its peak, were observed from 1960 to 1994 and this work continues (Shulman *et al.*, 1994). Interannual variability in the lipid content, though small, is real, with a coefficient of variation of 12–15%, as estimated at certain points of this period of time.

Observations on the fatness of Black Sea anchovy in the autumn post-feeding period were made during the period 1956–1973 (Figure 40) (Danilevsky, 1964; Dobrovalov, 1970; Danilevsky *et al.*, 1979; Shulman and Dobrovalov, 1979). The year-to-year variations in fatness are much more pronounced than in sprat (coefficient of variation > 20%). This is probably because anchovy need warm water and sprat cold, so changes in the temperature regime of the Black Sea elicit more stress in the anchovy. For much of the time, sprats keep drifting in waters lying over the thermocline, where the temperature is as low as 7–13°C (Gusar and Getmantsev, 1985), while

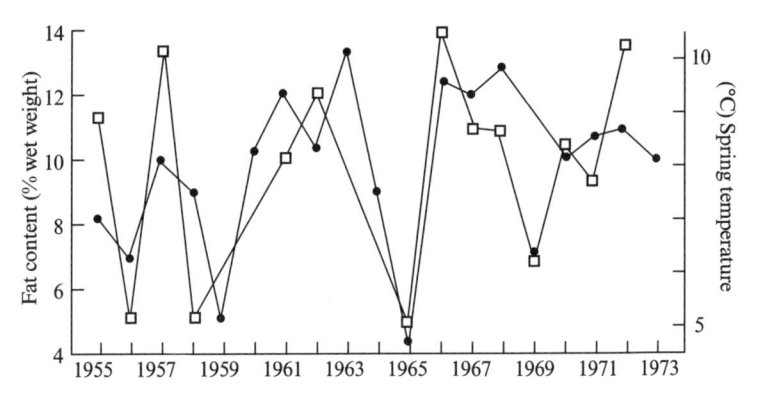

Figure 40   The lipid content of Black Sea anchovy following the feeding period, in relation to the average temperature of the water in spring. Temperature data from published sources. ●, lipid content; □, temperature. (After Shulman and Dobrovalov, 1979.)

anchovy are exposed to temperatures which change dramatically and which may differ by several degrees from year to year in the upper water layers. Variations in the temperature of sub-surface waters greatly affect the growth of plankton, which, in the long term, determines the rate and degree of lipid accumulation in the anchovy; note the similarity between the curves for temperature and lipid accumulation in Figure 40.

Studies on the fatness of Azov Sea anchovy made between 1953 and 1974 have yielded further information (Shulman, 1960b; Danilevsky, 1964; Taranenko, 1964; Shulman, 1972a; Dubrovin *et al.*, 1973; Luts and Rogov, 1978). Between 1953 and 1966, the curves describing the fatness of Azov and Black Sea anchovy stocks were very similar (Shulman, 1974). The Azov and Black Seas are located in the same climatic zone, and their hydrological and hydrochemical conditions vary in a similar way, inducing similar fluctuations in the nutritive base and so the lipid contents of these stocks. After 1966 it is difficult to compare the data because of differences in the sampling and the manner of treating the samples.

4.4.2. Carbohydrates

The lipid resources of a fish are more stable than are the carbohydrates, which in muscle can be reduced by half within 15 s of intense activity (Love, 1980). Since over the longer term the carbohydrate levels are maintained by gluconeogenesis from lipids and proteins, it might seem that no meaningful information can be obtained from carbohydrate analyses, especially since, in the very act of sampling, the muscle glycogen level will fall to minute values. In actual fact, the 'resting' level of muscle glycogen can very easily be assessed, and further knowledge can be gained on the energy dynamics of fish, as studied in Atlantic cod.

The measurement of glycogen in the liver presents no problems, since there can be as much as 200 mg of it in the liver of a fish weighing 1 kg (Black and Love, 1986). Measurement in the muscle depends on the fact that, during muscular effort, the glycogen is broken down to lactic acid, which lowers the pH (reviewed by Burt, 1969). Some of the glycogen remaining in the muscle after the death of the fish is also converted to lactic acid via the Embden–Meyerhof–Parnas pathway, while some is depolymerized to glucose by the action of amylase (Burt, 1969). The glucose does not, of course, lower the pH, but, if there is any change in the proportion of glucose to lactic acid, it must be very small, because the concentration of free glucose 24 h after death correlates positively with the initial glycogen concentration (Black, 1983). In the exercised muscle of living mammals, the lactic acid formed is rapidly released into the blood, but in fish it is virtually all retained (Dando, 1969). Wardle (1978) showed that it is *stressful* exercise that causes the fish muscle to retain its lactic acid, probably under the influence of the catecholamine hormones.

The lactic acid produced from glycogen by violent exercise, such as that induced by capture and sampling, therefore, supplements that produced from glycogen after death, with the end result that the pH, 24 h after death, should be the same whether the fish struggled at death or not. Such has been observed by Love and Muslemuddin (1972), who found that the post-mortem pH of cod killed almost instantly was the same as that of those previously exercised stressfully. Finally, a relationship between the pH of the white muscle 24 h after death and its initial glycogen content was established by Black and Love (1988), the 'initial' value being that obtained from rested fish caught in a hand-net, and stunned with a single blow on the head. The sample was rapidly excised and dropped into liquid nitrogen. Doubtless a real initial value, if it could be obtained, would show less scatter between samples and exhibit a correlation coefficient greater than the –0.67 shown in Figure 41.

There is a good correlation between the concentration of glycogen in the white muscle and that in the liver (Figure 42). Hence the pH of cod muscle 24 h after death (when the residual glycogen has reached its minimum value) gives a good working guide to the total carbohydrate resources of the fish. As a depleted fish begins to build up stocks of energy, the liver, virtually devoid of lipid, increases its store of glycogen before the lipids start to increase (Figure 43). The results appear to show that the priority in a fish severely starved is to build up some carbohydrate reserves to enable it to swim and capture prey, rather than lipid reserves which are not so quickly available.

The carbohydrate resources of cod, during experimental refeeding after severe starvation, increase to values many times higher than in cod fed continuously throughout the experiment; the 'overshoot phenomenon'

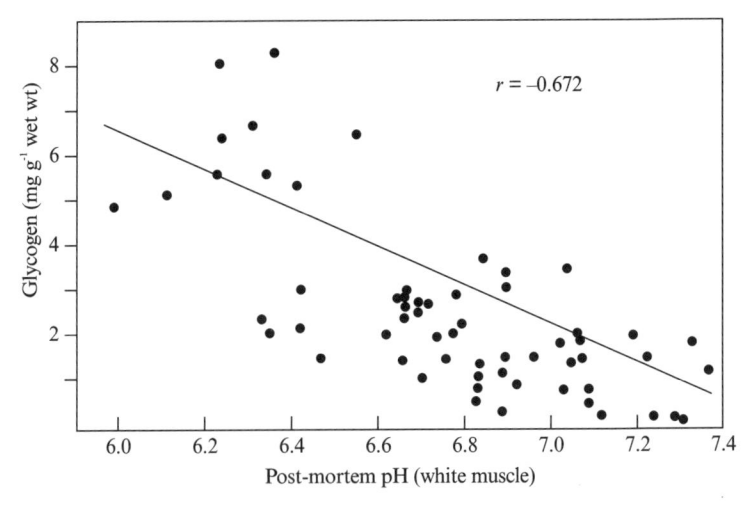

Figure 41 The relationship between the initial glycogen concentration in the white muscle of rapidly killed, rested cod and the pH of the muscle 24 h after death. (After Black and Love, 1988.)

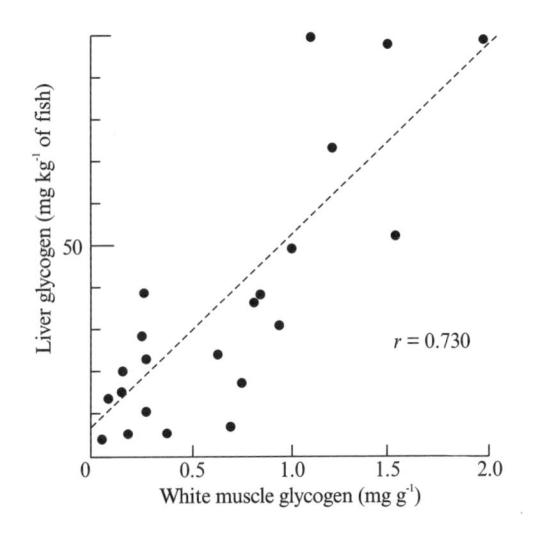

Figure 42 The glycogens of the liver and white muscle of Atlantic cod. (After Black and Love, 1986.)

(Love, 1979; Black and Love, 1986). Values have been found to reach a maximum about 100 days after the start of refeeding, and then to drop quite sharply to normal, even though feeding continued. This spontaneous decrease within the red and white muscle and the liver is quite different from the behaviour of lipid, which, as we have seen, continues to rise as

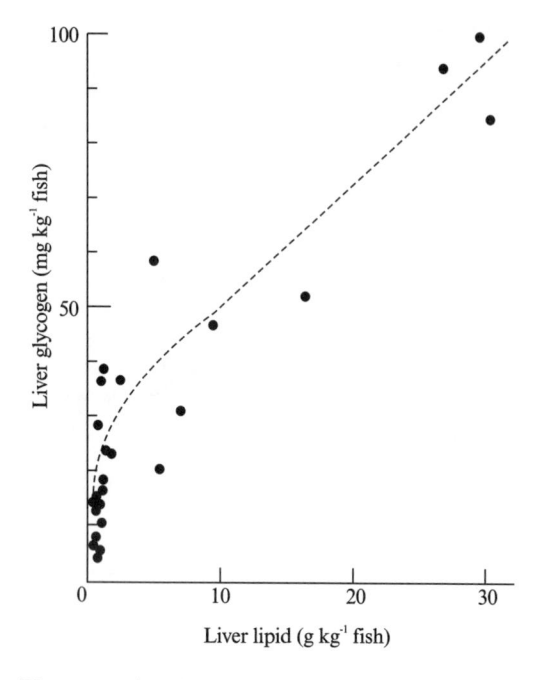

Figure 43   The proportional restoration of energy resources in the liver of starved Atlantic cod during refeeding. (After Black and Love, 1986.)

long as the feeding remains plentiful. Maybe the high glycogen levels cancel the mechanisms of gluconeogenesis. Whatever the purpose of the phenomenon, it is potentially relevant to the study of interannual fluctuations. It has also been well documented in cod from the ocean (Love, 1979) and in cultured salmon (Lavéty *et al.*, 1988). The post-mortem pH of wild cod is 6.6 or more for most of the year, but around June there is a fall, corresponding with the high values of glycogen (Figure 44). When the pH was measured 24 h after death in batches of cod caught at regular intervals from 1969 to 1977 (apart from 1970) at approximately the same locality, it was found that the dates and spread of low values (i.e. of maximum glycogen) varied considerably from year to year. Variation is to be expected between individuals of any batch of cod that are free to swim long distances, even though they are taken from the same fishing ground over the years. Despite this, the findings illustrated pose an interesting question.

High values of glycogen were observed in more fish caught from April to July 1971 than in other years. Were such fish more depleted the previous winter? Conversely, the high glycogen values were almost non-existent in fish caught in 1974. Was the food so plentiful the previous winter that endogenous feeding was unnecessary?

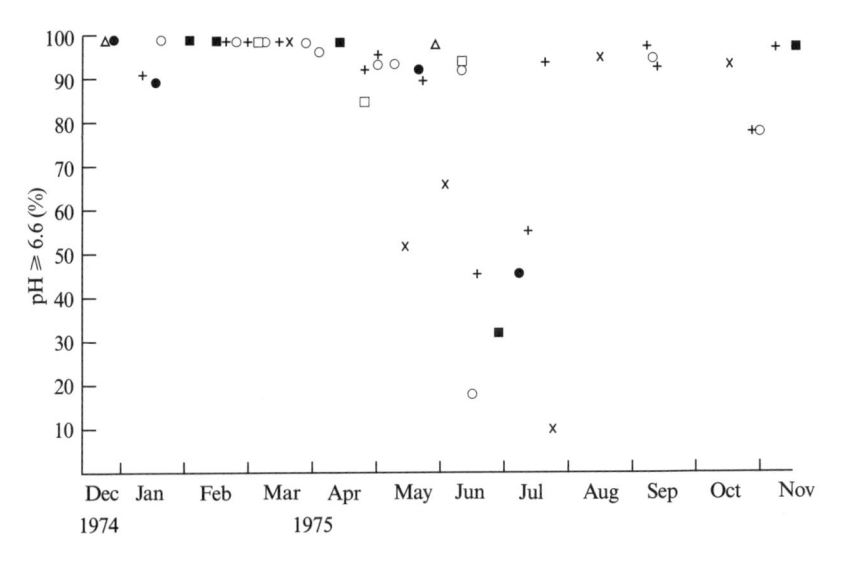

Figure 44   The percentage of batches of 20 cod which exhibited post-mortem pH values of 6.6 or over during a year. The fish were obtained from several grounds, and the results show that the phenomenon of low pH in summer is widespread. (After Love, 1979.) Grounds: ●, Shetland; ○, Faroe; +, North Sea; ×, West Scotland; △, Iceland; ■, North Norway; □, South Norway.

The glycogen phenomenon probably gives no information about the quality of summer feeding, since the high level of glycogen is not maintained throughout the feeding period, but it might well give useful information about the ecological situation in earlier months.

4.5.   DYNAMICS OF ABUNDANCE, BEHAVIOUR AND DISTRIBUTION PATTERN

Metabolic rhythms underlie population dynamics, behaviour and distribution, despite their complicated nature. As a corollary, the influences of abundance and complex behaviour patterns on metabolism are equally important. The relationship between metabolism and population dynamics appears to be simple. Nikolsky (1974) regarded recruitment, adult stock, total mortality and food supply as interdependent parts of the whole.

Recruitment of a species depends on the abundance of the brood. Abundance, in its turn, is influenced by the fertility of the population and by the quality of the reproductive products. Such quality comprises specific features of the genome and the content of plastic (proteins, phospholipids and cholesterol) and energy (triacyl-glycerols, wax esters and glycogen) substrates. The higher the contents of these substances in oocytes and spermatozoa, the better the fertility and development of the eggs and larvae (Kim, 1974; Shatunovsky, 1980; Eldridge *et al.*, 1981; Konovalov, 1984, 1989; Gosh, 1985; Zhukinsky, 1986; Zhukinsky and Gosh, 1988). In addition, any increased content of polar lipids engenders better metabolic activity of cell membranes, the increased content of proteins intensifies biosynthesis, and the increased content of reserve substrates furnishes the biosynthesis with energy. Beside all this, a greater reserve of energy is beneficial to the eggs and larvae, permitting survival in less favourable environments, such as temperatures outside the optimum range and hazardous concentrations of pollutants, shortage of food and the presence of predators. As pointed out earlier, food deficiency may dramatically affect fish larvae during their transition to the exogenous mode of feeding, inducing mass mortality. Energy stored in greater quantities than the norm would then save many lives.

Polyenoic fatty acids play an important part in the proper development of eggs and larvae (Sargent and Henderson, 1980; Walton and Cowey, 1982; Yuneva *et al.*, 1990). An increased content of docosohexaenoic acid may influence the development of sense organs and higher nervous activity of the larvae which, coupled with an increased content of neutral lipids, would promote better escape from predators. There is a close relationship between the levels of plastic and energy components contained in reproductive products and those in the bodies of the brood fish, primarily females (Shatunovsky, 1963, 1980; Krivobok, 1964; Shulman, 1974; Holdway and Beamish, 1985; Shatunovsky and Rychagova, 1990). The proteins, lipids and glycogen stored in the muscle and other tissues of the fully grown fish are transformed in the liver and then transferred to the reproductive products.

We can therefore sketch the following series: quality of brood fish – quality of reproductive products – quality of eggs and larvae – size of recruitment. The scheme is represented in Figure 45, where the content of docosohexaenoic acid present in the white muscle of female pink salmon is directly proportional to the survival rate of their eggs and larvae. The quality of brood fish has both a direct – through the quality of the reproductive products – and an indirect influence on the quantity of recruitment. 'Quality' implies not only increased content of biologically significant substances in the body, but concomitant accelerated protein synthesis to provide higher growth rate. The resulting increased body weight is directly related to rising fecundity, which again contributes to recruitment.

An increased stock of recruits, resulting from the effects of a better food supply on the parent fish, will deplete the food supply available to the population. This poorer food supply will reverse the entire process, reducing energy stores, growth rate of the fish, quality and quantity of reproductive products and, finally, the total numbers of the population. The reduced numbers of fish, assuming that the food supply remains more or less stable, are then able to eat

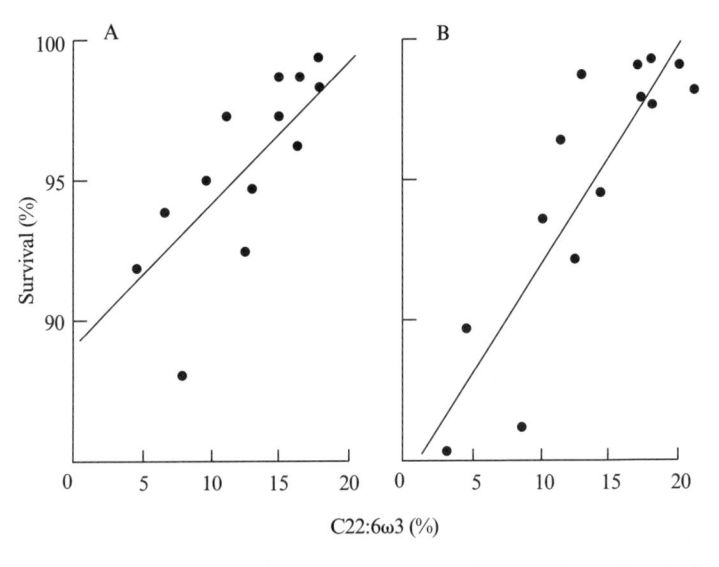

Figure 45   Survival of eggs (A) and larvae (B) of pink salmon in relation to the concentration of docosohexaenoic acid in the triacyl-glycerols of white muscle of female brood stock. (Yuneva *et al.*, 1990.)

more and the pendulum swings back. The whole cycle settles into a fluctuating routine (Nikolsky, 1974). This is an over-simplification: one can hardly assume that the nutritive base (available food) would be more stable than the number of consumers feeding on it. Besides, internal and external factors may affect the survival rates of eggs and larvae so strongly that they will die no matter how perfect their 'quality' is. For instance, the survival rate of eggs and larvae of pelagic fish in the Black Sea depends on temperature conditions at spawning time (Dekhnik, 1979). In contrast, biosynthesis proceeding in the brood fish is governed by the substrates transported from organs and tissues which were enriched through intensive feeding in the pre-spawning period.

Nevertheless, the principle of population self-regulation that entails interrelation between the environment and the metabolism can be admitted as a working hypothesis. It is applicable also to other organisms which evince significant fluctuations in their numbers and occasional population explosions, as in insects, rodents and others (Elton, 1958; Shilov, 1985). Hormonal (catecholamine) stress can also limit numbers in rats (Christian, 1950; Slonim, 1971; Shilov, 1977) and the post-spawning mortality of Pacific salmon could be a combination of abnormal hormonal balance and total exhaustion (Black *et al.*, 1961; Idler and Truscott, 1972; Ardashev *et al.*, 1975; Ando, 1986). Maksimovich (1988) stressed that overabundance of salmon in the spawning grounds ruins their breeding capacity (Chebanov *et al.*, 1983). Moreover, abnormally high losses of juvenile fish of the salmon family are also linked to the raised content of catecholamines in their bodies which might be the result of excessive fish density (Podlesnykh and Ardashev, 1990). Robertson *et al.* (1963) demonstrated that the excessive cortisol was the lethal agent, and that its presence in high concentrations in the dying fish was not just coincidence. Pellets of cortical hormone were implanted in immature rainbow trout so that cortisol was released continuously into the blood stream. The fish lost weight, developed infections of the skin and died within 5–9 weeks. Death occurred much more rapidly at higher water temperatures.

Food supply evidently underlies all regulatory processes: the endocrine mechanism that is engaged when the population becomes excessive occurs simultaneously with a dramatic reduction in food supply. Numbers in the population may also be influenced by wintering, when the consumption of food either drops markedly or ceases altogether. The low temperature itself may be important as well as the shortage of food, since whatever food is available is digested more slowly (Pearse and Achtenberg, 1917). At such a time, the fish have to resort to endogenous feeding, mobilizing reserves and breaking down actual muscle cells to supply energy. Many species initiate gametogenesis at the same time, so that by the end of winter many fish have become dangerously depleted. Black and Love (1986) showed that in Atlantic cod, and presumably many other species, the different substrates are used up in a definite order. Lipids from the liver and glycogen from liver and white muscle start to be mobilized at the outset of starvation, liver lipids being reduced to minimal values first. Before any of these constituents have reached their lowest limits, the proteins of white muscle and dark muscle and the glycogen of red muscle begin to be used, muscle glycogen reaching limiting values before the other two substrates, which continue to be used until the fish dies.

Different species adapt in different ways. The glycogen levels in Atlantic cod drop to very low values, and, far from creating more of it from protein or lipid precursors, the gluconeogenic capacity of cod actually diminishes during severe starvation. In the liver, the enzymes phosphoenol pyruvate carboxykinase, fructose diphosphatase and alanine aminotransferase were shown by Love and Black (1990) to decrease significantly after 22 weeks without food. The species adapts to the low glycogen levels by virtually ceasing to swim. In contrast, rainbow trout, a more active fish, does exhibit gluconeogenesis and increases these enzymes during starvation for 8 weeks (same authors). Carp retain considerable quantities of glycogen during long periods of starvation (Love, 1970), and starving eels retain lipids (J.A. Lovern, personal communication). There are also differences in the withdrawal of substrates from different organs, brain and heart being virtually unaffected by starvation (Love, 1958, 1970).

Excessive depletion of lipids, adenyl nucleotides and proteins during wintering may result in mass mortality, especially if the temperature is very low. Many workers have reported a relationship between the energy reserves at the beginning of winter and the subsequent survival rate of the fish (Berman, 1956; Kirpichnikov, 1958; Mukhina, 1958; Polyakov, 1958; Higashi *et al.*, 1964; Zhidenko *et al.*, 1994). Some species suffer heavy losses during winter migrations also, as the intensity of feeding is reduced. The anchovy from the Azov Sea can die out from lipid deficiency.

The energy reserves of fish are almost completely drained, not only through being channelled into reproductive materials, but also through supplying energy for the spawning behaviour and the act of spawning itself. However, death does not eliminate the entire population, but mainly the senior age groups and the most severely exhausted fish of any age (Morawa, 1955; Shatunovsky, 1980). Males of the Azov round goby do not feed for a long time while watching over the eggs, depleting their lipid reserves until they die (Shulman, 1967). Here again, not all the fish are eliminated, and some restore their reserves and spawn again the following year (Rashcheperin, 1967). Low energy reserves decrease the immunity of the fish, increasing its mortality through disease (Mikryakov, 1978). Thus, over the total annual cycle, metabolic processes underlie population dynamics, being regulatory factors that foster a relatively stable existence, their rates being influenced by the abundance of the population.

Each period of the annual cycle (pre-spawning migration, feeding, etc.) has been given a name which reveals its biological basis, and manifests itself as a definite behavioural pattern. Moreover, the population can progress from one period to the next only if the entire set of metabolic processes supporting the new period have been set in motion. Otherwise the population or part of the population would die, through being locked into the previous period. We have already offered examples of this type of mortality as occurring during wintering and wintering migration. We would not group post-spawning mortality with the other types, since fish dying at that stage have accomplished the vital task of reproduction. However, in the Aberdeen work on cod, it did appear that depletion continued for a short while after the completion of spawning, perhaps from repair work after shedding eggs or because food was still scarce. It is not noticeably fatal for the brood fish if they are unable to complete gametogenesis during the pre-spawning period. Among responses to deterioration of the environment are interruption of vitellogenesis and trophoplasmatic growth, or the resorption of reproductive products. Such an event, however, may gravely endanger the population if it occurs on a large scale, by reducing recruitment. In recent years, resorption of reproductive products in fish has been found to have been provoked by increase in the concentrations of toxicants entering the water (Lukyanenko, 1987).

In the section devoted to daily rhythms, we introduced the idea of a 'biological clock'. Here we deal with a 'biological calendar' over the annual period. It is based on an ecological trigger which launches each stage, the key factor here being the photoperiod. Any population, be it of fish, birds or other animals, adjusts its biological calendar according to this factor. Climatic conditions, particularly temperature, are not constant and vary between years and seasons. They may turn adverse: for instance, a sudden drop of temperature may produce a lethal effect or retard metabolic activity in a fish population. For this reason the constant factor (photoperiod), which does not depend on climatic fluctuation, has been taken as a reference point.

The effect of photoperiod on metabolism, behaviour and endocrine secretion has been studied in birds (King et al., 1963; Dolnik, 1965) and fish (Gerbilsky, 1941; Fontaine, 1948; Hoar, 1953; Polenov, 1968; Barannikova, 1975; Maksimovich, 1988, 1989). Fish chosen as objects of study were mainly diadromous; see, for example, the studies on smoltification (Barannikova, 1975; Thorpe et al., 1980, 1982; Krayushkina, 1983; Klyashtorin and Smirnov, 1990; Varnavsky, 1990a,b). However, a similar scheme is valid for marine fish also, although the effect is smaller. The most important influences in the sequence of changes are the neuroendocrine and endocrine systems (hypothalamus, hypophysis, thyroid, adrenal cortex and gonads) which disseminate the effects of the 'light impulse' into every aspect of metabolic activity. This in turn results in the shaping of complex behaviour patterns in the population.

An increased light period was found to stimulate the development of reproductive products in birds and warm-water fish, while reduction was found to retard spawning and to trigger pre-winter feeding. The ways in which this induces changes in trends and the rates of metabolism have been discussed in section 4.2. Exogenous and endogenous factors in the periods of the annual cycle should be regarded as indissolubly interrelated. Photoperiod is the primary exogenous factor, while implementation of the sequence of metabolic processes is the endogenous factor without which none of the annual cycle would be possible. The endogenous factors are strictly specific for each period. For example, the pre-spawning period entails the biosynthesis of generative tissue, the pre-wintering period the formation of lipid reserves to support normal living during the winter, and the wintering period is characterized by the maintenance of plastic and energy metabolism at a certain level. The pituitary hormones somatotrophin and prolactin play important roles in regulating the protein and lipid metabolism (Lee and Meier, 1967; MacKeown et al., 1975; Sautin and Romanenko, 1982; Semenkova, 1984; Sautin, 1985, 1989; Trenkler and Semenkova, 1990). The behaviour of the population is rigidly regulated and determined by endogenous factors. The pre-spawning period is associated with intensive feeding or, in migratory fish, with moving to the spawning grounds. During spawning time, migratory fish spread over these sites and many species display a tendency towards decreasing their feeding rate. Both the post-spawning and pre-wintering periods are distinguished by high feeding rates; when satisfactory lipid reserves have been accumulated by the fish, their activity in seeking food is reduced and they tend to shoal. In migratory forms, this may result in crowded stocks of fish. During wintering, the fish gather into tight, sluggish shoals that ascend or descend slightly within the water depth.

The inverse relationship between the level of accumulated lipid and the feeding activity of the fish can be observed from changes in the vigour by which they are attracted to artificial light. The sprat from cold waters of the Black Sea develop the most intense photoreaction in winter, when their lipid content drops to a minimum (these fish do not have a wintering period in their annual cycle). As the lipid reserve accumulates, this reaction weakens and finally disappears in the summer when the lipid content of the fish is maximal. The content of neutral lipids is lower in sprats caught with the aid of artificial lights than in those caught by trawl fishing, when both catches were taken at the same time, demonstrating again that it is the hungry fish that are most attracted by light.

Manteifel (1980) and Gusar and Getmantsev (1985) suggest that fish which dash to the source of artificial light are simply showing curiosity (Pavlovian reflex 'What is this?'), which is closely related to preying and feeding. The reaction disappears once the fish are sated, while the shoaling behaviour increases at the same time and migratory fish tend to migrate. That lipid accumulation is an important factor in developing proper migratory condition in the pre-wintering period has been demonstrated in studies of the Azov anchovy (Lebedev, 1940; Vorobyev, 1945; Shulman, 1957, 1960b, 1972a; Taranenko, 1964; Luts and Rogov, 1978). Lipids stored by the fish are the basic source of energy supporting winter survival. While feeding during August and September, the anchovy are gathering in more and more concentrated shoals, moving on to the Kerch Strait, and in October they overcrowd water areas adjacent to it. The lipid reserve has been found to be equally important in achieving the correct pre-migratory condition in other species – Black Sea anchovy (Danilevsky, 1964; Chashchin, 1985), Caspian kilka (Rychagova, 1989), West Pacific sardine, ivasi (Shvydky, 1986), Baltic herring (Krivobok, 1964), Atlantic herring (Shubnikov, 1959) and some other species.

All these facts are cogent evidence of the importance of endogenous metabolic factors in bringing about specific behaviour patterns and implementing the annual cycles. Studies on birds have given similar results (Dolnik, 1965). However, internal impulses alone do not trigger the transition from one period of the annual cycle to another, in particular to initiate the wintering migration. An external stimulus is required. For example, Caspian sturgeon begin spawning migration in response to increasing turbidity of the water flowing in from the Volga river (Gerbilsky, 1956). Juveniles of some families, such as carp, salmon and sturgeon, may respond in the same way to a sudden decrease in available food (Kizevetter, 1948; Krivobok, 1953; Lovern, 1964; Yarzhombek, 1964; Malikova, 1967; Akulin et al., 1969). The premature (August–September) migration of fry of the Azov anchovy into the Black Sea,

migration. Hence, a population with maximum lipid reserves will leave the Azov Sea through the Kerch Strait in the last days of September, those moderately fat in October and the relatively lean residue only in November. As regards behaviour, the fattest anchovy gather into the densest shoals while passing through the Strait and their run is heavy and coordinated (Shulman, 1972a, 1974); the leaner anchovy 'ooze' through the Strait in small groups, giving a poor run.

All the above remarks about behaviour apply to distribution. Locomotion whilst on the winter ground, run of the migrating stock, spawning dispersion, gathering into the feeding or migrating shoal – each of these characteristics of distribution is closely associated with the metabolic pattern of the population.

Black Sea sprat, unlike anchovy, does not make lengthy migrations. Populations concentrate at certain sites, and their migratory activity is limited to comparatively short trips offshore in autumn for spawning and inshore for foraging in spring and summer. However, as in the anchovy, this limited migration is closely linked to lipid accumulation. When food is abundant and the lipid reserves high, sprats gather into vast shoals, which come as near as possible to the shore. The phenomenon may be seen all over the shelf waters of the Black Sea, especially the northwestern part. In years when the food supply is poorer, the stocks are less dense and remain in deeper waters, avoiding the coast.

The fatness of the sprat taken in the summer on completing their feeding period has been found to relate closely to the catch per unit of commercial fishing effort (Gusar et al., 1987). This again shows the importance of lipid reserves in the formation of shoals even when the fish do not migrate. The connection between lipid reserve and the distribution of the fish over a large area has been determined for populations of Black Sea anchovy (Danilevsky, 1964; Chashchin and Akselev, 1990), Caspian kilka (Rychagova, 1989) and Pacific species: ivasi, yellowfish (Vdovin and Shvydky, 1993) and walleye pollock (Shvydky and Vdovin, 1991).

In exploring the dependence of complex behaviour patterns of fish on their metabolism, one should keep in mind that any population is a complex genetic, morphological and, most importantly, physiological-biochemical system. In such a system, a considerable proportion of the fish is far from satisfying the norms required for the development of certain behaviour patterns. When they are swimming in a hydrodynamic flume, fish may easily be differentiated according to their ability to withstand continuous high-speed flow. Among them are leaders capable of swimming for many hours – sometimes for more than a day. There are also relatively numerous 'recessive' fish, those poor wrecks that have been pressed hard against the grating by the flow in the first hour or, occasionally, the first few minutes. It was found in experiments on horse-mackerel that 'recessive' fish could make 16% of the total number (Shulman et al., 1978). A similar proportion was deduced in experiments conducted with red mullet. This may not be mere coincidence, but indicate a more or less constant percentage of enfeebled individuals in a population.

Our investigations showed that these weakened fish had the lowest content of triacyl-glycerols – 55 mg% in red muscle and 32 mg% in white. In fish with normal swimming performance, these estimates were 110 mg% and 87 mg%, respectively (Shulman et al., 1978). The content of essential lipid components such as docosohexaenoic acid in phosphatidyl ethanolamine of red and white muscle is also lower in the 'recessive' fish (Yuneva et al., 1992). We have discussed the heterogeneity of physiology and biochemistry in spawning groups of pink salmon in Chapter 3. Similar experiments performed in the flume also revealed the same divergence of swimming capabilities and lipid composition in silver salmon (Zaporozhets, 1991). Fingerling fry of the cichlid *Aequides pulcher* with enhanced catecholamine content (L-dopa and dopamine) in the brain exhibited more of the 'searching' reaction than did those with lower levels (Nechaev, 1989; Nechaev et al., 1991). The same group of fry dominated the others and settled over the water body with greater activity. The haematological characteristics of enfeebled fish identified by poor performance in the flume were below the norm (Belokopytin and Rakitskaya, 1981)

The social and hierarchical structure of sockeye salmon populations has been thoroughly studied by Chebanov et al. (1983) and Semenchenko (1988). These workers found that the quality and quantity of the spawned sex products was inferior in the subdominant (stressed) group. The distinct difference in growth rates amongst different groups of one-finned greenling fry shows which fish will develop into leaders and which into outsiders (Vdovin and Shvydky, 1993). There is a high level of elimination of the latter groups. The hierarchical relationships known among higher terrestrial animals have much in common with those of fish (Shilov, 1977, 1985; Thorpe et al., 1980).

So far, we have considered the dynamics of lipid resources and their importance to the migration of anchovy and sprat. Studies of a different kind link the dynamics of red muscle biochemistry to migration in Atlantic cod. Since red muscle is concerned with continuous swimming as distinct from 'burst' activity, there is more of it in active species of fish than in sluggish (illustrated in Love, 1970, p. 26). Its biochemistry and physiology are designed for steady, slow activity, and its relatively high energy stores, vascularity, and content of haem pigments are designed to supply and transport energy to the contractile tissues as efficiently as possible. Although there is as yet no evidence of a marked change in the relative quantity of red muscle according to swimming requirements of the fish,[*] it has been shown that the level of haem pigments, which assist oxygen transport, is dynamic and changes according to need. A survey of cod caught in September from ten fishing grounds, ranging from Aberdeen (Scotland) to West Spitzbergen (Love et al., 1974) showed that the

---

*Greer-Walker and Pull (1973) did in fact demonstrate a small *absolute* increase in the red muscle of saithe that had been forced to swim for 42 days, but the white muscle increased much more, reducing the relative content of red.

mean proportion of red muscle in the total musculature of the caudal peduncle varied from ground to ground only within the narrow range of 31.3% to 36.1%, but that the pigmentation of the red muscle was considerably deeper in fish from South Spitzbergen and West Spitzbergen than in the cod from any other ground.

Most stocks of cod are 'local', that is, they remain in one area all their lives. In particular, the Faroe Bank cod, as already noted, are confined to a relatively small mound in the ocean, isolated by deep water on all sides. The cod caught off Spitzbergen, however, swim for many hundreds of miles each year, feeding there in the summer-time then swimming southwards, past Bear Island to the North Cape of Norway and further south to Lofoten, where they spawn. The stock is called 'skrei' in Norwegian, and is distinguished by having heads that are narrower than those of other cod stocks. Several other species of fish are known to exist as lean river-forms or stouter lake-forms.

The red muscle from cod from all the fishing grounds in the survey, apart from the two Spitzbergen grounds, was a pale 'dusky' brown, while the Spitzbergen cod exhibited a dark, chocolate-coloured lateral band of tissue. The colours of extracts of the red muscle were almost twice as intense as in the extracts from the other cod stocks, while that of the Faroe Bank cod was the palest (not significantly). A visit to Tromsø in February and examination of the red muscle of cod caught nearby off the Lofoten Islands confirmed the very dark colour of the same stock, now at the other end of its run.

The distribution of colour within a batch of 50 fish told an interesting story. In all the fish stocks apart from those from Spitzbergen, the colour showed a Gaussian distribution, that is, the largest number of individuals exhibited a particular value while progressively smaller numbers possessed red muscle of progressively stronger, or progressively weaker, intensity. Figure 47 shows the typical pattern of this kind in fish from the Faroe Bank. No such distribution was exhibited by fish caught off Spitzbergen, in which the values showed a wide horizontal spread without a clear peak, one specimen having a colour some five times as intense as that from a fish at the other end of the scale (Figure 48). The latter illustration relates to the fish caught at the end of their northern migration at Isafjorden (West Spitzbergen), and a similar pattern was observed in fish from South Spitzbergen.

A plausible interpretation of the results (Figure 47) is that the fish with the most strongly coloured red muscle had just arrived from their long migration from Norway, the haem pigmentation being necessary to transport large quantities of oxygen to the contractile cells. The pale red muscle from the same batch may have represented fish which had been at the feeding ground for some time, the pigmentation having faded because of reduced swimming activity. To test this theory, live, freshly caught cod were exercised in a circular channel in which they were chased by a spot of light which they preferred to avoid (stress-free exercise). Only fish that were consistently swimming were kept in the

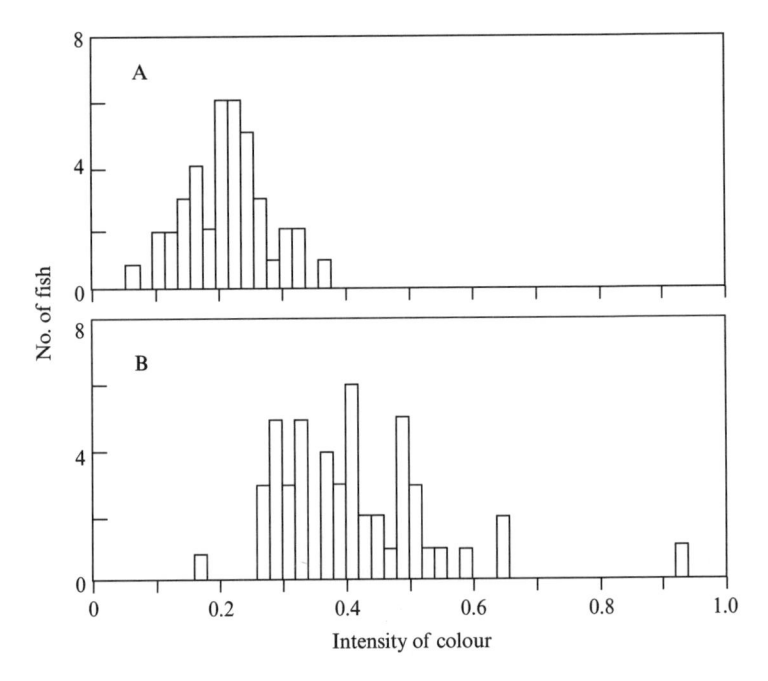

Figure 47   The distribution of intensity of pigmentation of red muscle among a batch of 50 cod caught on (A) the Faroe Bank and (B) West Spitzbergen. Faroe Bank fish are a stationary stock and Spitzbergen fish are migratory. (After Love *et al.*, 1977.)

experiment. After 28 days, the colour of the red muscle was significantly darker than that in the fish sampled at the point of capture, while the red muscle of those rested in a darkened tank of still water was paler than at the point of capture. Thus the red muscle was being pigmented according to swimming requirement (Love *et al.*, 1977).

Having established this fact, it was possible to monitor the average activity of cod during a year (Figure 48). These fish, caught on Aberdeen Bank (N.E. Scotland) gradually increased their swimming activity from April towards a maximum at the end of August, after which there was a continuous, steady reduction in activity until the following spring time. It seems most likely that activity rises with increased temperature and also, perhaps, with increased availability of food.

Further information was obtained from the fish caught on different grounds. It had been intended on the voyage to take the trawled fish immediately after capture and place them in an aquarium on the ship, then to exercise them to exhaustion (for a different experiment). In the catches south of Spitzbergen, capture in the trawl had virtually killed all the fish, and when the most active fish were placed in the aquarium they would not swim. In contrast, the fish from the two Spitzbergen grounds were vigorous and

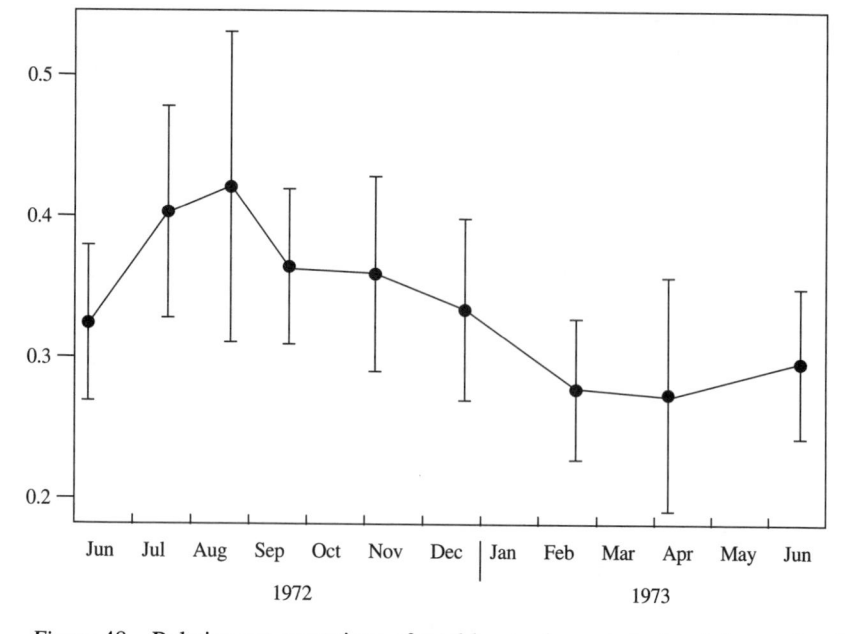

Figure 48   Relative concentrations of total haem pigments in the dark muscle of cod during a 12-month period. The ordinate is the mean optical density at 512 nm of a 4-cm light path.

flipped furiously about on the deck after capture. They could not be exercised to exhaustion – indeed, the man detailed off to chase them round the tank became exhausted first.

In this work there were no indications of the source of this extra vigour. The concentration of liver lipid was similar in the cod from all the grounds, but variations in total lipid between grounds are better reflected in the size of the liver. The concentrations of glycogen in the livers was much greater in cod from the Faroe Plateau than in those from any other ground, probably reflecting the rich nutrition of this stock. The pH value of the muscle 24 h after death, the indicator of the concentration of glycogen in the muscle of the living fish before capture, also showed nothing remarkable in the Spitzbergen fish.

The most likely explanation seems to lie in more efficient transfer of energy and oxygen to the contractile sites in the migratory fish. Poston *et al.* (1969) found that, in brook trout, exercise for 20 weeks in a flume improved the stamina of the fish, so that vigorous exercise caused less reduction in muscle glycogen in the muscle and produced less lactate in the blood and muscle, compared to fish which had not been 'trained' in this way. The glycogen in the muscle also recovered more quickly in the trained group during a period of rest after exercise. The reason for the long migration of this single stock of cod is not known. The study has done nothing to reveal the factors that trigger the migration, but shows that the migration itself causes biochemical changes in the fish.

Not every fish within a population will develop migratory and spawning capacity each year. The activity is limited to fish that meet the physiological requirements, the 'leaders'. The rest of the population are 'recessive' and have nothing to do but follow. A similar division has been reported by Shvydky (1986) in Pacific sardine, and in birds by Dolnik (1965), and it explains why a great many birds die during migration. Death certainly takes away, recessive, individuals which are incapable of making long migrations, for one or another reason, in fish as in birds. 'Recessives' compete for food and make easy prey for predators.

5. The Metabolic Basis of Productivity and the Balance of Substance and Energy

5.1. DEFINING PRODUCTION AND ENERGY

Production, usually defined as biomass produced per unit of time, is a subject that has been explored for many decades because of the economic importance of fish. It is investigated by estimating the intensity and efficiency of transformation of matter and energy.

The Black Sea provides sufficient data for assessing production and the balance of populations, calculated by the following equations:

$$C = Q + P + F \tag{8}$$

or
$$C = Q_b + Q_a + P_s + P_g + F \tag{9}$$

where C is the energy or weight of food consumed, Q is the total metabolism or total consumption of energy, Q_b is basal metabolism, Q_a is active metabolism, P is the energy or weight of the total production, P_s is the energy or weight of somatic production, P_g is the energy or weight of generative production, and F is the energy or weight of unassimilated food. This notion has been used in studies of the physiology of higher animals since the 1920s (e.g. Terroine and Wurmser, 1922; Ginetzinsky and Lebedinsky, 1956).

Sometimes a supplement is needed for equation 8:

$$C - F = Q + P = A \tag{10}$$

where A is the energy or weight of assimilated food, the flow of energy through the system being examined. From the balance equation one can calculate the share contributed by the production to the total balance of matter and energy in the population. Efficiency of production is estimated from coefficients of utilization of food consumed (K_1) and assimilated for growth and production (K_2), following Ivlev (1939) and Brett and Groves (1979):

$$K_1 = P/C \tag{11}$$

and
$$K_2 = P/A = P/(Q + P) \tag{12}$$

In studying productive processes in living systems, it is essential to combine the substance and energy aspects. This reveals means and forms of energy accumulation, transformation and utilization by individuals and populations and also evaluates energy demands for processes such as protein growth and lipid accumulation. The integrated approach adds complications; in addition to the assessment of energy potential (calorie content, fresh- and dry-weight of fish in a population), it is necessary also to estimate the content of protein, lipid, glycogen and total mineral substances in the fish body. The balance equation 8 is therefore modified with due regard for biochemical components:

$$
\begin{array}{cccc}
C & = Q & + P & + F \\
\downarrow & \downarrow & \downarrow & \downarrow \\
C_p & = Q_p & + P_p & + F_p \\
+ & + & + & + \\
C_f & Q_f & P_f & F_f \\
+ & + & + & + \\
C_g & Q_g & P_g & F_g \\
+ & + & + & + \\
C_m & Q_m & P_m & F_m
\end{array}
\tag{13}
$$

where C_p, Q_p, P_p and F_p are protein consumption, expenditure for metabolism, growth (production) and unassimilated. Similarly, suffix f relates to fat, g to glycogen and m to minerals.

This integrated approach differs from the traditional one whereby energy and substance balances were studied separately. Attempts to combine the two most often come down to the study of energy balance with elements of plastic metabolism included, disregarding the balance between different biochemical substrates. Only some studies on marine fish include energy and substance approaches. Examples of species studied in this way include cod, haddock, Baltic herring and flounder of the northern seas (Shatunovsky, 1978, 1980), plaice (Dawson and Grimm, 1980) and cod of Northern Norway (Eliassen and Vahl, 1982), Pacific sardine (Lasker, 1970), American plaice (MacKinnon, 1973).

Until recently, studies of the balance and production of fish have covered fairly brief time spans. However, productive processes as well as substance and energy mobilization proceed for the best part of a year, so the study must include an entire annual cycle of a population. Only by knowing the seasonal rhythms of accumulation, transformation and utilization of substances and energy can one reliably assess annual population production. Results obtained during long-term studies of the seasonal dynamics of chemical composition (Shulman, 1972a) were the basis for research on production in Black Sea populations (Chapter 4), but knowledge of the dynamics of biochemical substances alone is not sufficient for calculating production. One must know the values of absolute changes in these substances, protein increments and accumulation of lipids being the most important; it is possible only when data on the increase in weight are available for all age groups of fish examined. Such estimates are available for Black Sea fish (Shulman and Urdenko, 1989).

It is necessary to distinguish between somatic and generative components, and determine the quantitative ratio between them. It is not easy to use this approach when dealing with species of the Black Sea because of the protracted spawning period (Oven, 1976). Having taken that into account, however, we can assess the share of the total production that represents generative synthesis (Shulman and Urdenko, 1989).

Special emphasis should be put on the evaluation of energy consumed for metabolism. In assessing total, routine, active, standard and basal metabolism in Black Sea species, the data of Belokopytin (1978) and others were used, although they had been obtained from small numbers of fish. It has also been necessary to study the locomotory activity of different species in laboratory experiments and in the sea (Belokopytin, 1990, 1993). This has permitted the calculation of the average daily rate of swimming in different periods of the annual cycle under both sets of conditions. From the rate of oxygen uptake at different swimming speeds, it is possible to calculate the level of total energy metabolism and compare it with the level of standard metabolism, i.e. the level observed under experimental conditions with limited mobility. Having then determined the relationship between the swimming rates of individual fish or small schools recorded in experiments and in nature, the corresponding energy expenditures can be estimated.

In studying seasonal dynamics of energy metabolism, it is essential to discover how the rate of oxygen consumption changes with ambient temperature. We have used data of Ivleva (1981) and the results of experiments on oxygen uptake at different temperatures and Q_{10} coefficients from Belokopytin and Shulman (1987). Records of the temperatures at which the fish live enabled calculation of their energy metabolism during each season. To transfer from energy expenditure to consumption of specific biochemical substrates, experiments were performed to measure the roles of lipids, proteins and glycogen in supporting metabolism under different modes of life, inherent mobility and functional condition (Muravskaya and Belokopytin, 1975; Muravskaya, 1978). The results were scaled up to population level, using data on size-and-weight and age structures of given populations, as well as the sizes and identities of prey animals appropriate to the different sizes of the predators – the qualitative and quantitative compositions of their food.

Data on the numbers and biomass of the total stock of species examined in the Black Sea provided a basis for computing their production and balance. Efficient fishing devices and echo-sounder records were employed. The study allowed the determination of productivity of fish and of the effect of substance and energy consumption on population level over the whole annual cycle and for each species. It was also possible to evaluate the share contributed by the total stock of abundant fish species to the total cycle of matter and energy in the whole of the Black Sea.

5.2. PRODUCTION OF SUBSTANCE AND ACCUMULATION OF ENERGY

Methods for assessing the production of a population were developed some time ago (Vinberg, 1968, 1979; Mann, 1969; Zaika, 1972, 1983). Production (P) contributed during time span t is calculated using the following equation:

$$P_t = \Delta B + B_e \tag{14}$$

where B is the increment of population biomass or energy equivalent of the biomass and B_e the elimination of biomass by mortality, feeding and fishery.

The object of our present consideration is net production, that is, the increment of biomass. Gross production is the total of substances synthesized, including material subsequently catabolized. Production is usually quoted as 'per day', although in fact the assessment is often carried out over a considerable period such as a month or a year.

The production of a population is described with a differential equation, as follows:

$$P'(t) = \frac{dB}{dt} + \frac{dB_e}{dt} \tag{15}$$

Hence production contributed over a specific time span (t_1, t_2) is

$$P(t_1,t_2) = \int_{t_1}^{t_2} P' \, (t \mathrm{d}t) \qquad (16)$$

For a population heterogeneous in size or weight groups, the following equation is valid:

$$P'(t) = \int_{W_o}^{W_m} \mathrm{d}w/\mathrm{d}t \; N \mathrm{d}w \qquad (17)$$

where w is the weight group, W_m is the final weight of a single weight group, W_o is initial weight and N is the density of the numbers distribution by weight. Calculation of the production per unit weight of the population gives the specific production:

$$P/B = P' \, (t_{1,2})/B \qquad (18)$$

The specific daily rate of production of the population has dimension, namely the weight of production per unit of weight of the population per day. Specific production (P/B) is converted to total production (P):

$$P/B \cdot B = P \qquad (19)$$

Data from studies over several years (Shulman and Urdenko, 1989) were used to calculate production. The increments of weight of anchovies throughout the year are shown in Figure 49. Figure 50 shows how increments of fresh weight can be converted into increments of dry matter, protein, lipid and caloric values. In Figure 51, the monthly increments of protein in anchovy are separated into somatic and generative components. To convert from the increase in weight of the fish to production it is necessary to compute the specific daily rate (intensity) of increments of all examined age groups per month by applying the equation:

$$\triangle W/W = \frac{2(W_n - W_o)}{n(W_n + W_o)} \qquad (20)$$

where W_n is the final weight of the fish, W_o the initial weight of the fish and n = 30 days.

Table 10 shows seasonal variation in the age composition of Black Sea anchovy and sprat. The weight of an average fish in a population is evaluated from the age structure of the population and the share of the total biomass contributed by each age group. In horse-mackerel, pickerel and whiting, the proportion of the different age groups keeps stable all the year round, while in

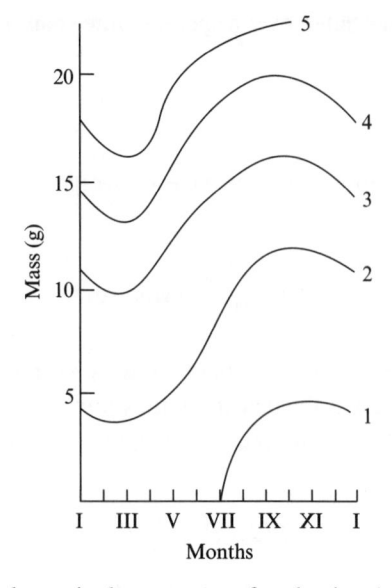

Figure 49   Annual change in the wet mass of anchovies: 1, born this year; 2, born last year; 3, two years old; 4, three years old; and 5, four years old.

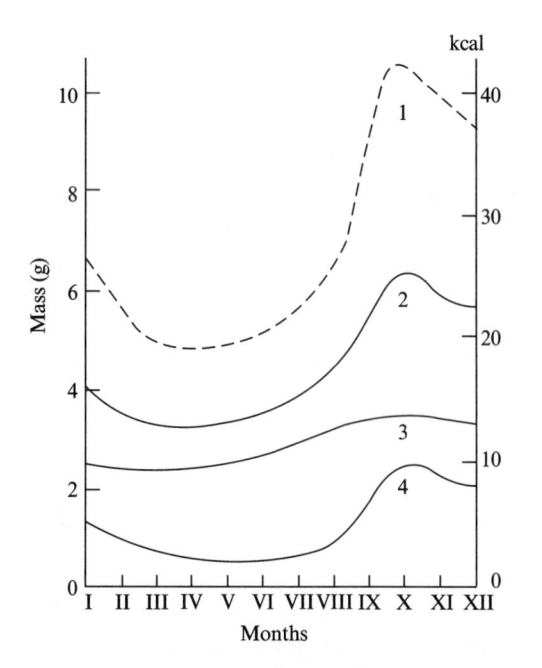

Figure 50   Somatic production of anchovy 3 years old. Right-hand scale: 1, calorific value (broken line). Left-hand scale: 2, lipids; 3, proteins; 4, dry matter.

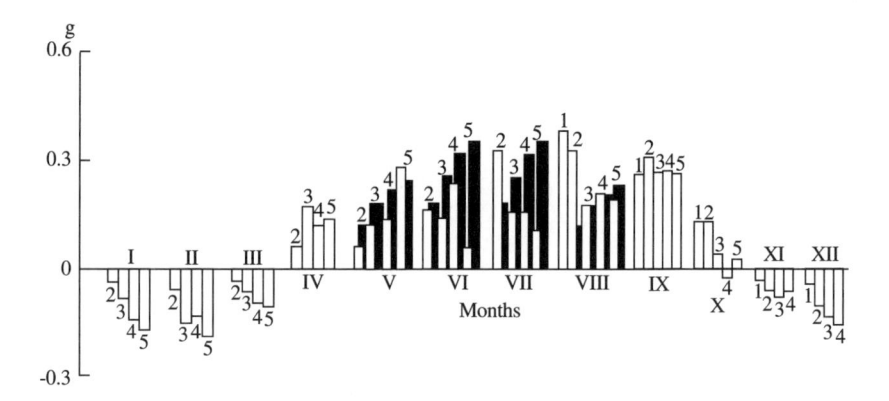

Figure 51   Monthly somatic and generative growth increases in anchovy (g protein g^{-1} per month): 1–5, age groups; I–XII, months of the year. Open columns, somatic growth; filled columns, generative growth.

Table 10   Average age composition of Black Sea anchovy and sprat (% of biomass).

	Age groups				
	0	1	2	3	4
Anchovy					
January–July	—	40	40	15	5
August	20	40	30	10	—
September	40	30	20	10	—
October–December	50	30	15	5	—
Sprat					
January	—	55	30	15	—
September	10	55	35	—	—
October–December	10	55	35	—	—

anchovy, sprat and red mullet it varies significantly. The annual somatic and generative increments in fresh and dry matter, protein, lipid and the energy equivalent of weight can then be calculated for an average fish in the population. Examples of the kind of results obtained are shown in Figure 52. Values obtained conform with those of the specific somatic and generative production (P_s/B and P_g/B) – see equation 18. All the characteristics are expressed in milligrams and calories per gram weight of fish per day.

In populations of these species, the ratio between female and male fish is usually 2:1. The generative increase observed in male fish is small compared with its total increment and is only 10% of the corresponding increase in the female – and only 5% of the total population when the sex ratio is taken into account. No sex difference in the somatic increment appeared in the six

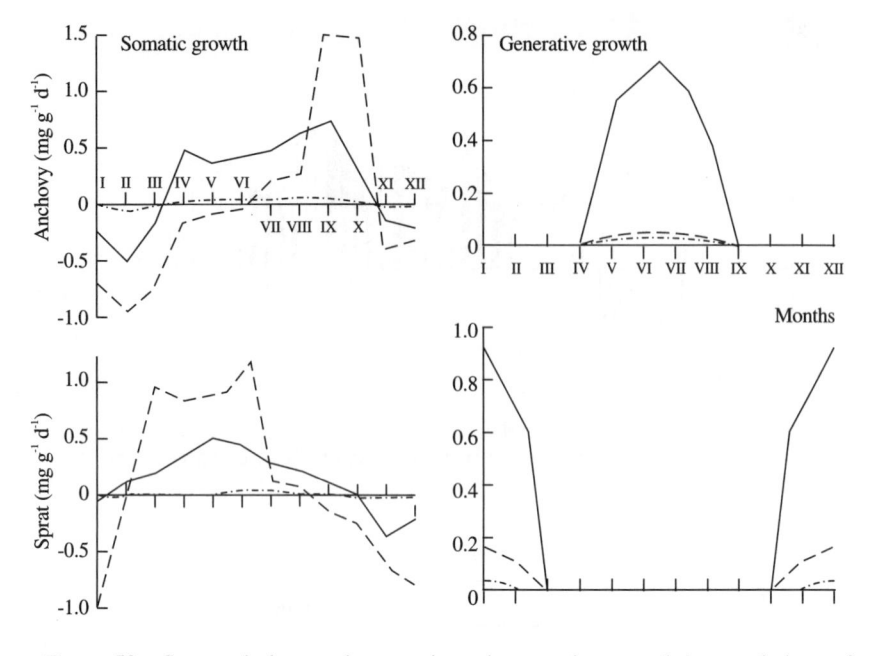

Figure 52   Seasonal changes in somatic and generative growth in populations of anchovy and sprat (mg g⁻¹ fresh tissue): solid line, proteins; broken line, lipids; dotted and broken line, mineral matter and glycogen.

species, with the exception of red mullet; thus, no correction for the somatic production of male fish is needed in populations of the other five species. Being as little as 10%, the correction can be ignored in assessing total production. In the case of red mullet, despite the sexual dimorphism in the incremental rate, the correction can also be ignored.

When the production of different Black Sea fish is compared, the six species investigated divide roughly into two groups, one comprising anchovy and sprat and the other, pickerel, whiting, horse-mackerel and red mullet. The sizes and biomasses of fish within a group are similar, as are some biological features.

Anchovy and sprat live in pelagic waters and live for a comparatively short time – 4 years in anchovy and 6 years in sprat, although 5- and 6-year-old sprats are rare. Both live on plankton. The average sizes of the senior age groups of anchovy and sprat are 12 cm and 11 cm, and their weights 23 g and 20 g, respectively. Both have an extended spawning season. Their temperature preferences differ, in that anchovy live in water at 12–25°C, sprat at 7–10°C. Because of this temperature difference, vitellogenesis occurs in spring and spawning occurs from June to August in anchovy, while the corresponding periods in sprat are autumn and winter – December to March (Oven, 1976). The densest gatherings of anchovy are found in winter, those of sprat in summer. Unlike anchovy, sprat do not make extensive migrations around the Black Sea, but follow the thermocline, ascending from the depths into shallow areas and returning as the temperature changes.

As regards production, somatic and generative production proceed at the same time in anchovy during the warm season when a plentiful food supply makes this possible. However, generative production takes only 4 months, while somatic production lasts for 7 months. This is because of intensive body-growth in the post-spawning period (September–October) of mature fish and of juvenile yearlings (August–October) which contributes as much as 30% of the total somatic production of the population.

In contrast, somatic and generative production in sprat occur at different times: somatic production takes place in the spring and summer, generative during autumn and winter. Sprats are known to spawn at water temperatures of 7–8°C in the Baltic and North Seas, and this same temperature characterizes the waters of the Black Sea where the sprat are found in winter. Even in summer these fish do not cross the bounds of the temperature gradient and so live at a low temperature throughout the year. The separation of somatic and generative production seems to relate to the inadequate food supply suffered by sprat inhabiting the northern extremity of their range (Biryukov, 1980; Ipatov and Kondratyeva, 1986).

The fish in the second group (four species) are also important components of the ecosystem of the shelf, differing among themselves in their ecology. Horse-mackerel, which are active predators, require warm water both near the bottom and the surface. Red mullet are less active, feeding on the sea bottom and also requiring warm water. Whiting require cold water and are near-bottom predatory fish displaying relatively low mobility, while pickerel is a near-bottom fish, intermediate in its characteristics. These four species are comparable in size and weight, older age groups having body lengths of 17–21 cm and weights of 60–90 g.

In horse-mackerel, somatic and generative production occur in the warm season as in anchovy. However, the formation of genital products begins and ends 2 months earlier than somatic growth. It seems as though generative production is supported both by the consumption of food and also the utilization of material from the body of the fish itself. Evidence for this is provided by negative values for somatic production during the first half of the year. The periods of somatic and generative production coincide in red mullet and horse-mackerel, suggesting identical temperature preference in these fish. In the spring, somatic and generative productions do not coincide in time.

The generative and somatic productions of whiting, as in sprat, occur in different seasons, the former in cold months and the latter in spring, summer and autumn, the spring peak relating to the growth of immature yearlings: whiting and sprat require cold water, while horse-mackerel, red mullet and anchovy need warmth. In terms of the creation of somatic and generative tissues, pickerel is intermediate between the warm- and cold-loving fish of the Black Sea. The discrepancy in time separating the two processes in pickerel is not as marked as in whiting and sprat, but as in horse-mackerel the act of creating genital products is accompanied by consumption of the body tissues.

Temperature is therefore a determining factor influencing the timing and character of both somatic and generative production in Black Sea fish. All the species examined form a series, which is related to their temperature preference: anchovy – horse-mackerel – red mullet – pickerel – whiting – sprat. In the first species the somatic and generative productions coincide, while in the last there is a large space of time between them. In horse-mackerel, red mullet and pickerel there is a partial overlap.

The principal materials encompassed by the word 'production' in fish are proteins and lipids (Figure 52), the proportion contributed by glycogen and inorganic substances being no more than 10% of the other materials. Proteins contribute mostly to body structure and lipids to energy. The two productions may not coincide in warm-water fish, where protein production begins and ends earlier than that of lipid. In addition, peaks of protein production and lipid accumulation do not coincide (Shulman, 1972a). This is because the increase in length and weight coincides with generative development, while lipid accumulation, necessary for the fish to survive winter, occurs later in these species. In sprat and whiting, which require cold water, the proteins and lipids are produced simultaneously. It is not easy to distinguish sharply between the functional importance of proteins and of lipids, since proteins, like lipids, do supply a modicum of energy to the organism. In the cold season, when zooplankton become scarce, anchovy and sprat consume lipids and some proteins from their stores, but in anchovy this process coincides with wintering and in sprat with maturing and spawning. Like anchovy, horse-mackerel and red mullet consume lipids and proteins in January–March, but proteins are not vital energy reserves in any of the four species since their lipid reserves are very high. Protein is more important in Black Sea whiting, goby and scorpion fish, which possess only small lipid reserves. Protein is an important source of reserve energy in all members of the cod family, scorpion fish, pike, perch and some carp, being mobilized after lipid reserves have been depleted.

Pickerel consume relatively low amounts of their reserves during wintering, perhaps because of intensive feeding prior to winter, or because they inhabit colder waters and so are less affected by the seasonal temperature drop. The utilization of reserves during the winter has been studied in several species (Mukhina, 1958; Polyakov, 1958; Shcherbina, 1989). In addition to triacyl-glycerols, phospholipids can also be used as endogenous sources of feeding (Dambergs, 1964; Love et al., 1975; Ross and Love, 1979; Zagorskich and Kirsipuu, 1990), despite their role in the structure of muscular tissue. On the other hand, normal levels of glycogen are maintained, through gluconeogenesis, in at least some species when endogenous feeding is in progress during wintering.

Results of analytical studies on white muscle in winter and summer months are given in Figures 53 and 54; the patterns for red muscle and liver are quite similar. As stated earlier, the main fractions are triacyl-glycerols and phospholipids, but only the former increase and decrease during the annual cycle. The highest concentrations of triacyl-glycerols are found in red muscle and liver, but these organs account for a mere 7% or less and 5% or less of the body weight, respectively, in the species under consideration. On the other hand, white muscle comprises up to 50% of the body weight, so in fact contains the major part of the triacyl-glycerols.

Phospholipids, essential components of all cell membranes, range from 2% to 50% of the total lipids of different tissues of Black Sea fish. However, what appears to be a large range is mostly the result of variation in triacyl-glycerols, changes in phospholipid content being small. Like proteins, they are 'building materials' and are components of relatively stable structures. Although the phospholipid contents of the livers of scorpion fish, round goby and whiting are high, the content in the whole fish can be as low as 0.5–1%, comparable to its glycogen content.

The Atlantic cod offers an example (in a non-fatty fish) of extreme proportions of phospholipids and triacyl-glycerols. As much as 88% of the lipids of the white muscle (total lipid content 0.5% to 0.6%) consists of phospholipids and only about 1% is triacyl-glycerols. When the fish are starved, the minute proportion of triacyl-glycerols does not change, so in this situation a

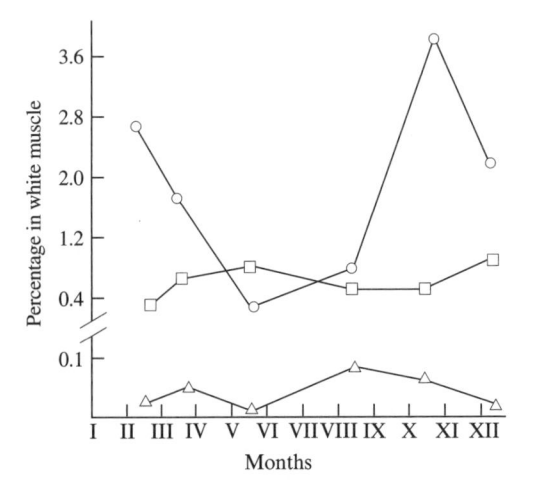

Figure 53   Seasonal changes in the lipid fractions of the white muscle of anchovy: ○, triacyl-glycerols; □, phospholipids; △, non-esterified fatty acids.

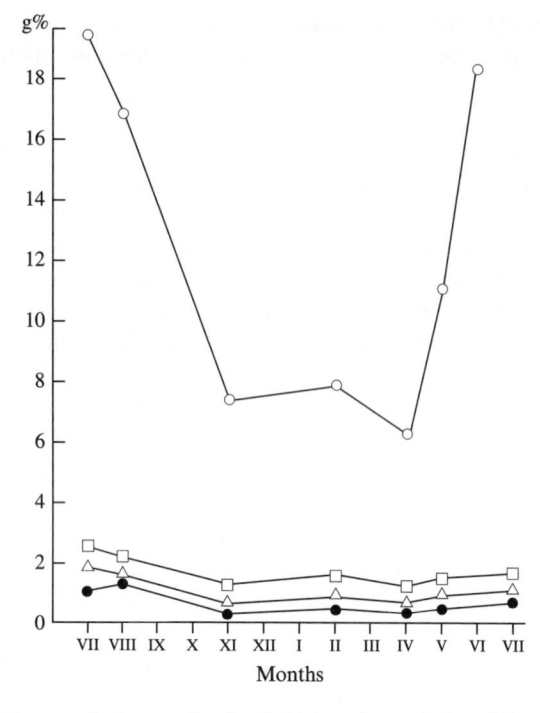

Figure 54   Seasonal changes in the lipid fractions of the white muscle of sprat:
○, triacyl-glycerols; □, phospholipids; △ , cholesterol; ●, non-esterified fatty acids.

proportion of them, at least, appears to be structural rather than a store of energy. When extreme starvation causes the breakdown of some of the white muscle cells, the proportion of phospholipid decreases to about 78%, and a corresponding 'rise' in cholesterol indicates that this component is retained even under such conditions (Ross, 1977).

Phospholipids are most actively employed in the creation of somatic and generative production. Gonad maturation is accompanied by a sharp increase in the phospholipid concentration of the liver and blood (Shchepkin, 1972), indicating enhanced synthesis and transport into the gonads. The phospholipid concentration is greater in the gonads than in somatic tissues.

A considerable quantity of body proteins are also involved in generative synthesis, evidenced by substantial consumption of these proteins during the pre-spawning period by horse-mackerel, red mullet, whiting and pickerel. This would not have occurred if the food proteins had been adequate for the purpose. As horse-mackerel, red mullet and scorpion fish mature, the transport of serum proteins (albumin, α- and β-globulins) proceeds more intensively (Golovko, 1964; Kondratyeva, 1977). The triacyl-glycerols supply energy for the generative synthesis and for the accumulation of reserves in the gonads themselves (Shchepkin, 1979).

By using ^{14}C tracers, the rates of protein, phospholipid and triacyl-glycerol metabolism have been measured in pickerel and scorpion fish (Khotkevich, 1974). It was found that during gonad maturation the anabolic and catabolic rates of these substances sharply accelerate in the liver and in white and red muscle. Here lies the explanation for the discrepancy between somatic and generative production processes in the three species studied. Vitellogenesis causes a redistribution of plastic material and energy from muscle into gonad via the liver and blood. The endogenous feeding is observed not only during winter, when it compensates for food shortage (Love, 1970, 1980; Cowey and Sargent, 1979; Sorvachev, 1982), but also during pre-spawning foraging when food is consumed at a high rate. Maturation requires a great deal of material.

Data are available on the transport and transformation as distinct from their redistribution of endogenous proteins, lipoproteins, phospholipids and triacyl-glycerols into developing gonads, (Krivobok, 1964; Wang et al., 1964; Love, 1970, 1980; Lapin, 1973; Medford and Mackay, 1978; Shackley and King, 1978; Lizenko et al., 1980; Shatunovsky, 1980; Dabrowsky, 1982; Eliassen and Vahl, 1982; Henderson et al., 1984). Increases in the lipid and protein contents (and overall weight) of the liver is a characteristic of maturation in many species of fish (Krivobok, 1964; Kozlov, 1975; Shevchenko, 1977; Shatunovsky, 1980; Ando, 1986). It is in the liver that the reproductive materials are synthesized, as observed earlier in higher animals (Fontaine, 1969). Cytological studies have demonstrated a dramatic increase in the number of hepatocytes that synthesize proteins (Love, 1970, 1980; Neifakh and Timofeeva, 1977; Shatunovsky, 1980). The spawned eggs are rich in proteins, phospholipids and triacyl-glycerols, which provide for the proper structural and functional development of the embryos and larvae until the latter become capable of exogenous feeding (Lasker, 1962; Wang et al., 1964; Love, 1970, 1980; Vanstone et al., 1970; Healy, 1972; Hayes et al., 1973; Neifakh and Timofeeva, 1977; Clarke et al., 1978; Woo et al., 1978; Wootton, 1979; Constantz, 1980; Lizenko et al., 1980; Shatunovsky, 1980; Cetta and Capuzzo, 1982; Kamler et al., 1982; Lapina and Lapin, 1982; Chepurnov and Tkachenko, 1983; Dabrovsky et al., 1984).

There is a clear interrelation between the metabolism of proteins, phospholipids and triacyl-glycerols during the formation of both somatic and generative tissues. In the blood flow this interrelation appears as the continuous formation of lipoprotein complexes (Kulikova, 1967; Lapin and Shatunovsky, 1981; Sautin, 1985; Gershanovich et al., 1991). In lipoproteins of very low density, triacyl-glycerols comprise 20–65%, the remainder being proteins. Low-density lipoproteins comprise about 50% each of proteins and lipid substances, while those of high density comprise only 30–50% of lipids. The transport of non-esterified fatty acids and triacyl-glycerols is achieved in association with albumins, while phospholipids and cholesterols are transported together with α-globulins.

The fatty acid composition of phospholipids and triacyl-glycerols has been studied in anchovy and sprat throughout their annual cycles (Yuneva, 1990). The saturated fatty acids 14:0, 16:0, 18:0, monoenoic 14:1, 16:1, 18:1, 20:1, 21:1 and polyenoic 18:2, 18:3, 20:2, 20:4ω6, 20:4ω3, 20:5ω3, 22:5ω6, 22:5V3 and 22:6ω3 were identified, but the nine most quantitatively significant are 14:0, 16:0, 18:0, 16:1, 18:1, 20:1, 22:1, 20:5ω3 and 22:6ω3. The qualitative pattern of fatty acids described here is common to marine fish the world over (Viviani, 1968; Rzhavskaya, 1976; Ackman, 1983; Sidorov, 1983; Henderson *et al.*, 1984; Bell *et al.*, 1986).

Of the saturated fatty acids in winter-caught anchovy it was found that 16:0 predominated in the phospholipids, while 14:0 was the most important in the triacyl-glycerols. The proportions of 16:1 and 18:1 were also high in the triacyl-glycerols, but in the phospholipids it was the polyenoic acids, principally 20:5ω3 and 22:6ω3 that predominated. Because the phospholipids are a relatively small fraction in anchovy, the fatty acid composition of the total lipids is fairly close to that of the triacyl-glycerols alone. These specific features indicate a particular structure and function of the two main lipid fractions. The triacyl-glycerols, containing mostly saturated and monoenoic acids, are the source of energy for lipolysis. On the other hand, high metabolic activity in the polyunsaturated fatty acids of phospholipids indicates the adaptation of cell membranes to changing conditions.

In the anchovy, the total saturated fatty acids content in the reserve and structural lipid fractions is maximal in the autumn and minimal during the rest of the year. However, the polyenoic acids in these two fractions are maximal in winter, spring and summer, but minimal in autumn. During spring and summer, a substantial reduction is observed in the quantity of 16:0, 18:1 and 22:6 in the triacyl-glycerols, perhaps because of transfer to sexual products in which they accumulate as stored energy (saturated and monoenoic acids) or as structural elements (polyenoic acids). They are of similar importance in this context in capelin, horse-mackerel, cod, Pacific saury, eelpout and trout (Jeffries, 1972; Dobrusin, 1978; Ackman, 1983; Henderson *et al.*, 1984).

In Atlantic cod, the effects of maturation are different from those of simple starvation (Takama *et al.,* 1985). Two groups of cod were kept without food. The gonads of one group were maturing during the period, while the fish in the other had been castrated. No difference was observed in the lipid compositions of the flesh, but it was found that in the group which was maturing as well as starving (imitating the situation found in the wild) there was more depletion of 22:6ω3 from the liver. This fatty acid was found to predominate in the gonads of either sex. There were also indications that 18:1 was being removed preferentially from the livers of the maturing group, this fatty acid also being an important constituent of gonad lipids.

Sprat, which lives in colder waters than anchovy, exhibits different dynamics of fatty acid composition. Phospholipids display a single maximum in summer and minimum in winter of their constituent polyenoic acids, mostly 22:6. The cycle seems to relate to varying functional activity, which is high in summer and low in winter (Yuneva *et al.*, 1986). A similar picture characterizes the triacyl-glycerols. It seems that the growing unsaturation of reserve lipids observed in summer sprat results from the need for a more rapid mobilization to feed the more rapid metabolism. The basic energy accumulators are saturated and monoenoic fatty acids, and their mobilization leaves the remaining lipids more unsaturated. The sprat contains more of the fatty acids 20:1 and 22:1 than the anchovy, these fatty acids characterizing the planktonic crustaceans from colder water (Sargent, 1978; Ackman, 1983). In the protein of sprat and other fish (Love, 1970, 1980; Kleymenov, 1971; Cowey and Sargent, 1979; Ostroumova, 1983; Sidorov, 1983) glutamic and aspartic acids as well as valine, leucine, lysine, isoleucine, phenylalanine, threonine and methionine are present in large amounts. The amino acid composition remains unchanged throughout the annual cycle.

The activity of alkaline phosphatase in fish scales is relevant to the study of the biochemical characteristics of production (Senkevich, 1967: red mullet, round goby). This enzyme promotes calcification of bones and scales, and its activity correlates well with the linear growth of the fish (Figure 55), so it is most intense during the warm season. Data from cold-water fish are not available.

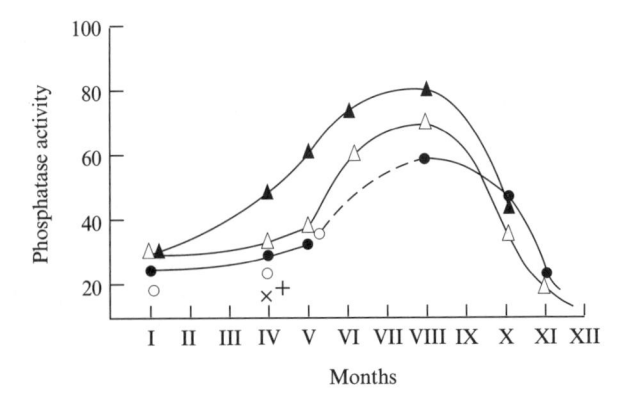

Figure 55   Activity of alkaline phosphatase in scales of red mullet during the annual cycle. (After Senkevich, 1967.) Age (years): ▲, 1; △, 2; ●, 3; ○, 4; ×, 5–6.

Annual structural and functional characteristics of populations are of special interest. The annual specific production $(P_\Sigma/\overline{B})$ is estimated by using the following equation:

$$P_\Sigma/\overline{B} = \int_0^{12} (P/B)\, dt \qquad (21)$$

or
$$P_\Sigma/\overline{B} = \sum_{i=1}^{12} (P/B)\, \Delta t \qquad (22)$$

where P/B is the monthly production. Equation 22 is valid for steady biomass, while variable biomass, which is our interest, should be calculated as follows:

$$P_\Sigma/\overline{B} = \frac{\displaystyle\sum_{i=1}^{12} P_i}{\dfrac{1}{12}\displaystyle\sum_{i=1}^{12} B_i} \qquad (23)$$

The annual specific production is described by means of a P/B (production to biomass) ratio, which is widely employed in ecological studies. It is possible to calculate this coefficient even without data on the numbers and biomass of the population, using instead the figures for daily or monthly specific production contributed by a 'common member' of this population. Data of numbers and biomass are of course essential when estimating absolute production (equation 19).

It is not difficult to assess the average biomass of a fish stock over many years, using data on the numbers and biomass of samples. In Black Sea anchovy, it is 284 kt in May. By November the figure doubles as a result of replenishment and of somatic increase. The average annual biomass (B) amounts to 426 kt for anchovy, 280 kt for sprat, about 36 kt for horse-mackerel, 6 kt for red mullet and 9 kt for whiting (no data are available for pickerel). Results of the studies on annual specific and absolute production in populations of the fish examined here and P_Σ/B coefficients are given in Table 11.

From equations 21 and 22 it follows that the specific annual production is determined by adding the estimates for specific production obtained in each month of a year, regardless of the fact that negative estimates emerge over about half that time (Figures 53 and 54). From available literature it is clear that values for specific and absolute annual production are most often assessed solely from positive estimates of the constituent parts (Zaika, 1972, 1983), but it is our belief that both positive and negative estimates should be taken into account. Negative increments have been reported also by Wootton et al. (1980) and Shvydky and Vdovin (1991). As regards a living organism or a population, the negative

Table 11   Annual average production of total fish populations.

	Anchovy	Sprat	Horse-mackerel	Red mullet	Pickerel	Whiting
B Biomass						
Wet mass	4.26×10^{11}	2.80×10^{11}	3.57×10^{10}	6.00×10^{9}	—	9.00×10^{9}
Dry mass	1.11×10^{11}	0.70×10^{11}	1.07×10^{10}	1.86×10^{9}	—	1.71×10^{9}
Protein	0.72×10^{11}	0.41×10^{11}	0.62×10^{10}	1.14×10^{9}	—	1.26×10^{9}
kcal	6.82×10^{11}	4.48×10^{11}	6.10×10^{10}	12.00×10^{9}	—	9.90×10^{9}
P_s Wet mass	1.92×10^{11}	1.00×10^{11}	1.10×10^{10}	3.30×10^{9}	—	4.41×10^{9}
	(0.45)	(0.36)	(0.31)	(0.55)	(0.63)	(0.49)
P_g	1.16×10^{11}	1.10×10^{11}	1.54×10^{10}	4.32×10^{9}	—	2.79×10^{9}
	(0.27)	(0.39)	(0.43)	(0.72)	(0.20)	(0.31)
P_s Dry mass	0.46×10^{11}	0.25×10^{11}	0.28×10^{10}	0.78×10^{9}	—	0.81×10^{9}
	(0.11)	(0.09)	(0.08)	(0.13)	(0.18)	(0.09)
P_g	0.28×10^{11}	0.34×10^{11}	0.51×10^{10}	0.96×10^{9}	—	0.54×10^{9}
	(0.07)	(0.12)	(0.14)	(0.16)	(0.07)	(0.06)
P_s Protein	0.32×10^{11}	0.14×10^{11}	0.19×10^{10}	0.60×10^{9}	—	0.63×10^{9}
	(0.08)	(0.05)	(0.05)	(0.10)	(0.13)	(0.07)
P_g	0.23×10^{11}	0.26×10^{11}	0.34×10^{10}	0.60×10^{9}	—	0.45×10^{9}
	(0.05)	(0.09)	(0.10)	(0.10)	(0.05)	(0.05)
P_s Lipid	0.13×10^{11}	0.11×10^{11}	0.11×10^{10}	0.18×10^{9}	—	0.18×10^{9}
	(0.03)	(0.04)	(0.03)	(0.03)	(0.05)	(0.02)
P_g	0.08×10^{11}	0.08×10^{11}	0.14×10^{10}	0.36×10^{9}	—	0.09×10^{9}
	(0.02)	(0.03)	(0.04)	(0.06)	(0.02)	(0.01)
P_s kcal	2.68×10^{11}	1.69×10^{11}	1.70×10^{10}	4.56×10^{9}	—	4.77×10^{9}
	(0.63)	(0.60)	(0.48)	(0.76)	(1.04)	(0.53)
P_g	1.56×10^{11}	1.89×10^{11}	3.25×10^{10}	6.00×10^{9}	—	3.15×10^{9}
	(0.37)	(0.68)	(0.91)	(1.00)	(0.46)	(0.35)
P Wet mass	3.09×10^{11}	2.10×10^{11}	2.64×10^{10}	7.62×10^{9}	—	7.20×10^{9}
	(0.73)	(0.75)	(0.74)	(1.27)	(0.83)	(0.80)
Dry mass	0.74×10^{11}	0.53×10^{11}	0.79×10^{10}	1.74×10^{9}	—	1.44×10^{9}
	(0.17)	(0.21)	(0.22)	(0.29)	(0.25)	(0.16)
Protein	0.55×10^{11}	0.40×10^{11}	0.53×10^{10}	1.20×10^{9}	—	1.08×10^{9}
	(0.13)	(0.14)	(0.15)	(0.20)	(0.19)	(0.12)
Lipid	0.21×10^{11}	0.20×10^{11}	0.25×10^{10}	0.54×10^{9}	—	0.27×10^{9}
	(0.05)	(0.07)	(0.07)	(0.09)	(0.07)	(0.03)
kcal	4.24×10^{11}	3.58×10^{11}	4.95×10^{10}	10.56×10^{9}	—	7.92×10^{9}
	(1.00)	(1.28)	(1.39)	(1.76)	(1.49)	(0.88)
P_g/P_s Wet mass	0.60	1.10	1.40	1.31	0.31	0.63
Dry mass	0.61	1.36	1.82	1.23	0.41	0.68
Protein	0.72	1.86	1.79	1.00	0.40	0.72
Lipid	0.61	0.73	1.27	2.00	0.40	0.50
kcal	0.58	1.12	1.91	1.32	0.44	0.66
P_Σ/B Wet mass	0.72	0.75	0.74	1.27	0.83	0.80
Dry mass	0.67	0.84	0.74	0.93	0.86	0.82
Protein	0.76	0.97	0.85	1.05	0.90	0.85
kcal	0.62	0.80	0.81	1.88	0.87	0.81

Notes: wet and dry mass, protein and lipid are given in grams; specific production (in brackets) is given per 1 g of wet mass of population.
P_Σ/B is given per 1g of wet mass, dry mass, protein or 1 kcal.

increase in production is not a mere abstraction but an indication of real consumption of substance and energy. For the half year mentioned above, the population consumes, rather than contributes to, its biomass, biochemical substrates and energy potential, the consumption subtracting as much as 10–30% from the peak production value (Shulman, 1972a) during wintering and maturation. The computed annual production is therefore less than that estimated by routine techniques. For anchovy, P_Σ calculated from fresh weight is 21% less, from dry weight 41% less, from protein estimate 21% less and from energy equivalent 51% less than estimates obtained by routine procedures.

Assuming that the annual increase in each age-group of the population is as follows: (1) W_1-w_0, W_2-w_1 ... W_n-w_{n-1}; the resulting values are overstated owing to the difference that has been ignored: W_1-w_1, W_2-w_2 ... W_n-w_n. It is more correct to take into consideration the annual increase in weight: (2) w_1-w_0, w_2-w_1 ... w_n-w_{n-1} or (3) W_1-W_0, W_2-W_1 ... W_n-W_{n-1}. Applying this last method (3) the first difference is overestimated while the last is underestimated, giving a balance between the two for a population (Figure 56).

Having tested this method, many workers will still want to asses annual production from positive values only. If such is the case, Table 12 can be used to determine coefficients of conversion for annual production estimates for all components or for positive values only. Estimates are highest for anchovy, somewhat less in red mullet, horse-mackerel, pickerel and sprat, while there is no difference in whiting. This is the ranking of the species studied for differences between somatic and generative production. It has

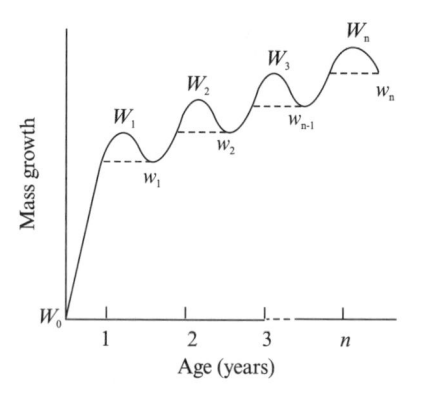

Figure 56   Scheme of mass growth of fish. W, mass of fish at the end of the feeding period; w, mass at the end of wintering.

Table 12   Percentage reduction in the computed figures for total annual production of fish when **all** monthly values, positive and negative, are taken into account, rather than when only positive values are used. In whiting there is no reduction, the two figures are the same.

	Anchovy	Sprat	Horse-mackerel	Red mullet	Pickerel	Whiting
Wet mass	21	1	18	14	3	0
Dry mass	41	9	27	31	15	0
Protein	21	0	12	20	2	0
Energy equivalent	51	26	31	36	10	0

been suggested previously that the differences may result from individual species preferring warm or cold water. A similar cause is responsible for the differences shown in Table 12. The higher the ambient temperature preferred by a species, the more pronounced is the negative component of annual production. This negative component indicates the consumption of energy and reserves during over-wintering and in reproductive development. At higher preferred temperatures, more reliance is placed on endogenous sources during the unfavourable cold season. Whiting are truly cold-water fish and can feed all the year round, so they have no negative component of production.

When the energy equivalent of weight is applied in these calculations the difference increases. This is related to utilization of lipid reserves in winter; weight for weight lipid yields more energy than protein, so it is used preferentially. In all species examined, 70% of dry matter production is protein, the remainder being presumably lipid. If glycogen and inorganic matter are omitted, the annual percentage production of protein and lipid is, respectively: anchovy, 74.3 and 25.7; sprat 67.8 and 32.2; horse-mackerel 67.1 and 32.9; red mullet 69 and 31; pickerel 76 and 24; and whiting 75 and 25. Comparing these values with Table 11, we see the total absolute production (P_Σ) of the anchovy population is 1.5 times that of the sprat on a wet weight basis; 1.3 times higher on a dry weight basis; 1.4 times higher on a protein basis and 1.2 times higher on an energy basis. This difference results from the higher numbers and biomass of the anchovy population. However, P_Σ/B coefficients that describe the annual specific production are higher in sprat because of greater generative production; the absolute quantities of dry matter, protein, lipid and caloric contents of the reproductive products of sprat are very much higher than in anchovy (Shulman et al., 1983). The organic content of fish eggs developing in low temperatures is known to be relatively high (Kozlov, 1975; Shatunovsky, 1980). The cold water sprat and the warm water anchovy obviously have different production strategies. The extensive mode shown by the anchovy implies an increase based on high initial biomass and numbers, with relatively low production efficiency. The abundance of anchovy in the Black Sea may reflect their wider distribution than sprat, which are confined to the deeper cool waters below the thermocline.

Annual horse-mackerel production is an order of magnitude less than that of sprat and anchovy, but still an order of magnitude greater than red mullet and whiting. The annual P_Σ/B coefficients of all six species are of the same order, although the estimates differ. The coefficient is highest in red mullet, followed by pickerel, whiting and horse-mackerel, while that in anchovy is the least. Because of their wider range of distribution and more diverse nutritive base, horse-mackerel increase in numbers at the expense of demersal and demersal–pelagic species that inhabit a narrow belt of water above the in-shore shelf.

There is an inverse relationship between the absolute and the specific production of anchovy, sprat, horse-mackerel and red mullet (Figure 57). High absolute production is a result of the extensive mode (high initial abundance), while high specific production is related to the intensive mode with a more efficient increase of the biomass.

In the species discussed here, the annual P/B coefficients are 0.72–1.27 (fresh weight), which, converted to the energy equivalent, become 0.62–0.88. On a dry weight basis, the values are intermediate between these two ranges. These coefficients are very close to those reported elsewhere for the same Black Sea species: 0.75 for mackerel and sprat, and 0.69 for anchovy (Danilevsky *et al.*, 1979). The actual annual specific production values for Black Sea fish can be estimated from these figures. The range of annual P/B coefficients of other fish species cited in the literature is 0.45–1.18 (Mann, 1965; Backiel, 1973; Penczak *et al.*, 1976).

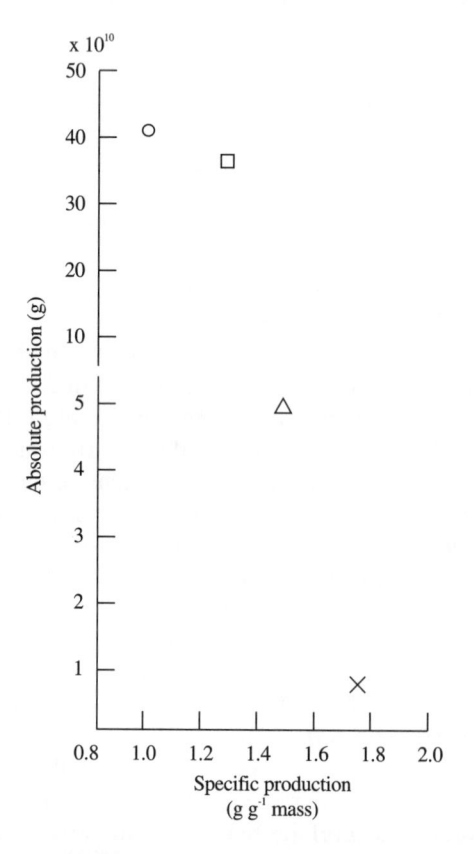

Figure 57   Relationship between absolute production (g) and specific production (g g⁻¹ mass): ○, anchovy; □, sprat; △, horse-mackerel; ×, red mullet.

If it is assumed that P/B coefficients are about 1 for adult Black Sea fish, data on larvae and fry have been ignored in the population calculations (Sorokin, 1982). The latter author supposes that production contributed by this section of the population, although later much reduced by natural mortality and predation, is as much as 50% of the total population. Certainly, elimination of larvae and fry of Black Sea pelagic fish is very high (Dekhnik, 1979), in the absence of which, Sorokin concludes, the total annual production of the population would double and the annual P/B coefficients would be about 2. However, Sorokin admitted that his calculations were rather approximate. Zaika (1983) has also pointed out that fry mortality has been disregarded in production calculations. All the cases considered here therefore refer to the adult part of the population, into which grown-up yearlings are also placed. Annual P/B ratios converted into fresh weights are 0.67 for Azov anchovy, 1.13 for Azov kilka ('Tyulka') and 1.46 for round goby (Greze, 1979). The large supply of food in the Azov Sea fosters better production of fish when compared with the Black Sea; production is shown in Table 12, disregarding negative values.

Since averages of biomass of a population estimated over many years are relatively stable, the annual P_Σ and B_Σ should be equal. In anchovy, the natural mortality is approximately 70% of the average annual biomass, average losses from commercial fishing and predators being 30% and 10%, respectively (Danilevsky and Mayorova, 1979). The effect of parasites on production is small enough to be ignored (Shchepkina, 1980a,b), as is the effect of detachment of scales or slime. The total annual elimination of the populations of anchovy amounts to as much as 110% of the average annual biomass, that is, 470 kt. B_Σ is therefore 1.5 times greater than P_Σ. The degree of congruence of these two estimates is fairly good, the discrepancy probably arising from inaccuracies in the estimate of elimination. Production can be calculated from measurable characteristics such as the increase in weight, age structure and numbers in a population, but the components of elimination, apart from losses through commercial fishery, are not actually measurable in practice.

Despite this, data obtained from sprat show better comparability than those from anchovy. Natural mortality in the sprat population, observed over many years, was 57%, losses from predators 10% and from commercial fishery 3%. The total elimination is therefore 70% of the total biomass, or 200 kt. The difference between P_Σ (210 kt) and B_Σ is only 5%.

Ivanov (1983) reported that the natural mortality of horse-mackerel in the Black Sea was 45% and the removal by commercial fishery 13%, making 58% elimination (losses from predators being disregarded) or 21 kt. However, our calculations showed that the total annual production of this fish was 26 kt, a difference of 20%. This value may be reduced if the losses from predators are taken into account. Assuming such losses to be 10%, as in the case of sprat, the total elimination of horse-mackerel would be 68% or about 25 kt. The difference between production and elimination is then only 4%.

From the data of Danilevsky *et al.* (1979), the natural mortality and losses from the fishery of red mullet can be reckoned as 60% and 30%, respectively. Assuming that predators remove 10% of the stock, the total elimination would amount to 100% of the biomass or 6 kt. However, production yielded by red mullet is 8 kt – the difference is 25%. Data on the natural mortality of whiting are lacking. These comparisons therefore show a reasonable correspondence between the two basic characteristics of stocks of Black Sea fish – production and elimination.

The assessment of generative production of fish with an extended spawning period (multiple spawning) implies exact knowledge of the proportions of eggs spawned during the whole period and the total weight of fully grown oocytes. As regards Black Sea fish, the question was settled by Oven (1976). A comparison between generative production of these fish and the weights of their bodies after spawning showed that females (apart from those of pickerel) produced E-stage oocytes of combined weights equal to 20–65% of the corresponding body weights over the whole period (Shulman *et al.*, 1983).

The generative production of multiportion fish in the Black Sea amounts to 37–67% of the total production (Tables 11 and 13), so the ratio between generative and somatic productions ranges from 0.58 to 1.91, apart from the 'small-portion' pickerel, where the figures are 24–31% and 0.31–0.44, respectively. The generative production is therefore comparable to somatic production. According to Tseitlin (1988), the generative production yielded by polycyclic fish amounts to 70% of the total. Pickerel, anchovy and whiting give more generative than somatic production, while sprat, red mullet and horse-mackerel give less. The weight of reproductive products produced within the life spans of Black Sea fish is comparable with the ultimate weight of their bodies. As the proportion of males in the populations is 1/3 and their generative production is 1/10 of that yielded by females, the share contributed by males to the total generative production of the population is a mere 5%.

Fish are not unique in the animal kingdom in their multiportion spawning. In organisms as diverse as worms, molluscs, crustaceans, insects and birds the yield of generative production is not merely comparable with somatic, but is sometimes far greater (Shpet, 1971).

Table 13   Generative production of Black Sea fish as a percentage of total production.

	Anchovy	Sprat	Horse-mackerel	Red mullet	Pickerel	Whiting
Wet mass	37	2	58	57	24	39
Dry mass	41	57	64	55	28	38
Protein	38	64	67	50	26	42
Energy equivalent	37	53	65	57	31	40

In these Black Sea species, apart from pickerel, a complete generative synthesis followed by spawning can occur as frequently as once every 10 days. Somatic production, in contrast, accumulates over the lifetime of the fish but also allows regular self-regeneration (i.e. replacement of all the dry matter of the body). Annual somatic production occurs over a 6-month period of favourable feeding and temperature conditions, during which the organic material in the body of an anchovy regenerates ten times. If generative and somatic production (extra tissue) are more or less equal in these multi-spawning fish, the actual somatic biosynthesis greatly exceeds generative biosynthesis because of this replacement factor. It must be taken into account when assessing the biosynthetic activity of fish.

The share of total production contributed by different age groups varies between species (Table 14). In anchovy 70% of the total production (fresh weight) is contributed by yearlings and 2-year-old fish; in sprat the value is 80%, and in red mullet over 80%. Three age groups (yearlings, 2- and 3-year-old fish)

Table 14   The production of different age groups as a percentage of the total production of the year.

	Age group	Anchovy	Sprat	Horse-mackerel	Red mullet	Pickerel	Whiting
P_Σ wet substance	0+	17	*	5	4	35	*
	1+	38	53	25	48	32	25
	2+	31	28	25	48	15	25
	3+	12	19	25	—	12	20
	4+	2	—	10	—	6	20
	5+	—	—	10	—	—	10
P_Σ dry substance	0+	17	—	5	8	35	—
	1+	38	55	25	46	32	25
	2+	31	28	25	46	15	25
	3+	12	17	25	—	12	20
	4+	2	—	10	—	6	20
	5+	—	—	10	—	—	10
P_Σ protein	0+	16	—	5	4	35	—
	1+	38	56	25	48	32	25
	2+	32	31	25	48	15	25
	3+	12	13	25	—	12	20
	4+	2	—	10	—	6	20
	5+	—	—	10	—	—	—
P_Σ calories	0+	17	—	5	12	35	—
	1+	38	54	25	44	32	25
	2+	31	28	25	44	15	25
	3+	12	18	25	—	12	20
	4+	2	—	10	—	6	20
	5+	—	—	10	—	—	10

* Since the first group of sprat and whiting appears in winter and spring, not in summer as in other species, the starting point is 1+, not 0+.

yield 75% of the total production of horse-mackerel. The contribution from fry or fingerlings is relatively high (35%) only in pickerel, in which, together with yearlings, this age group contributes as much as 80% of the total production. In whiting, the four age groups from fingerlings to 3-year-olds each contribute similar amounts to production, the share contributed by 4- and 5-year-olds being of minor importance. Similar pictures emerge whether one is estimating protein, dry matter or energy. As stated earlier, the contribution of those larvae and fry that had died or been eaten is usually disregarded in estimating total production. However, Nikolsky et al. (1988) have calculated the contribution made to production by 'eliminated' recruit anchovy. Individual weights and elimination rate of larvae not older than 18 days were taken into consideration (Dekhnik, 1979) and extrapolated to fry not older than 3 months (Table 14). The elimination of larvae and fry accounted for 30–50% of total production, so the annual P/B (fresh weight) should be increased from 0.72 to 0.96 in anchovy. Similar corrections appear to be justified in other Black Sea species.

5.3. SUBSTANCE AND ENERGY EXPENDITURE

Appreciation of substance and energy expended on metabolism and growth is essential to the understanding of production in living systems. Expenditure and production have been successfully considered as an integrated whole in several investigations on fish (Ivlev, 1939, 1960; Gerking, 1952, 1966, 1972; Vinberg, 1956).

Expenditure in individual fish can be assessed by measuring excretion of nitrogen (Gerking, 1952; Karzinkin, 1962; Mann, 1965, 1969), phosphorus (Gandzyura, 1985) and compounds labelled with ^{14}C (Karzinkin, 1962; Fedorova, 1974; Romanenko et al., 1980; Belyaev et al., 1983; Shekhanova, 1983; Yarzhombek et al., 1983). It can also be assessed from losses of biochemical components under experimental muscular loading (Pentegov et al., 1928; Idler and Clemens, 1953; Idler and Bitners, 1958; Job and Gerald, 1976; Moore and Potter, 1976; Boëtius and Boëtius, 1980; Gleebe and Leggett, 1981; Beamish and Legrow, 1983) and prolonged fasting (MacKinnon, 1973; Gerald, 1976; Muravskaya, 1978; Stevens and Dizon, 1982; Yarzhombek et al., 1983). Respiratory quotient (Morris, 1967; Kutty, 1968; Fischer, 1970; Jedryczkowski and Fischer, 1973; Alikin, 1975; Kutty and Mohamed, 1975) and ammonia quotient (Stroganov, 1962; Jobling, 1980; Kutty, 1968; Kutty and Mohamed, 1975; Sukumaran and Kutty, 1977) have also been used. The weights of lost tissue can be converted to energy-equivalent by the use of a bomb calorimeter (Webb, 1975; Gerald, 1976; Diana and McKay, 1979; Dawson and Grimm, 1980; Beamish, 1981), but this equipment is inadequate to deal with the material from a large fish.

Oxygen uptake is not very useful in studies of energy expenditure, since 30–40% of it is not used for phosphorylated oxidation or ATP resynthesis, but for other varieties of oxidation such as peroxic and enzymatic oxidation which results in the production of heat (Khaskin, 1981). In any case, the efficiency of oxygen uptake as measured from the P/O coefficient varies according to various eco-physiological factors (Verzhbinskaya, 1968; Arsan et al., 1984; Savina, 1992). The heat produced is dissipated into the environment, except in the case of certain very active fish that possess a heat-exchange mechanism (Stevens et al., 1974) permitting the body temperature to rise and thereby increase muscle power (Love, 1980, p.321).

Anaerobic pathways may be of more importance, for then oxygen is either not consumed at all, or its consumption is little involved in substrate exchange (Morris, 1967; Kutty, 1968; Hochachka, 1969; Kutty and Mohamed, 1975; Hochachka and Somero, 1977; Johnston, 1977; Sukumaran and Kutty, 1977; Duthie, 1982; Klyashtorin, 1982). This circumstance is of special significance as regards aquatic animals which can excrete unoxidized or insufficiently oxidized products of anaerobic metabolism through the body surface (gills, skin) into the water. In these cases, in addition to oxygen consumption, other parameters must be measured to understand energy metabolism. These include enzymatic reaction rates and the dynamics of proteins, lipids, glycogen, macroergic phosphates, and their metabolic products (amino- and fatty acids, ADP and AMP, glucose, pyruvate and lactate) and the turnover of labelled compounds.

Oxygen consumption data have been combined with biochemical observations, using data on oxygen uptake by Black Sea fish living under optimum conditions. Such conditions are known to stimulate maximum efficiency of energy utilization, reducing the errors that usually arise when the evaluation is based on oxygen consumption. As mentioned in Chapter 3, $Q = a W^k$, where Q is the level of energy metabolism, W is the weight, and a and k are coefficients. The coefficient 'a' is the rate of oxygen uptake in an organism of body weight 1 g. It is convenient to use this when the metabolic rates of animals of widely different sizes and species are being compared under different living conditions. Oxygen consumption was measured by Belokopytin (1968, 1978) in six species of Black Sea fish (Table 3, page 60). To these figures we have added data from Shulman et al. (1987) and Stolbov (1992).

The range of temperatures experienced during the annual cycle is 17–25°C in anchovy, 12–20°C in pickerel and 9–20°C in whiting, and the relationship between ambient temperature and oxygen consumption was determined in the three species (Belokopytin and Shulman, 1987; Skazkina and Danilevsky, 1976). To understand the general character of the relationship between temperature and oxygen uptake, we also used data from Skazkina (1972), Klovach (1983) and Shekk (1983). Our calculations show that the Q_{10} is 2.2 on average for the species studied (Belokopytin and Shulman, 1987), and this agrees with that suggested by Vinberg (1983) for a whole range of aquatic poikilotherms. Similar Q_{10} values [standard metabolism: total and active metabolism measurements by Fry (1957, 1971), Brett and Groves (1979), Beamish (1981), Klyashtorin (1982) and Matyukin *et al.* (1984) may display quite different temperature dependence] have been reported for many species of fish (Beamish, 1968, 1978; Brett, 1970; Daan, 1975; Klyashtorin, 1982; Klyashtorin and Smirnov, 1983).

Table 15 allows determination of '*a*' coefficients for six species of fish at environmental temperatures (Figure 58). Extrapolating energy metabolism from single organisms to populations poses great problems, because the rate of oxygen uptake is usually measured in isolated individuals with artificially restricted mobility. Fish in a shoal display a 'group effect' in which the metabolic rate is reduced. On the other hand, their mobility in the wild is higher

Table 15   Environmental temperatures experienced by Black Sea fish during the annual cycle (°C).

	Jan	Feb	Mar	Apr	May	Jun	July	Aug	Sep	Oct	Nov	Dec
Anchovy, horse-mackerel and red mullet	8	7	9	11	13	20	23	23	20	16	12	10
Pickerel	8	7	9	10	12	18	20	20	16	14	12	10
Sprat	7	7	8	9	10	11	11	12	13	11	9	8
Whiting	7	7	7	8	8	8	9	9	9	8	8	8

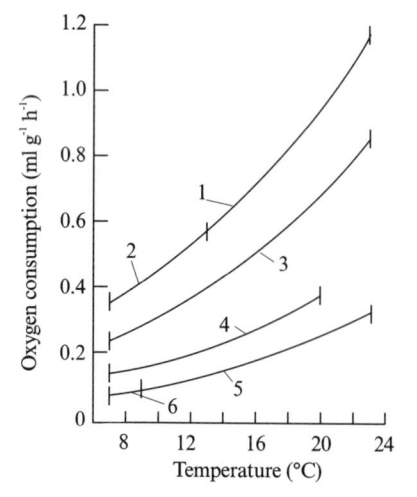

Figure 58   Relationship between the oxygen consumption of different species and the temperature of the habitat in which they were caught: 1, anchovy; 2, sprat; 3, horse-mackerel; 4, pickerel; 5, red mullet; 6, whiting.

than that under experimental conditions. Finally, the metabolic rate under natural conditions is strongly influenced by feeding – the specific dynamic effect of food (SDEF) – whereas unfed fish are most often used in experiments on standard metabolism. According to Alekseeva (1959, 1978), the group effect reduces oxygen consumption of grey mullet, horse-mackerel, red mullet, pickerel, whiting and scorpion fish which have shoaled under experimental conditions also. The average decrease of uptake per fish in a shoal was 30% when compared with the value obtained in isolated individuals. Interestingly, the oxygen consumption increases by 30% during feeding, so the opposite effects balance each other. The latter value can also be obtained by averaging the mass of data accumulated on SDEF (Lasker, 1962; Paloheimo and Dickie, 1966; Brett, 1970; Smith, 1973; Beamish, 1974; Pierce and Wissing, 1974; Timokhina, 1974; Ware, 1975; Elliott, 1976; Miura et al., 1976; Kitchell et al., 1977; Brett and Groves, 1979; Tandler and Beamish, 1979; Flowerdew and Grove, 1980; Holeton, 1980; Jobling and Davies, 1980; Smith et al., 1980; Jobling, 1981; Cho et al., 1982; Johnston, 1982; Kerr, 1982; Klyashtorin, 1982; Diana, 1983; Hamada and Maeda, 1983; Klyashtorin and Smirnov, 1983; Turetzky, 1983; Yarzhombek et al., 1983; Amineva and Yarzhombek, 1984; Karpevich, 1985a,b; Khakimullin, 1988; Karamushko, 1993).

The levels of metabolism of Black Sea fish in relation to their natural mobility were determined by Belokopytin (1990). First, the relationship between swimming rate and oxygen consumption in a Brett-type tunnel respirometer was determined, using horse-mackerel, pickerel, anchovy and annular bream. The relationship is described by the equation:

$$Q = Q_0 b^v \tag{24}$$

where Q is the total metabolism, Q_0 is the basal metabolism, i.e. metabolism at 'zero' rate, v is the swimming velocity and b is a coefficient. Belokopytin (1978, 1993) gives the values obtained by using this equation on different species. A similar picture has emerged from many other species (Brett, 1970; Webb, 1971; Matyukhin, 1973; Beamish, 1974; Yarzhombek and Klyashtorin, 1974; Ware, 1975; Kiceniuk and Jones, 1977; Brett and Groves, 1979; Duthie, 1982; Klyashtorin, 1982; Stevens and Dizon, 1982; Matyukhin et al., 1984). Secondly, the natural daily mobility of fish was examined experimentally during different seasons (Vyskrebentsev and Savchenko, 1970; Belokopytin, 1993). The experiments were conducted in spacious tanks where fish were less restricted in movement than in the study of standard metabolism. The average distance covered and average daily swimming rates were also deduced. Stage three involved underwater observations, by means of cinematography, of fish shoaling under natural conditions (Belokopytin, 1993).

Applying equation 24, the levels of total energy metabolism were determined in fish under natural and laboratory conditions (in the second stage described above, metabolism has been named 'routine' because of its closer similarity to natural as compared with strictly 'standard' metabolism). These levels were then compared with those known in standard metabolism of Black Sea fish 'population coefficients' (Table 3, page 60), i.e. the ratio between total metabolism recorded in nature and standard metabolism (Table 16). These ratios are 2.3 for highly mobile fish (anchovy, horse-mackerel), 1.4 for moderately mobile (red mullet and pickerel), and 1.0 for the inactive whiting and scorpion fish. Klyashtorin (1982) cites similar coefficients for fish from other seas. We found these quotients very helpful when converting rates of energy metabolism from individual fish to the level of a population. The quantitative assessment of total metabolism of fish in nature and the energy balance of a population both depend on the proper extrapolation of data from experimental to population level. The extrapolation is indeed the most important element in balance calculations.

The population coefficients given above for highly mobile fish are similar to the coefficients of conversion from standard (Q_{st}) to total metabolism (introduced by Vinberg, 1956) and are about 2. Similar values of Q/Q_{st} were reported for cod (Kerr, 1982) and grey mullet (Flowerdew and Grove, 1980). Brett and Groves (1979) suggested that this coefficient be named after G.G.Vinberg as 'Vinberg-2', 'Vinberg-1' being applied in standard metabolism, where it should be increased from 2 to 3–4 (Mann, 1965). Experiments on sockeye salmon gave a value for Q/Q_{st} of 3.5 (Brett and Glass, 1973), and 3.0 for bleak (Ivlev, 1962). Ware (1975) concluded that the coefficient was more than 2 in pelagic and juvenile fish.

The population coefficients of moderately mobile, as distinct from highly mobile, Black Sea fish decrease to 1.4, and low mobile to 1, i.e. to the approximate values known for standard metabolism. These facts lend emphasis to the importance of conducting experiments in order to substantiate the conversion calculations. It is incorrect to extrapolate experimental data to total metabolism in fish without making direct observations on their motor activity in the natural state (Paloheimo and Dickie, 1966; Beamish, 1968; Brett, 1970; Healy, 1972; Elliott, 1976; Tytler, 1978; Kerr, 1982; Diana,

Table 16　Population coefficients (Q_Σ/Q_{st}) of Black Sea fish.

Anchovy	2.3
Sprat	2.3
Horse-mackerel	2.3
Red mullet	1.4
Pickerel	1.4
Whiting	1.0

1983). Lack of telemetry (measurement from a distance) under natural conditions is, as Brett and Groves (1979) believe, the bottleneck in studies of fish bioenergy. The same opinion was expressed by Klyashtorin (1982), who found that the average daily oxygen consumption of fish in nature exceeded their standard metabolism by anything from 20% to 300%, depending on the mobility of the species.

We have already stated that total metabolism (Q) is the sum of basal (Q_b) and active (Q_a) metabolism:

$$Q = Q_b + Q_a \tag{25}$$

As Belokopytin (1968, 1978) reported, the percentage of basal metabolism (Q_b) estimated from standard (Q_{st}) was 40% in highly mobile horse-mackerel and anchovy, 75% in moderately mobile red mullet and about 100% in the sluggish scorpion fish. Sometimes basal metabolism is named 'supporting' metabolism, which is incorrect and leads to misunderstanding. Metabolism and diet support the balanced equilibrium, $P = 0$.

Using the population coefficient, the ratio of Q_b to Q is 17% for highly active fish, 54% and 100% for moderately mobile and sluggish, respectively. Then Q_a^* is 84%, 46% and 0%, respectively. Similar estimates were given by Yarzhombek and Klyashtorin (1974), who found that, on average, the energy consumed in swimming could amount to 90% of the basal metabolism in active fish and 30% in sluggish species.

To determine the rate of oxygen uptake in each age-group during the annual cycle, one should calculate the 'a' coefficients at given temperatures, Q_{st} from the weight of the fish (W), using $Q = AW^k$, and total metabolism (Q) by means of Q_{st} multiplied by the 'population coefficient'. Figure 59 shows the results of our computations for anchovy. In order to make comparisons of estimates of production and other characteristics of balance more convenient, the energy metabolism of the fish has been assessed in kilocalories. The oxycaloric quotient (Q_{ox}) of 4.8 cal ml^{-1} of oxygen was used in estimating oxygen consumption (Brody, 1945; Vinberg, 1956; Ivlev, 1960; Brett, 1973; Matyukhin, 1973; Beamish et al., 1979; Klyashtorin, 1982). This quotient corresponds with the functional norm, when fish are swimming smoothly, oxygen debt not being found and the respiratory quotient (RQ) 0.8, implying complete oxidation of catabolized substrates (Beamish, 1974) apart from protein. Some investigators (Brett and Groves, 1979; Elliott and Davison, 1975) suggest reducing the value of Q_{ox} to 4.64 cal ml^{-1} O$_2$ in predatory fish.

---

*In non-Russian literature, active metabolism is denoted by Q, not Q_a, and implies total metabolism during intensive swimming (Fry, 1957; Prosser, 1969; Brett, 1970; Webb, 1971; Brett and Glass, 1973; Elliott, 1976; Brett and Groves, 1979; Beamish, 1981; Duthie, 1982; Kerr, 1982). The term 'scope of activity', used to define Q_a, is rather awkward. In Russian literature, by 'active metabolism' we mean energy consumed in locomotion (Vinberg, 1956; Ivlev, 1962; Klyashtorin, 1982).

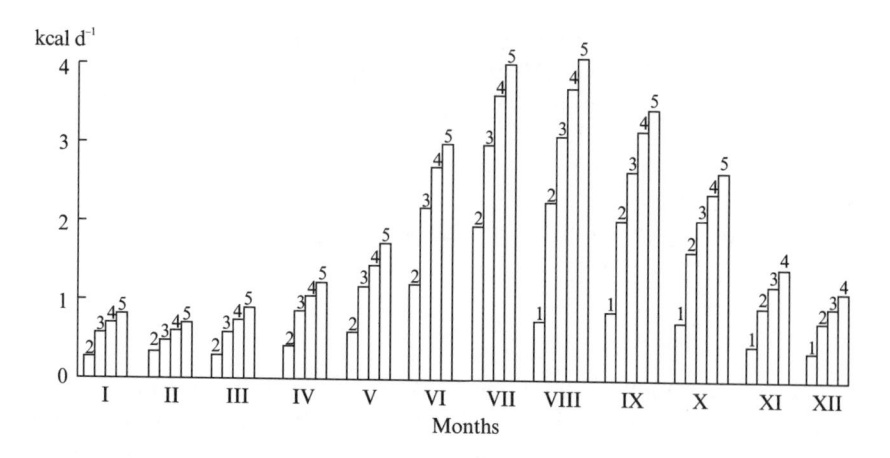

Figure 59   Energy expended on total metabolism by anchovy of different year-classes throughout the year.

The rate of oxygen uptake of an average fish in a population can be calculated as follows: firstly, a series of Q_s should be determined from average weights according to the age structure of the population, as in the estimation of production. Values obtained are then divided by the weight of the 'average fish', and the result represents the intensity (specific rate) of total energy metabolism in the population, Q/W (Figure 60). There are therefore similarities and contrasts in the curves that describe changes in the specific rate of energy metabolism in populations of the six species during their annual cycles. The similarity arises largely from the temperature effect; in all these fish, apart from whiting, the metabolic rate increases in the summer to a maximum in September when the sea is warmest, minimal oxygen being consumed in winter. Therefore, the curves describing changes in metabolic rate usually have one peak, corresponding with the temperature curve.

The specific rate of energy metabolism in a population also varies between species, depending upon the innate mobility rate and size of the fish, as well as the temperature; the three factors can produce a combined effect. The specific metabolic rate is considerably higher in the mobile anchovy, sprat and horse-mackerel than in the less active red mullet, pickerel and whiting. The estimates were highest in the population of anchovy (high mobility, small size and high preferred temperature). The high specific metabolic rate in sprat is partly the result of their small size, but in horse-mackerel the result of the high preferred temperature. The seasonal rhythm of metabolic rate is more pronounced in the warm-water horse-mackerel than in cold-water sprat. In whiting, the metabolic rate is remarkably low because of the combined effect of the three factors cited above. In populations of red mullet and pickerel, which are relatively inactive and require

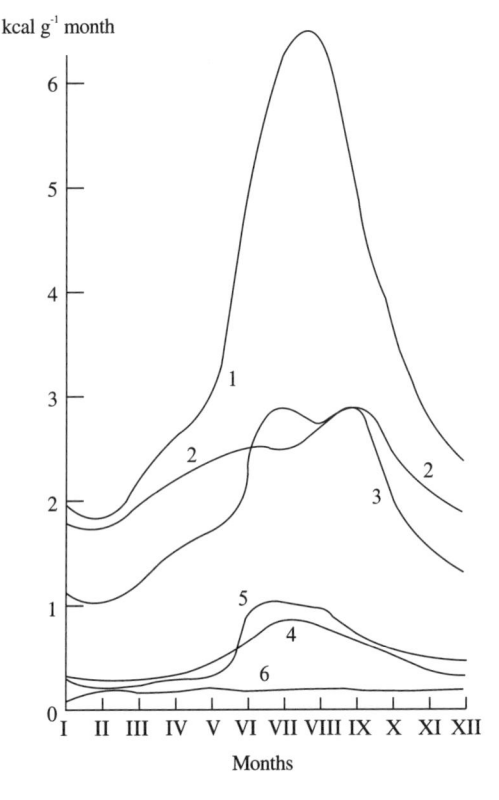

kcal g^{-1} month

Figure 60   Expenditure of energy on total metabolism by populations of: 1, anchovy; 2, sprat; 3, horse-mackerel; 4, pickerel; 5, red mullet; 6, whiting.

warm water, a distinct seasonal rhythm of metabolic rate can be traced, while in the whiting, a resident of waters lying under the thermocline, such a rhythm is not found.

The utilization of energy substrates has been studied in experiments in which the fish are starved. There is a dramatic loss in weight (Suppes *et al.*, 1967; Savitz, 1969; Love, 1970, 1980; Shulman, 1972a; MacKinnon, 1973; Gerald, 1976; Caulton, 1978b; Muravskaya, 1978; Jobling, 1980; Sorvachev, 1982; Stevens and Dizon, 1982; Yarzhombek *et al.*, 1983; Boëtius and Boëtius, 1985), but since it is the lot of many species to undergo extreme depletion under natural conditions, the disorganization of swimming muscle and some other organs seems to cause no permanent damage. Provided that the depletion does not actually kill the fish, complete recovery is usually possible after food again becomes available. In 'lean' fish, the loss of protein appears to prevail (Solomon and Brafield, 1972; Shulman, 1974; Storozhuk, 1975; Shevchenko, 1977; Yakovleva and Shulman, 1977; Medford and Mackay, 1978;

Muravskaya, 1978; Diana and Mackay, 1979; Shatunovsky, 1980). However, in all starved fish it is the lipid that is mobilized first, except possibly in the eel. In fatty fish, much lipid is used from the flesh, while in lean fish it is used from the liver. In both types of fish, muscle protein is mobilized only when the lipid resources fall below a critical level. Black and Love (1986) showed that energy substrates are mobilized in a definite sequence, white muscle protein, for example, being metabolized at an earlier stage than red muscle protein, while, on refeeding, the latter is replenished before the former.

Carbohydrate (glycogen) losses are often negligible. Under fasting, the glycogen content can be kept at a stable level by gluconeogenesis, derived from lipids and proteins. Similar results are known from the studies of winter fasting of fish in nature (Mukhina, 1958; Polyakov, 1958; Shcherbina, 1989; Shekk et al., 1990). Gluconeogenesis does not assist all species during starvation, however. In fasting cod, for example, the activities of the gluconeogenetic enzymes phosphoenol pyruvate carboxykinase (PEPCK) and fructose diphosphatase (FDPase) actually decline in the liver and muscle during prolonged starvation. Alanine aminotransferase also decreases in the liver. It really seems that cod reject this facility, and the glycogen levels steadily decline in red and white muscle, and the liver. In contrast, the same enzymes increase in starving rainbow trout, in which the carbohydrate levels are maintained. The strategy employed to deal with glycogen lack in cod seems to be in controlling the swimming activity. At later stages of starvation, cod virtually cease to swim, but spend most of their time completely stationary on the bottom of the aquarium. They maintain their balance and show slow opercular movement. Rainbow trout, on the other hand, continue to swim quite vigorously throughout starvation (Love and Black, 1990). The reason for this species difference may relate to their activity in differing habitats – relatively stationary water on the ocean floors for cod and running streams for trout.

Embryos and larvae of fish not capable of exogenous feeding follow the same mode of substrate utilization during their catabolism as adult fasting fish (Fessler and Wagner, 1969; Atchison, 1975; Woo et al., 1978; Dabrovsky et al., 1984; Gosh, 1985; Pionetti et al., 1986).

The data available on the loss of biochemical substrates permits estimation of the rate of energy metabolism in starving fish. The technique is often applied in experiments made on fish embryos and larvae (Lasker, 1962; MacKinnon, 1973; Boëtius and Boëtius, 1980; Constantz, 1980; Cetta and Capuzzo, 1982). The technique is laborious, so is rarely used in studies on adult fish; we employed it in studying squid (Stenoteuthis oualaniensis) from the Indian Ocean (Shulman and Nigmatullin, 1981). The supply of biochemical substrates for locomotion is particularly important for active fish, where swimming can account for as much as 80% of the total metabolism.

Using the ammonia quotients calculated by Muravskaya and Belokopytin (1975), one can deduce that the swimming energy of active fish is 20% derived from protein and 80% from lipids. At rest (standard metabolism), the share contributed to the energy supply by protein increases 1.5-fold, the ratio of protein to lipids now being 30:70. In the sluggish scorpion fish, the ratio between protein and lipids is 50:50. These figures agree with those obtained from other species (Kutty and Mohamed, 1975; Jobling, 1980).

Biochemical studies therefore yield results essential to the quantitative assessment of energy expenditure during swimming. Interest in the problem began about seven decades ago, when the spawning migration of chum salmon into the Amur River was studied by Pentegov et al. (1928). They calculated the losses of lipids, proteins and the energy equivalent of weight in fish that had migrated 2000 km from the mouth of the Amur to their spawning grounds, and also the average energy used per kilometre during migration. Similar work was performed by Idler and Clemens (1953) on sockeye salmon from the Fraser river in Canada. Their investigations are often cited in the literature as an example of the correct approach to determining the energy consumed by locomotion of fish (Vinberg, 1956; Ivlev, 1962; Brett and Groves, 1979). However, we can hardly share this enthusiastic approval. During the spawning migration of salmon, energy is used as reproductive products develop in addition to that used for swimming. Where it has been used to provide generative synthesis, energy does not lend itself to quantitative assessment. Gleebe and Leggett (1981) made a similar error in their calculations of energy consumed by migrating and maturing shad in American rivers. It would be better that the portions of energy used in locomotion, maturation and basal metabolism are taken together in evaluating the situation during spawning migration, as has been done in experiments on lampreys (Moore and Potter, 1976; Beamish, 1978; Beamish et al., 1979).

Energy expenditure has been experimentally studied and assessed through the losses of lipids and fatty acids in coho salmon during prolonged swimming in a flume (Krueger et al., 1968). We used this approach with horse-mackerel and red mullet (Shulman et al., 1978). This mode of calculation of energy expenditure in the course of uptake of biochemical substrates has been used very successfully in field and experimental studies of birds (King et al., 1963; Dolnik, 1969) and insects (Kinsella, 1966). It is also essential, however, to know the basal and active metabolism through the amount of oxygen consumed. The method is quicker to carry out and, more to the point, it is often impossible to make representative extracts for biochemical analyses, especially when the subjects are large fish. Working out the total loss of substrates sustained is no easy task. It is also technically impossible to estimate the energy equivalent of the organism by burning it whole in a bomb calorimeter.

It should not be forgotten that estimates of energy consumed, calculated from losses of substances, may not agree with those from the oxygen uptake because of the formation of under-oxidized products of protein and lipid catabolism (Brett, 1973; Klyashtorin, 1982). These products can be produced even during cruising, despite the current notion about their complete oxidation.

In Chapter 3 it was stated that the ratio between succinate and lactate dehydrogenase activity measured in muscle and liver increases in summer when the water temperature rises and locomotor activity increases (Emeretli, 1981a,b), falling again in winter. The energy potential (situation which encourages high metabolic rates) was found to increase in the summer-time through the increased production of erythrocytes and higher haemoglobin levels in the blood (Kotov, 1979; Rakitskaya, 1982) and also higher myoglobin levels in the red muscle (Figure 48). In addition, the oxygen-binding capacity of the blood increases (Klyashtorin, 1982). Putman and Freel (1978) confirmed that a relationship exists between blood characteristics and locomotor activity.

It seems likely that the balance between aerobic and anaerobic energy metabolism, and with it the balance of utilization of different substrates, alters during the annual cycle. Data are somewhat scanty, so this hypothesis should not yet be accepted as proved. Using available estimates of the rates of energy metabolism in different age groups of anchovy (Figure 59), one can assess the amounts of energy consumed in the form of protein and of lipid. Dividing the value obtained by the corresponding caloric coefficients, 4.1 and 9.3, one can estimate the weights of each of these two basic substrates that have been used (Figure 61). In evaluating the quantity of protein mobilized, the physiological caloric coefficient (4.1) is used rather than the chemical one (5.6). We applied the latter when calculating the energy equivalent of the body during our study of production (section 5.1).

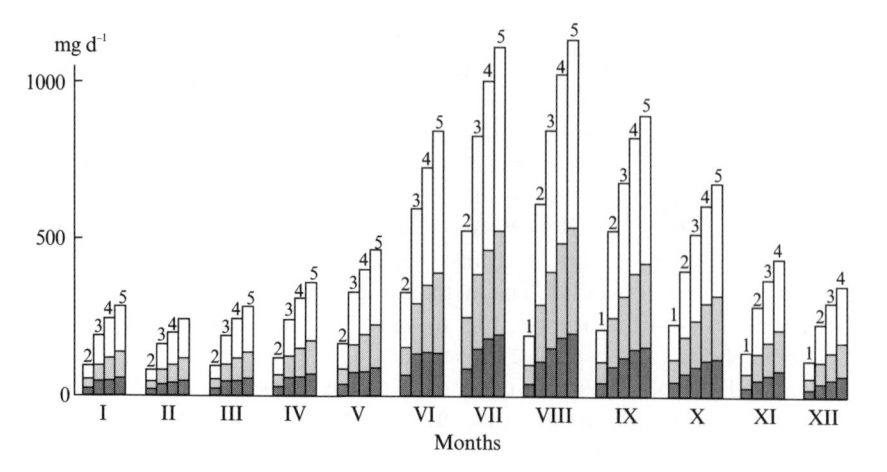

Figure 61 Expenditure of substances on total metabolism during a year by anchovy. Whole column: wet matter; light stipple: lipids; heavy stipple: proteins. The small numbers represent age groups.

Energy consumed by fish can be conveniently regarded as the amount of protein and lipid used up, glycogen being only a small and variable fraction. Figure 62 shows the average expenditure of the two main constituents in populations of Black Sea fish over the annual cycle. The curves are similar to those that describe energy expenditure (Figure 60), indicating that the processes of substance and energy utilization have much in common. In the flesh of fatty fish such as anchovy, sprat and horse-mackerel, lipid, in preference to protein, is expended, while in non-fatty fish like red mullet and whiting the expenditure of flesh proteins is the most obvious feature.

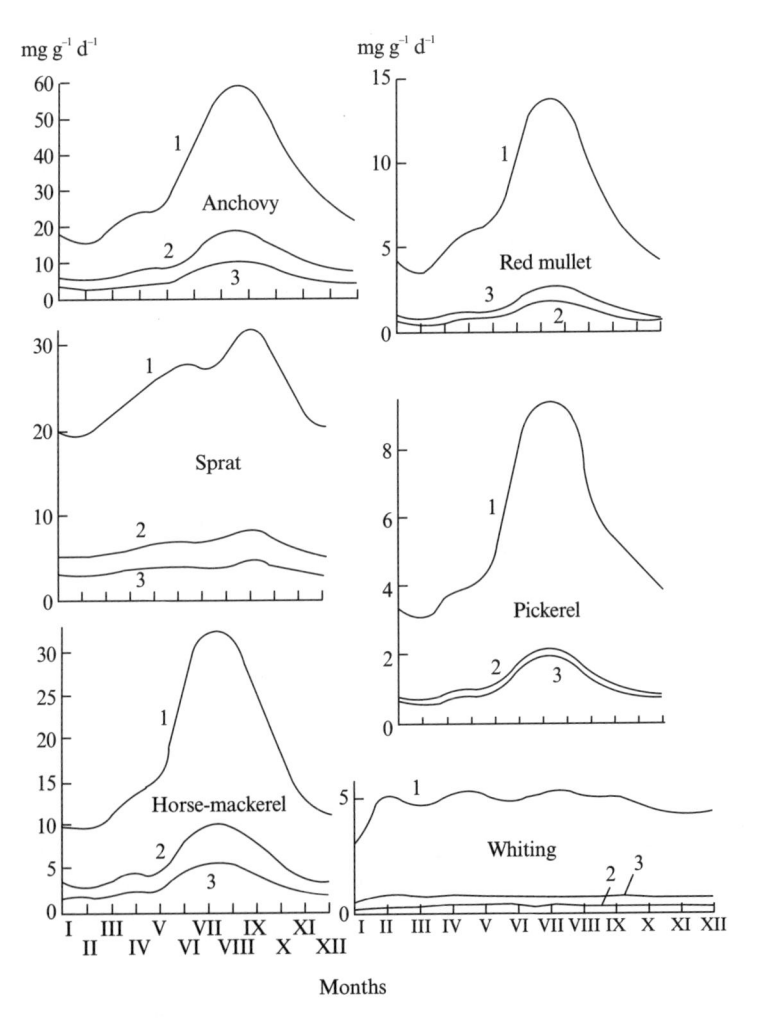

Figure 62   Expenditure by populations of different species on total metabolism during a year: 1, wet tissue; 2, lipids; 3, proteins.

However, the sequence of mobilization must always be borne in mind, lipids being mobilized before the contractile muscle is 'raided'; in many species of non-fatty fish, large resources of lipid are in fact stored in the liver. Pickerel take up an intermediate position, lipid and protein being removed equally from the flesh.

The annual expenditure of substance and energy (Q_Σ) in populations has been assessed using the equation:

$$Q_\Sigma = \sum_{i=1}^{12} Q\Delta t \qquad (26)$$

Table 17 shows the results obtained.

The estimates are about 1–2 orders of magnitude greater than the corresponding production values (Table 11), which represent the proportion between the constructive processes and the total metabolism of populations. The proportion of energy expended by each age group in relation to the expenditure of the total population is identical to that of the total production (Table 14).

5.4. CONSUMPTION, TRANSFORMATION AND UTILIZATION OF SUBSTANCE AND ENERGY

Consumption of food (nutrition), growth (production) and metabolic expenditure are the constituents of a balance which underlies the functioning of heterotrophic organisms and populations. The most important component of this triad is nutrition. Food is the link between the organism and its environment, because it is food alone that supplies organic matter and energy to higher organisms.

Table 17   Annual expenditure of substance (g) and energy (cal) per total population in the Black Sea.

Species	Energy	Protein	Lipid
Anchovy	173.98×10^{11} (90.84)	8.44×10^{11} (1.98)	14.95×10^{11} (3.51)
Sprat	74.68×10^{11} (26.67)	3.63×10^{11} (1.30)	6.41×10^{11} (2.29)
Horse-mackerel	83.72×10^{10} (23.45)	4.10×10^{10} (1.15)	7.21×10^{10} (2.02)
Red mullet	31.38×10^{9} (5.23)	3.03×10^{9} (0.51)	2.04×10^{9} (0.34)
Pickerel	— (5.75)	— (0.42)	— (0.43)
Whiting	17.82×10^{9} (1.98)	2.16×10^{9} (0.24)	0.99×10^{9} (0.11)

Figures in brackets are intensity of metabolism, i.e. per 1 g of wet matter of population over the year.

Various approaches have been developed for the study of substance and energy in living organisms. Attention is drawn to only one approach, that involving the concept of balance. Equation 13 ($C = Q + P + F$) is used to convert assessment of substance production, energy accumulation and their expenditure in biological systems to assessment of consumption, transformation and utilization of matter and energy. The total biosynthesis and metabolism of a population relate closely to trophology. The 'balance' approach then acquires special value as a key in solving problems of ecology. It allows a detailed description of the condition of a population as interrelated to the other components of the ecosystem. The description includes quantitative assessment of the flow of substance and energy (Lindemann, 1942), determining not only the intensity but also the efficiency of production (Ivlev, 1939), food requirements and the trophic importance of the population in the biotic cycle (Vinberg, 1962).

In studying the balance of matter and energy, approaches that determine each part of the triad are especially important. However, because of problems with methods, it is not yet possible to obtain reliable estimates of food consumption in field studies, and existing estimates are not satisfactory. The indirect approach to studying the nutrition of fish is better.

Knowing the components of the balance equation, one can calculate the levels of food consumption in individual organisms and in populations without having to resort to laborious and unreliable methods, especially in the field (Meyen et al., 1937; Ivlev, 1939, 1962; Yablonskaya, 1951; Karzinkin, 1952; Krivobok, 1953; Vinberg, 1956; Malyarevskaya, 1959; Shulman, 1962; Gerking, 1966; Skazkina and Kostyuchenko, 1968; Lasker, 1970; Melnichuk, 1970; Ryzkov, 1976; Chekunova and Naumov, 1982; Chekunova, 1983; Shekk, 1983; Yarzhombek et al., 1983; Karpevich, 1985a,b). Yet, although most of these authors studied problems of fish culture, very few considered populations of fish in nature.

Principles and rhythms of food consumption revealed in the course of these researches provide a basis for calculating food requirements and diets widely applicable in hydrobiology and fish farming. However, the balance approach is not free from drawbacks, in particular the extrapolation of data obtained with individual organisms to natural populations. We discussed this problem in section of Chapter 5, when surveying difficulties in converting metabolic expenditure calculated from experiments to actual values in the field. Whether the investigator studies energy or nitrogen balance, he runs into similar difficulties. To resolve them, it is necessary to determine the correlation of energy and protein expenditures of fish occurring under laboratory and natural conditions. These ratios have already been computed as regards energy balance, but are not available for protein. It is known that protein metabolism is much more restrained than energy metabolism during changes in living conditions, and the state of the organism, particularly as regards stress, has a more pronounced effect on the metabolism of energy than of protein (Kaplansky, 1945). We therefore disagree with Mann (1965, 1969) in his criticism of the experiments on nitrogen balance performed by some researchers (Meyen et al., 1937; Karzinkin, 1952; Krivobok, 1953; Shulman, 1962). Handling stress was shown to have a negligible effect on the nitrogenous excretion of the fish used in the experiments. Vinberg (1956) demonstrated a similarity between the estimates of dietary intake obtained by applying the oxygen and nitrogen methods of computation. Further evidence was found in experiments carried out on squid that had been subjected to stress (Shulman et al., 1984). It is also pertinent to note that, with increased locomotor activity, the energy metabolism of Black Sea fish accelerates greatly, non-protein substrates being used (Muravskaya and Belokopytin, 1975). The balance-flow approach involves determining the food intake indirectly, by means of retrospective estimates of somatic and generative growth over a period (Bulgakova, 1993; Fedoseeva and Ovcharkina, 1993).

In many experiments and population studies, the consumption of food has been assessed from an index of stomach fullness (the weight of food in the stomach as a percentage of body weight) and also by measuring the rate of passage of food along the digestive tract (Novikova, 1949; Hatanaka et al., 1956; Healy, 1972; Backiel, 1973; Cameron et al., 1973; Daan, 1975; Mils and Forney, 1981; Kerr, 1982; Diana, 1983; Penczak, 1985). This method has long attracted criticism from physiologists and hydrobiologists (Karzinkin, 1952; Shulman et al., 1985). Indeed, the index of fullness characterizes static rather than dynamic aspects of food consumption, and swallowing is far more frequent than is complete emptying of the stomach and intestine. Besides, the rate of food movement and digestion depends on the temperature of the environment. These features show that this approach yields only an approximate value for feeding rate.

The balance method of estimating food consumption through oxygen uptake and excretion of nitrogen has also been suggested as an alternative to the index of fullness. It should be noted that it has given results 30–50% (occasionally up to 300%) higher than those obtained from indices of fullness. This appears to favour the balance approach, which yields good results not only for assessing the consumption of substances and energy but also in studies of growth (production) and metabolic expenditure. Data on food consumed at the height of the feeding season by Black Sea anchovy (Bulgakova, 1993), cod, wolf-fish and plaice in the Barents Sea (Karamushko and Shatunovsky, 1993), and yellowfin tuna (Olson and Boggs, 1986) coincide almost completely with balance estimates already known. Food must therefore be one of the known components of the balance equation (Paloheimo and Dickie, 1966; Brett, 1970; Zaika, 1972, 1983, 1985; Sergeev, 1979; Kerr, 1982; Stevens and Dizon, 1982).

We have emphasized the necessity of employing a combined substance-and-energy approach in studies of balance. Substance and energy uptake are better studied together, whether in fish culture or in basic studies of population ecology. A study of energy alone gives no indication of the biological value of the diet or whether it is balanced in its proportions of proteins, lipids, carbohydrates and minerals, or how available are the essential amino- and fatty acids, microelements and vitamins. These are key points in studies on fish farming and the physiology of nutrition (Malikova, 1962; Shcherbina, 1973; Cowey and Sargent, 1979; Sorvachev, 1982; Ostroumova, 1983). Unfortunately, this aspect is lacking in trophological investigations of natural populations of fish.

So far most attention has been given to consumption. It is now necessary to consider transformation and utilization. In the previous section, utilization and expenditure were discussed, disregarding the nature of the food eaten, but this factor will now be introduced.

When the indirect method of calculation is employed, the assimilated food is assessed prior to that consumed. Assimilated ('converted', 'transformed') food (Brett and Groves, 1979) is the sum of the total production and the total (both substance and energy) metabolic expenditure:

$$A = P + Q \tag{27}$$

where A is the principal characteristic in estimating the flow of matter and energy through a biological system and its production efficiency.

In order to convert from assimilated to consumed food, it is necessary to know the assimilation quotient (I):

$$I = A/C \tag{28}$$

and
$$C = A/I \tag{29}$$

The mean assimilation quotient as regards fish and other aquatic organisms is 0.8 (Vinberg, 1956). It was and still is widely used for calculating consumption or assimilation of food.

As $A = 0.8C$, then $C = 1.25A$. There has been much animated discussion about the numerical value of I (Brett, 1973; Elliott, 1976; Kitchell et al., 1977; Brett and Groves, 1979; Jobling, 1981; Shekk, 1983). It has been shown that the energy generated by assimilated food is not completely utilized in metabolism. Unlike lipids and carbohydrates, which yield all their energy during catabolism, proteins, when excreted as ammonia and urea, 'hold back' up to 10% of their energy potential unused. Vinberg believed that proteins were not completely oxidized, but he also believed that the energy used up by liquid excretion in aquatic animals never exceeded 3%. In computing the mean food assimilation quotient, he would add this estimate to that of solid excretion. At the present time, such a minor energy loss is thought to be an attribute of those fish in which proteins are an insignificant part of their diet. In carnivorous fish (such as all the fish considered in this survey), the energy of unassimilated food (both solid and liquid excretions) totals 27%, not 20%, so the assimilation quotient should be 0.73, not 0.8.

The increase by 1/3 found in the value of unassimilated food energy in predatory fish is so substantial that it was proposed to introduce a special term (U) into the balance equation, to denote liquid excretion. In the literature, a concept of 'metabolized energy' was adopted, implying assimilated energy (A) minus the liquid excretion energy (U) (Brody, 1945; Brett and Groves, 1979; Jobling, 1983). The basic equation would then be:

$$C = Q + P + U + F \qquad (30)$$

'U' is an inappropriate term, being the first letter of the word 'urine' – 80–90% of nitrogenous excretion takes place through the gills and skin, not through the kidneys (Kutty, 1968; Goldstein and Forster, 1970; Walton and Cowey, 1982). If that liquid excretion accounts for as much as 30% of the unassimilated energy, the corresponding correction to the amount of assimilated energy is only 10%.

Changing the value of U from 1.25 to 1.37 in equation 29 ($C = A/I$), and thereby increasing the estimate of food consumed by 10%, does not achieve much, particularly since this 10% is within the range of error admitted in the estimation of food consumption and the rest of the component parts of the balance equation. Therefore, in our subsequent calculations of the food consumed, we have retained the coefficient of 1.25 proposed by Vinberg.

Applying the material described in previous sections, it is possible to work out the consumption of fresh and dry matter and calorie uptake by every age group of the six species of Black Sea fish. Figure 63 illustrates the latter data obtained from the study of anchovy.

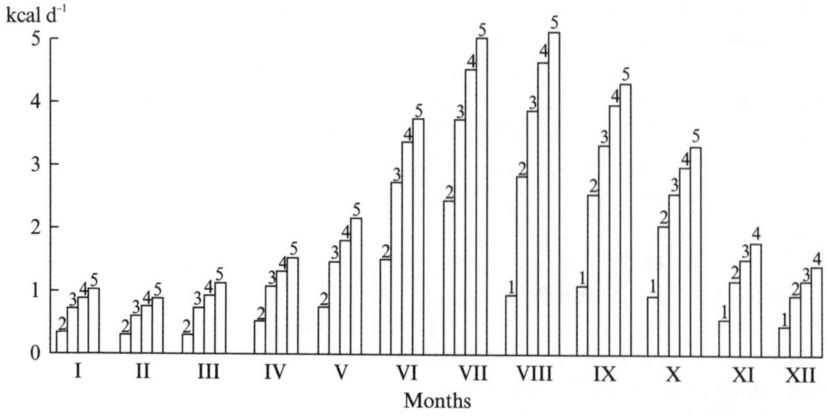

Figure 63   Consumption of calories by anchovy, year groups 1–5, during a year.

Food consumption entails the transformation of organic substances. It is therefore not possible to determine the consumption of proteins, lipids and glycogen merely by multiplying A by the quotient 1.25 as when dealing with fresh and dry matter and calorie uptake. To do that, one would have to know the chemical composition of the food.

Throughout the year, anchovy, sprat and pickerel feed on planktonic crustaceans (mainly copepods and arrow-worms), red mullet on molluscs, crustaceans and polychaetes, horse-mackerel and whiting on planktonic crustaceans during their first year and later on fish fry. The chemical composition of these food organisms from the Black Sea has been well studied (Vinogradova, 1967; Denisenko *et al.*, 1971; Kostylev, 1973, Trusevich, 1985; Table 18); however, the data in this table are averages and disregard seasonal variations in the content of certain compounds. Knowing the total food (converted into dry matter) eaten by Black Sea fish, one can define the proportions of proteins, lipids, glycogen and inorganic substances consumed (Figure 64).

In all the figures in this section, the food consumption has been estimated per whole fish. The basis of the level of food consumption is the food consumed per day per unit weight of the fish body. For wet and dry mass, it is given as a percentage of the wet or dry mass of the body; for protein, as a percentage of the protein mass; for energy, as a percentage of the energy equivalent of the body.

The values of daily rations are not identical when based on different equivalents. In anchovy, the smallest values were obtained on the basis of fresh matter, greater when converted to dry matter and protein, and maximal when calorie contents were considered. A similar picture emerged from studies on other species. The numerical values of the daily ration dropped as the age of

Table 18   Chemical composition of the food of Black Sea fish (% of dry matter).

	Proteins	Mineral substances	Lipids	Glycogen
Sprat	60	25	14	1
Anchovy	60	25	14	1
Red mullet	75	17	7	1
Pickerel	60	25	14	1
Whiting				
1 year old	55	25	15	5
2 years old	62	13	20	5
Average				
for population	60.25	16	18.75	5
Horse-mackerel				
1 year old	60	25	14	1
2–6 years old	62	13	20	5
Average for population	61.9	13.6	19.7	4.8

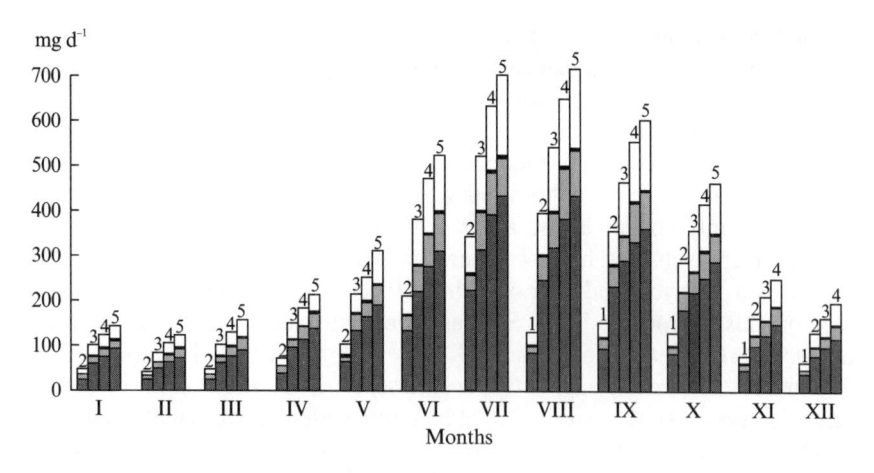

Figure 64   Food consumed by anchovy during a year, age groups 1–5. Total column, dry matter; open column, total mineral matter; solid black area, glycogen; light stipple, lipids; heavy stipple, proteins.

the fish increased. The food constituents consumed by 1 g of each population a day are shown in Figure 65. Protein was found to be the major material consumed, followed by mineral substances (anchovy, sprat, red mullet and pickerel) or lipids (horse-mackerel, whiting) and finally glycogen.

This series agrees with existing ideas about the nature of feeding in fish (Warman and Bettino, 1978; Brett and Groves, 1979; Cowey and Sargent, 1979; Love, 1980; Sorvachev, 1982; Walton and Cowey, 1982; Ostroumova, 1983). In the diet of the majority of fish, both carnivorous and herbivorous, protein-rich food greatly exceeds lipids and carbohydrates. Indeed, fish assimilate proteins well, lipids satisfactorily and carbohydrates poorly. The assimilation quotients of these three materials are, respectively, 0.9–0.98, 0.85 and 0.4 (Gerking, 1952; Karzinkin, 1952; Phillips, 1969; Fischer, 1970; Job and Gerald, 1976; Brett and Groves, 1979; Cho *et al.*, 1982; Jobling, 1983). The nutritional requirements of fish clearly differ from those of other vertebrates. The protein content of their food (60–70% dry weight) is 2–3 times greater than in mammals and birds, even birds of prey (Gerking, 1952; Walton and Cowey, 1982; Ostroumova, 1983).

Horse-mackerel and whiting differ from the remaining four species because of their diet of fish fry, while anchovy, sprat and pickerel are plank-tonivorous and red mullet are suspension feeders. The diet of the latter four species is richer in mineral substances than that of the two predator species, which is richer in lipids. The diet of Atlantic cod varies during the year. In the spring, when they are seriously depleted because of spawning and shortage of food, they feed voraciously on any kind of food, particularly fat-rich species such as sand-eels. As the summer progresses and their

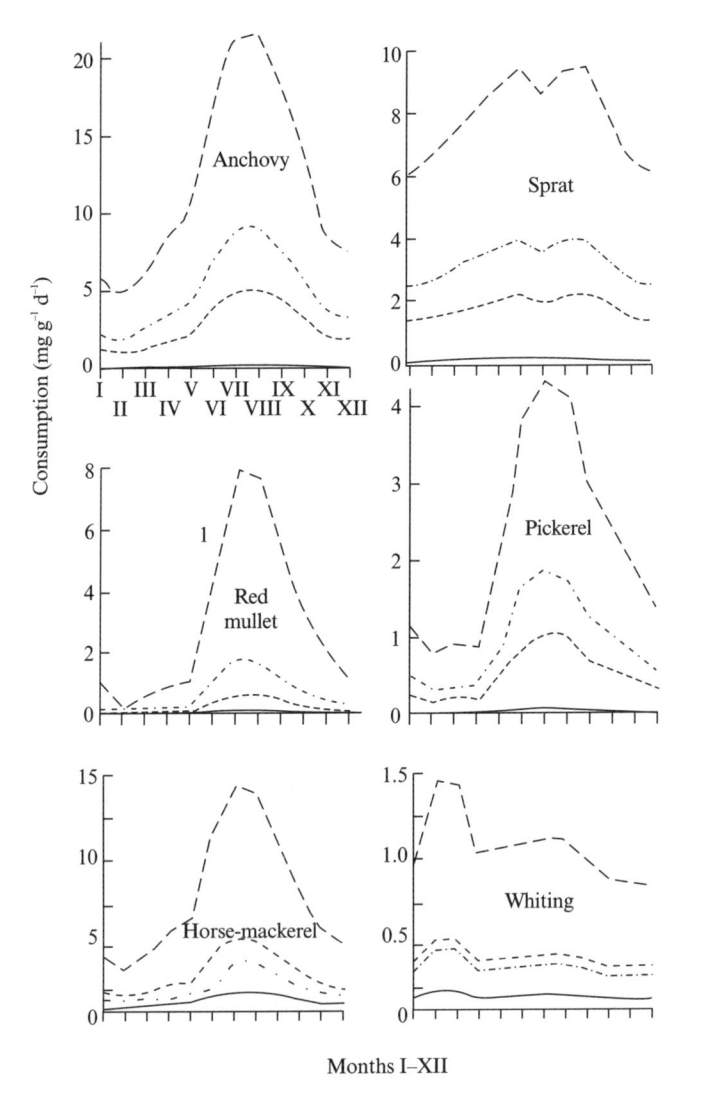

Figure 65   The consumption by populations of different species: coarse broken line, protein; fine broken line, lipids; broken and dotted line, total mineral matter; solid line, glycogen.

nutritional status improves, they become more selective and feed more on crustaceans (protein-rich).

It appears that the level of food consumption also varies in other species during the course of a year and differs according to the temperature preferences of the fish. As an example, during the intensive feeding period of summer, the rate of food consumption rises 1.5–1.7 times in sprat, while in whiting a similar increase is observed in winter.

The rate of consumption differs among species when their feeding rate reaches its maximum. Anchovy eat most (in summer), with horse-mackerel coming next. Third are sprats, which have a small body and develop high swimming activity to compensate for the low temperature of the water at the depth in which they live. Red mullet come fourth, and lastly come pickerel and whiting; the latter two engage in very little motor activity and consume food poor in calories. The relative levels of food consumption between species may be different when expressed in different ways: fresh matter, dry matter or calories.

In the cold season, mostly the winter time, the rate of feeding is highest in sprat (coinciding with their spawning), and also in whiting, although the relative rate in the latter species is not as high because of their greater body size, lower mobility and the composition of their food. Fish requiring warm water form the descending series anchovy, horse-mackerel, pickerel and red mullet. They eat moderately in the winter and do not cease completely even under severely cold conditions, but such feeding cannot altogether compensate for the expenditure of energy and plastic metabolism, so they lose proteins and lipids from their stores.

'Support' (maintenance) feeding is a common term used in the literature (Gerking, 1952; Karzinkin, 1952; Kitchell et al., 1977; Brett and Groves, 1979; Chekunova and Naumov, 1982; Elliott, 1982; Kerr, 1982), but the fish in the present survey are not really supported, merely spared the extremes of endogenous feeding. In the life cycles of natural fish populations, it is probable that true 'support feeding', which should provide a complete balance between input and expenditure of materials, is never found. Rather, at any one time, the balance is either positive (during growth) or negative (during starvation).

In commercial rearing of fish, it is essential to know how much foodstuff should be given for mere support (balancing metabolic output) and how much for actual growth. Present knowledge is insufficient for us to calculate the nutritional requirements for a steady state in a population, i.e. one which would not stimulate production but which would not reduce the number or weight of individuals.

The daily intake estimated in whole populations of fish conforms with the information above on the daily intake per unit weight of a population. It has been found to change from 2–5% to 9–22% in anchovy in winter and summer, respectively, from 1–2% to 5–12% in horse-mackerel, from 0.4% to 2–4.5% in pickerel and from zero to 4–5% in red mullet. In sprat and whiting, the daily intake is relatively stable, 2.8–7.5% and 0.5–1.3%, respectively.

Endogenous feeding supervenes if the balance between matter and output of energy turns negative (starvation). Yet even under intensive gonad development it is possible to evaluate the share contributed by endogenous feeding to the total flux of substance and energy, and to determine its ratio with exogenous feeding. Examples are to be found in the discrepancy between somatic and generative production yielded by horse-mackerel, red mullet, pickerel, whiting and sprat (Figure 52). Comparing the decline of somatic production observed in the pre-spawning period with the increase in generative production, one can assess the approximate share of generative tissue growth contributed by endogenous nutrition. The best results are obtained if estimates have been converted to protein, since lipid is used mostly for metabolic expenditure. Muscle proteins are converted by the liver into gonad proteins. In May, nearly all of the gonad proteins of red mullet are probably derived from body proteins, but in June it is exogenous food that prevails. In May, one-third to a half of the reproductive material produced by pickerel and horse-mackerel is derived from endogenous sources. In sprat (November) and whiting (December), the fraction is two-thirds, both of these species living in cold water. Sorvachev (1982) has paid attention to the problems of endogenous feeding in fish and Shcherbina (1989) has studied the effects of the winter environment.

Although much is known about daily or monthly consumption of food, data on annual food consumption are comparatively rare (Hatanaka *et al.*, 1956; Mann, 1965; Steele, 1965; Healy, 1972; Backiel, 1973; MacKinnon, 1973; Daan, 1975; Penczak *et al.*, 1976; Shatunovsky, 1978; Ochiai and Fuji, 1980; Chekunova and Naumov, 1982; Chekunova, 1983; Klovach, 1983). This is because many more difficulties are encountered in studies of an entire annual cycle compared with shorter time spans. Such studies are usually confined to the period of active feeding (May–October) in warm-water fish of temperate latitude. It is therefore not surprising that extrapolations to cover an entire year are often overestimated. Let us examine the daily rations of small fish – anchovy and sprat – and larger fish – horse-mackerel, red mullet, pickerel and whiting – averaged over a whole year. The estimates are 1.4–1.6 times greater in anchovy than in sprat, a surprising finding because anchovy are heavier than sprat, so should display a lower rate of food consumption, other conditions being equal. The distinctions in annual consumption of food between these two species result mostly from the different temperature preferences. Similarly, the whiting (from cold water) eat markedly less food than horse-mackerel and red mullet, which inhabit warm water, and pickerel, which are eurythermal.

Along with temperature, mobility is an important factor governing food intake. The active horse-mackerel consumes 2–3 times as much as red mullet, a less active fish. Yablonskaya (1951) found that verkhova (bleak), a highly active fish, consumed 20% more food than did carp of similar size. Karzinkin (1952) has also noted the relationship between food intake and mobility. Azov Sea anchovy (Shulman, 1962) and Black Sea anchovy show similar high intake of food during the summer months. Silverside, another pelagic species from the Black Sea, is another example (Klovach, 1983). In the six species of Black Sea fish it is clear that innate activity affects the feeding rate more than does the ambient temperature. The rate of food consumption is also influenced by body size. Small anchovy eat twice as intensively as the larger horse-mackerel.

Estimates of self-renewal of proteins are useful in defining the level of plastic metabolism. The process takes 13 and 167 days in populations of anchovy and whiting, respectively. However, estimates calculated for Black Sea fish contain errors. First, the calculations were based on the content of fat-free organic matter in the fish body, which material contains non-protein nitrogenous, and even nitrogen-free, components, in addition to 'real' proteins. Secondly, renewable proteins and the relatively stable collagenous structures were combined in making the assessment. Thirdly, it had been assumed that the rate of protein uptake relative to the total energy metabolism remained the same throughout the year, which is impossible. Even so, the values obtained are of the same order of magnitude as those available from studies of intermediate protein metabolism using isotope methods (Haschmeyer et al., 1979; Haschemeyer and Smith, 1979; Smith et al., 1980; Smith, 1981; Boëtius and Boëtius, 1985). Some workers have assessed the self-renewal of proteins on the basis of endogenous nitrogen excretion during long periods of starvation, but such excretion corresponds with that of basal metabolism and it would be preferable to use estimates determined on fish feeding normally.

It is essential to calculate the protein-to-calorie ratio of the food eaten to evaluate the balance of nutrients given to cultured fish, based on the energy and substance characteristics (Jobling, 1981). Sometimes the protein energy to total energy in the feed is preferred (Brett and Groves, 1979). It is best if calculated as for natural conditions, where it describes the role played by proteins in the energy balance of populations. It was found that 1 kcal is yielded by 82 mg of dietary protein in anchovy and sprat, 85 mg in horse-mackerel, 122 mg in red mullet, 90 mg in pickerel and 106 mg in whiting. In the same way, 1 g of food protein yields 12.3 kcal in anchovy and sprat, 17.8 kcal in horse-mackerel, 8.2 kcal in red mullet, 11.1 kcal in pickerel and 9.4 kcal in whiting. It is therefore evident that the highest protein content is found in the energy equivalent of the feed in red mullet, other species being successively less. These results correspond with the chemical compositions of the food eaten by the different species (Table 18). The protein content is greatest and high-calorie lipids lowest (75% and 7% of dry weight, respectively) in the diet of red mullet. The remaining species have a similar protein content (60–62%) but different proportions of other components.

In understanding such a subject as transformation of food, workers in aquaculture have decided advantages over those studying wild populations, having the ability to control the composition of the input. However, investigators of populations do in fact tackle the same problem, comparing the chemical composition of the food actually consumed in the natural habitat with that of the fish in the population. Therefore, in studying the process of matter transformation from input to output, one may assess the transformation of the biochemical constituents of food in the population as a whole, a goal unattainable without the sophisticated methods now available for studying intermediate metabolism.

The transformation of food is very similar in populations of anchovy and sprat (Figure 66). The total amount of lipid and protein assimilated into the body is also similar in the two species, but their proportions differ between the food and the fish body: the food is richer in protein and poorer in lipid when compared with the fish which consume it. Somewhat less than half of the body lipid appears to be formed from ingested protein ('liponeogenesis'), whereas all of the body protein originates in the diet. Some of the dietary lipid and, possibly, protein is transformed into glycogen. The mechanisms responsible

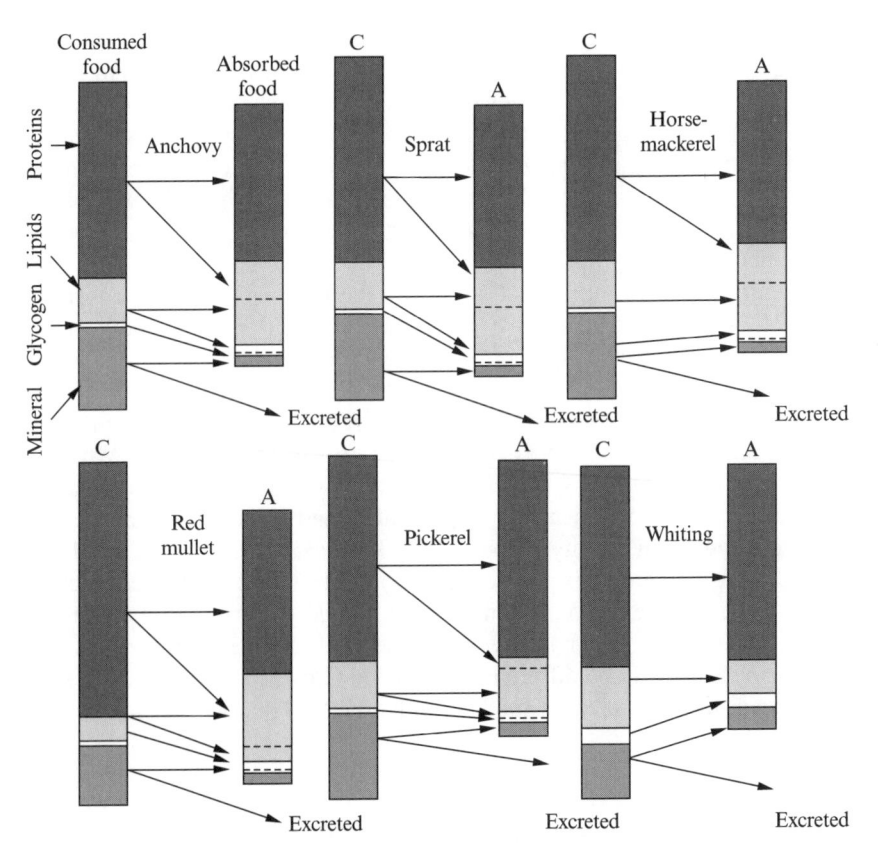

Figure 66   Scheme of the transformation of foodstuffs (relative quantities) in Black Sea fish. The key to the stipple is shown under 'Anchovy'.

for the transformation of all these substances have been well-studied in higher animals (Newsholme and Stort, 1973) and found to be similar in fish (Hochachka, 1969; Love, 1970, 1980; Cowey and Sargent, 1979; Sorvachev, 1982; Walton and Cowey, 1982).

Unlike organic substances, minerals in the food are assimilated with difficulty so that only about 80% of the total food is actually assimilated. In the sprat, the conversion of dietary organic matter into body tissues is more effective than in anchovy, but vice versa in the case of mineral components. Generally speaking, the same picture appears when comparing horse-mackerel and red mullet . In red mullet, however, only about 25% of the lipid deposited in the body originated in the food, 75% being produced from protein. Horse-mackerel, like anchovy and sprat, convert most of their dietary protein into body protein rather than body lipid. Pickerel differ from the other species in that virtually no body lipid is made from ingested protein and little of the protein is converted into glycogen. The body lipid of whiting all originates in the food, all dietary protein becoming body protein. In this species only, the dietary lipid exceeds requirements. Unlike the other four species, whiting and horse-mackerel consume more glycogen than is estimated to be in their bodies at any one time. Whiting assimilate the highest percentage of inorganic substances. Two patterns of transformation therefore emerge. In pickerel and whiting, ingested protein is converted almost exclusively into body protein, while in the other four species it is also an important precursor for body lipids, which supplement the dietary lipids.

A central point in nutritional studies is the destination of the consumed material – somatic and generative (or gonadal) production or metabolic expenditure. This leads to the concept of efficiency of production in populations. Two coefficients have been introduced to aid the understanding of food utilization in constructive processes: one represents food consumed for growth, and the other, food assimilated for growth (Terroine and Wurmser, 1922; Brody, 1945). These coefficients were first applied in studies of the ecology of aquatic organisms by Ivlev (1939), and are also known as coefficients of production activity (Brody, 1945; Karzinkin, 1952; Shpet, 1971).

At the beginning of this chapter, equations 11 and 12 were used in calculating the coefficients, K_1 ('consumed') and K_2 ('assimilated'). Russian researchers usually prefer to use K_2, while in western literature K_1 is used because K_2 is understood differently (Brett and Groves, 1979). The ratio K_1/K_2 is known as the coefficient of food assimilation, 'I' (equation 28), and

$$K_1 = IK_2 \qquad (31)$$

It is frequently claimed that K_1 describes the ecological efficiency of production, as it links two associated interpopulation trophic levels of a living system (Odum, 1959; Slobodkin, 1962; MacFadyen, 1963; Steele, 1965;

Mann, 1969; Shatunovsky, 1980; Penczak, 1985). By analogy, K_2 can be considered the physiological efficiency of production (the utilization of assimilated food), and V.S. Ivlev attempted to introduce K_3, the biochemical efficiency of production, but so far too little is known about the efficiency of constructive biosynthetic processes at tissue, cellular and molecular levels to make this concept feasible.

In recent western literature, K_1 is interpreted as 'gross conversion efficiency' and the western analogue of K_2 is 'net conversion efficiency' (Brett and Groves, 1979), but the biological sense of the two coefficients can be obscured by such generalized semantics. A better concept is 'efficiency of food consumption and assimilation for constructive processes' (Zaika, 1983).

The substance and energy utilization of Black Sea fish can be analysed using K_2. Figure 67 shows the dynamics of K_2 over the annual cycle of anchovy, on the basis of calories in fresh matter. Only positive values of the coefficients have been represented here; they can be zero during balanced equilibrium and negative if, during feeding, energy is expended rather than accumulated. Anchovy rely on endogenous feeding from November to March. If the coefficient of food assimilation is 0.8, then K_1 will be, according to equation 31, 80% of K_2. The difference between the two coefficients is small only in the case of protein, where assimilation is much higher.

Figure 67 shows that K_2 (assimilation of food) tends to decline with age in anchovy, a tendency that is even more pronounced in other Black Sea fish (Nagorny, 1940; Nikitin, 1982). Protein synthesis also wanes with age. The average K_2 (fresh matter and calories) differs between populations of each species (Figure 68), and K_2 calculated for fresh matter (hence, protein) differs from that calculated for calorie content (hence, total dry matter). The value of K_2 is usually greater on the basis of fresh matter than of calorie content. The coefficients determined in the context of food used for growth, each component being regarded separately, become closer together in value as food consumption diminishes (Vinberg, 1956). The value of K_2 varies considerably over the annual cycle of fish populations but many workers have regarded it as a stable entity for a species or an age group of a species, regardless of this

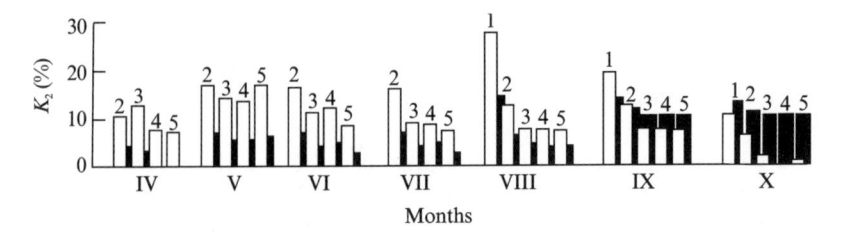

Figure 67 Efficiency of the utilization of assimilated food for production (K_2) in anchovy, age groups 1–5. White columns, wet matter; black columns, calories.

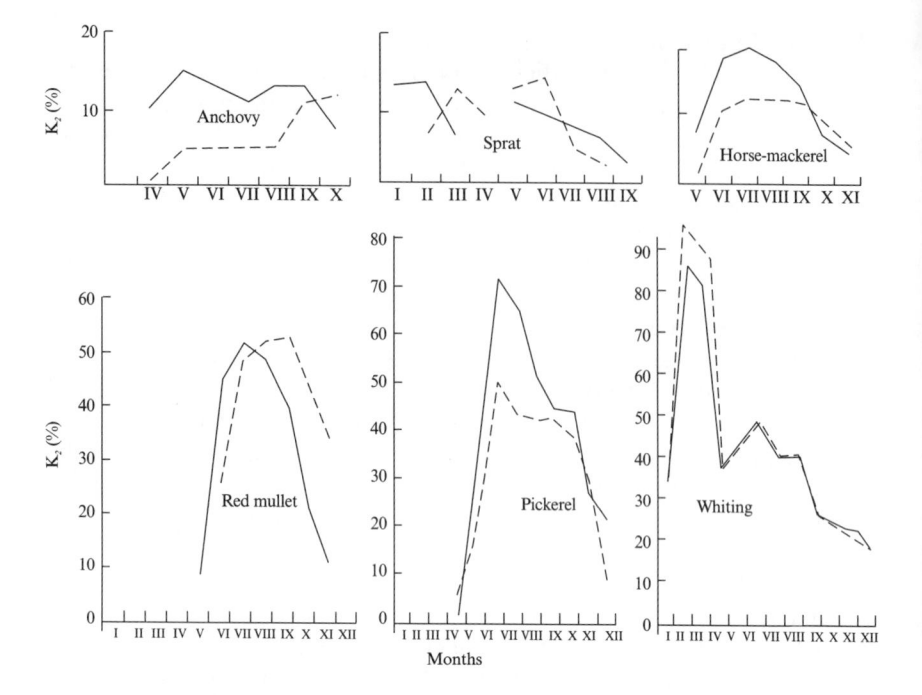

Figure 68   Efficiency of the utilization of assimilated food for production (K_2) in different populations: solid line, wet matter; broken line, calories. Note how the least active species use a larger proportion of food for production.

(Vinberg, 1956, 1962, 1986; Paloheimo and Dickie, 1966; MacKinnon, 1973; Greze, 1979). The last two authors supposed that because the coefficient was stable, the metabolic expenditure of an individual or a population could be evaluated from estimates of the food consumed and the resulting production, which can be calculated by the following equation:

$$P = Q\,\frac{K_2}{1-K_2} \qquad (32)$$

This method, regarded as 'physiological', is widely used for assessing production. However, from the arguments above and from present knowledge about the functioning of organisms, we cannot regard this method as physiological. There are, in fact, grounds for rejecting it altogether. During the annual cycle of a population and of individuals, the rate of food consumption, production of tissue, accumulation of energy and metabolic expenditure all vary greatly and regularly in accordance with the metabolic rhythm. Not being mutually coordinated, these processes induce substantial variation in the efficiency of food utilization for growth and production. Mann (1965) and

Gerking (1966) also criticized the 'physiological' approach to the evaluation of production based on the supposed invariability of K_2.

This approach can be advised only for assessing the production of a population or the weight increment in an individual over a whole year. One still has to calculate the annual K_2 beforehand, evaluating food consumption, metabolic expenditure and production, and annual estimates are based on monthly surveys, so are far from easy.

The high values of K_2 maxima recorded in red mullet (54% of the energy equivalent of weight), pickerel (50%) and whiting (90% – probably an overestimate) are of interest. In horse-mackerel, sprat and anchovy, these are much lower (13%, 14% and 12%, respectively). From the literature, the values of the coefficient range from 0 to 80% (Vinberg, 1962; Laurence, 1975; Calow, 1977; Brett and Groves, 1979; Holeton, 1980; Cetta and Capuzzo, 1982; Parry, 1983). No definite relationship was found between K_2 and the temperature preference of the fish (cold or warm). Red mullet (warm), whiting (cold) and pickerel (intermediate) use assimilated food equally efficiently for growth. Efficiency is lower in horse-mackerel, anchovy and sprat.

The relationship between the coefficient of food utilization and the ambient temperature is complex (Kitchell *et al.*, 1977; Brett and Groves, 1979). Some authors (Ivlev, 1938; Karzinkin, 1952; Vinberg, 1956, 1986; Shpet, 1971) claim that over the optimum temperature range of the particular species there is no such relationship. Maximum efficiency of food utilization for growth has been found at the optimum temperature (Shpet, 1971). Our data support this. Konstantinov *et al.* (1989), Konstantinov and Sholokhov (1990) and Konstantinov and Yakovchuk (1993) have demonstrated that K_2 increases considerably in juvenile fish exposed to variable temperatures. It is also possible to find a close (negative) relationship between K_2 and the natural mobility rate of the fish. Species with higher inbuilt rates of mobility consume more food, but have lower efficiency of food utilization for constructive processes – they direct much of the assimilated energy towards their high swimming performance.

To obtain a clear view of these relationships, annual, not maximum, estimates of the population coefficients, K_2, were computed for different species. The question arose as to whether only positive values should be used (covering the growth period), or all values over the 12-month period, which would allow the calculation of efficiency of production over the annual cycle. Annual coefficients have been estimated by the following equation:

$$K_2 = \frac{P_I + P_{II} \cdots + P_{XII}}{A_I + A_{II} \cdots A_{XII}} \tag{33}$$

In fact, the values for annual K_2 of the population obtained by using the two methods correlate well. K_2 calculated from the positive monthly estimates is 1.5–2.5 times greater than K_2 for the total annual cycle.

Analysis of a series of annual K_2 values confirms conclusions derived from their maxima recorded over the annual cycle (Figure 68). No relationship can be found between the efficiency of food utilization for growth (production) and the temperature of the habitat, but the efficiency of production does correlate with the natural mobility of the species, being higher in sluggish fish than in active. For the former, K_2 varies from 29% to 45% (21–33% using the second calculation technique), and for the latter from 6% to 15% (2.5–11.5%), respectively. Thus, food utilization for growth is lower in sardines, herrings, silversides, horse-mackerel, anchovy and the pelagic larvae of many species than it is in flounder, cod, haddock, goby, sole, annular bream, blenny, carp, roach, pike, perch and butterfish (Yablonskaya, 1951; Karzinkin, 1952; Krivobok, 1953; Steele, 1965; Everson, 1970; Fischer, 1970; Yoshida, 1970; Healy, 1972; Childress and Nygaard, 1973; MacKinnon, 1973; Smith, 1973; Daan, 1975; Laurence, 1975; Shevchenko, 1977; Shatunovsky, 1978; Childress et al., 1980; Ochiai and Fuji, 1980; Cetta and Capuzzo, 1982; Chekunova and Naumov, 1982; Sullivan and Smith, 1982; Chekunova, 1983; Houde and Schekter, 1983; Klovach, 1983; Tseitlin, 1983). Data from fish-breeding studies are not included here, since under intensive feeding the values of K_1 and K_2 increase, even in active fry of grey mullet and salmonids (Elliott, 1976; Flowerdew and Grove, 1980).

It has also been found that the efficiency of food utilization for growth varies in the same fish according to its (variable) activity. Trout swimming at low speed utilize more nutrients for growth than when swimming quickly (Davison and Goldspink, 1977). A similar situation was observed by Ware (1975) in roach and by Koch and Weiser (1983) in bleak. These observations explain why centrarchids held in a laboratory utilize food more efficiently than those in the wild (Gerking, 1966).

The inverse relationship between consumption and efficiency of utilization for constructive processes can be represented quantitatively by the equation:

$$K_2 = \frac{29 \cdot 1}{C} \qquad (34)$$

To convert into dry or fresh weights or calorie content, the same coefficients are used in the equation, while conversion into protein requires different coefficients. The inverse relationship between K_1 and K_2 on the one hand and the ration on the other has been pointed out by several workers (Paloheimo and Dickie, 1966; Kitchell et al., 1977; Paloheimo and Plowright, 1979).

When examined within the annual cycle of each species (Figure 68), K_2 is found to increase with increasing rate of feeding (intensive feeding is directed towards production). This has been demonstrated in production/ration curves (Gerking, 1952; Ivlev, 1955; Vinberg, 1956; Brett, 1970, 1979; Ware, 1975, 1980; Wootton, 1977; Ricker, 1979; Kerr, 1982; Zanuy and Carrillo, 1985). However, on the whole, it is not maximum but optimum ration that provides maximum efficiency of food utilization (Brett and Groves, 1979; Elliott, 1982). It can be concluded that the fish surveyed in the Black Sea differ in their strategies of adaptation to the nutritive base. In fish with high natural mobility, production results from high intensity of food consumption, while in sluggish fish it results from high efficiency of consumption.

Up to now, different aspects of substance and energy utilization for total production have been examined. It is necessary to look at somatic and generative production, protein retention and accumulation of lipid and, finally, the efficiency of utilization of assimilated food. The total K_2^{Σ} is the sum of somatic production K_2^s and generative production K_2^g:

$$K_2^{\Sigma} = K_2^s + K_2^g = \frac{P_s}{P+Q} + \frac{P_g}{P+Q} \tag{35}$$

Then $$K_2^s / K_2^g = P_s / P_g \tag{36}$$

So the ratio of efficiency of food assimilated for somatic and generative production is numerically equal to the ratio between somatic and generative production (Tables 11 and 13). The data show that the efficiency of food assimilated for somatic production per year in populations of different fish makes up 33–76% of the total efficiency, and that for generative production 24–67%. From these figures one can calculate K_2^s and K_2^g for annual estimates. K_2 obtained for the entire annual cycle underlies the calculations (second method); the coefficient of food assimilated for somatic production is then 2.0–27.3% and for generative production 0.9–18.3%. Proceeding on the assumptions made in section 5.2, in male fish the K_2 would be ten times less than that of females. This coefficient indicates the efficiency of substance and energy use for reproduction ('reproductive effort') (Mann, 1965; Healy, 1972; Backiel, 1973; Daan, 1975; Wootton, 1977, 1979; Woodhead, 1979; Constantz, 1980; Dawson and Grimm, 1980; Hirshfield, 1980; Ochiai and Fuji, 1980; Shatunovsky, 1980; Wootton et al., 1980; Chekunova and Naumov, 1982; Eliassen and Vahl, 1982; Kamler et al., 1982). As estimated by different authors in different species, coefficient K_1 can range over as much as 0.6–2.35% or more. The efficiency of food used for generative production, like total efficiency of production, was found to be lower in highly mobile forms. Values of K_2 will be considerably higher, provided that the estimates of generative production are correlated with those of food assimilated for this purpose. For example, converted into the energy equivalent of weight, such coefficients would be as follows: anchovy, 2.2%; sprat, 9.4; horse-mackerel, 7.1; red mullet, 24.5; pickerel, 25.9; and whiting, 48.1%. Converted into fresh weight, the values will be 11.7%, 13.7%, 17.9%, 48.2%, 33.8% and 40.8%, respectively. These estimates exceed those reported in the literature for species of comparable ecology of their reproductive period. When such values are found in in temperate and cold latitudes in fish with prolonged spawning periods, this may result from underestimated multiple generative production (Oven, 1976).

It has been noted that the assimilated food is composed mostly of protein and lipid. It is possible to assess how efficiently this food is used in the accumulation of protein and lipid stores in the fish. It is a question of total dry matter accumulation, not protein or lipid taken separately. Estimated as dry matter, K_2 is as follows:

$$K_2^{dry} \approx K_2^{protein} + K_2^{lipid} \approx \frac{P_{protein}}{P_{dry} + Q_{dry}} + \frac{P_{lipid}}{P_{dry} + Q_{dry}} \tag{37}$$

Then
$$K_2^{protein} / K_2^{lipid} \approx P_{protein} + P_{lipid} \tag{38}$$

Thus the correlation between the food assimilated for production of protein and accumulation of lipid can be numerically represented as the ratio between the two processes.

Table 19 demonstrates the results of the calculation, using data from Table 11. It can be seen that, depending on the species, from 2% to 24% of the dry food substance is assimilated for the total protein production and 0.7% to 8.0% for the total lipid production. Assuming that protein and lipid are adequately used for the somatic and generative productions, one can make relevant estimates (Table 19). The resulting data of K_2 of P_s of protein and lipid are of especial interest. They indicate the efficient assimilation of nutrients to provide somatic production and accumulation of lipid. The following ratios between food assimilated for protein and lipid production have been obtained for Black Sea fish: 2.6:1 in anchovy, 2:1 in sprat, 2.1:1 in horse-mackerel and red mullet, 2.7:1 in pickerel, and 4:1 in whiting.

Table 19   Efficiency of the utilization of assimilated dry matter for the production of protein and the accumulation of lipid (%).

Production	Anchovy	Sprat	Horse-mackerel	Red mullet	Pickerel	Whiting
P_Σ						
Proteins	2.3	3.7	4.4	17.7	17.3	24.0
Lipids	0.7	1.8	2.1	8.0	5.4	8.0
P_s						
Proteins	1.4	1.6	1.6	8.0	12.4	14.9
Lipids	0.4	0.8	0.7	3.6	3.9	4.9
P_g						
Proteins	0.9	2.1	2.8	9.7	4.9	19.1
Lipids	0.3	1.0	1.4	4.4	1.5	3.1

Applying the data of dry matter, protein and lipid of the assimilated food, it is possible to determine the biochemical (substance) expenditures for constructive processes in the populations. For this, one must calculate $1/K_2$ (Table 20). The biochemical 'cost' of a unit of weight of population indicates how much of the assimilated matter is used per unit. Similarly, as regards food consumed, this characteristic ($1/K_1$) is known as the food coefficient or feeding ratio, and describes the weight of food consumed for the production of a unit of weight of the population or organism. Data given in Table 19 are confined to the estimates of biochemical expenditure for the production of dry matter and protein from dry matter and protein, respectively. The biochemical cost of lipid produced from dry matter was estimated as 143.9 g for anchovy, 55.6 g for sprat, 47.6 g for horse-mackerel, 12.5 g for red mullet, 18.5 g for pickerel and 12.5 g for whiting. These estimates are far greater than those obtained for protein production.

The next step comprises the assessment of the biochemical cost of production of the energy equivalent of weight. It is found that, in order to produce 1 kcal, the anchovy assimilated 5.7 g of dry matter, sprat 3 g, horse-mackerel 2.4 g, red mullet 0.6 g, pickerel 0.7 g and whiting 0.6 g. It is also evident that energy expenditure on construction processes is much higher in mobile than in sluggish fish.

A further point of interest is how the assimilated food is used in metabolism (Q). The coefficient is as follows:

$$K_2^Q = \frac{Q}{P + Q} \qquad (39)$$

Where $A = P + Q$, it can be assumed that there is adequate knowledge of the quantities of substances and energy required for constructive processes, but it is not possible at present to evaluate the food expended in non-constructive metabolism.

Table 20   Annual biochemical cost ($1/K_2$) of production in populations (assimilated substance in g g^{-1} production).

Production	Anchovy	Sprat	Horse-mackerel	Red mullet	Pickerel	Whiting
P_Σ						
Dry matter	33.3	18.2	15.4	3.9	4.4	3.1
Protein	16.1	10.3	8.7	3.5	3.2	3.0
P_s						
Dry matter	55.6	41.7	43.5	8.6	6.1	5.1
Protein	26.3	28.6	26.3	7.1	4.3	5.2
P_g						
Dry matter	83.3	32.3	23.8	7.1	15.6	8.2
Protein	41.7	16.1	13.4	7.1	12.3	7.1

Energy entering the organism (enthalpy) is transformed in such processes as oxidation, phosphorylation (production of ATP), splitting of ATP and reactions that maintain biosynthesis (Khaskin, 1981). We never know precisely the proportions of the consumed oxygen directed to the different processes. The efficiency of oxidative phosphorylation when estimated at the cellular level in animals does not usually exceed 60%, but it has been reported to be as high as 80% in some cases (Calow, 1977; Parry, 1983). It may depend on the functional state and living conditions. It is probably lower in the whole organism compared with values at the cellular level. Up to 40% of assimilated energy is used for 'non-phosphorylated oxidation', the major part of the production of heat being dissipated into the surrounding water. According to Brett and Groves (1979), heat dissipation takes 12–30% of the energy derived from the food. Much of the dissipated energy (entropy) therefore never participates in useful action. Some authors (Mann, 1965; Backiel, 1973) name K_2^o the coefficient of energy dissipation. However, one should remember that useful, or productive, energy is also involved in heat production (Khaskin, 1981). It amounts to two-thirds of the total loss of energy by the organism (Brett and Groves, 1979) and usually relates to phosphorylation (Lehninger, 1972). However, peroxic, enzymatic and other non-phosphorylated oxidation is useful to the organism, so it is not unreasonable to suggest that all the energy metabolized is productive.

In populations of fish, the flow of assimilated food can be estimated from its utilization in basal metabolism (i.e. maintaining normal functioning of cells without physical activity), active metabolism (locomotion) and specific dynamic action (energy used in digesting and absorbing food).

Belokopytin (1968, 1978) studied the basal metabolism (Q_b) of Black Sea fish by applying two independent methods: (1) measuring oxygen consumption in anaesthetized fish; and (2) placing fish in a flume respirometer and varying their swimming speeds. In the latter method, the data were extrapolated to zero swimming speed. He found that basal metabolism was 17% of the total metabolism in fast-swimming fish (anchovy, sprat and horse-mackerel), 54% in the moderately mobile red mullet and pickerel, and approaching 100% in sluggish fish (whiting).

Active metabolism (Q_a) is defined as the oxygen uptake required to provide both muscular contraction and the operation of all the other systems which support locomotion (respiration, blood circulation, excretion, ion transport). Judging from published data, the energy cost of physiological action depends in the end on the systems that support locomotion. The functioning of the gills of a swimming fish takes up 5–10% of the oxygen consumed (Cameron et al., 1973; Holeton, 1980; Kramer and McClure, 1981; Klyashtorin, 1982; Klyashtorin and Smirnov, 1983; Furspan et al., 1984) and the cardiovascular system uses up to 40% (Kiceniuk and Jones, 1977). A considerable amount of energy is required for excretion and ion transport, including osmoregulation (Febry and Lutz, 1987). Based on these data, one can endeavour to assess the share contributed by each system to the active metabolism of fish, an exercise which would reveal the proportion of consumed oxygen directly expended in muscular work. Such a figure is claimed by Matyukhin *et al.* (1984) to amount to 80% of the total active metabolism. Available data are too few for a real evaluation of oxygen balance in actively swimming fish, but the reader is referred to papers by Kokshaysky (1974) and Klyashtorin (1982) for studies on the performance of fish during muscular contraction.

'Feeding metabolism' comprises actual eating, digestion and assimilation, and results in a rise in oxygen consumption. While the contribution of each stage is not known, it is believed that the combined effects of the transformation of food can amount to 10–40% or more of the initial level (Brett, 1970; Muir and Niimi, 1972; Smith, 1973; Beamish, 1974; Pierce and Wissing, 1974; Kitchell *et al.*, 1977; Brett and Groves, 1979; Tandler and Beamish, 1979; Flowerdew and Grove, 1980; Klyashtorin, 1982; Turetsky, 1983). Enzymic hydrolysis of food, enhanced peristalsis, transport through intestinal membranes and catabolism of assimilated substances (amino acids and proteins in particular) are all recognized as factors in feeding metabolism (Brody, 1945; Tandler and Beamish, 1979; Jobling and Davies, 1980; Jobling, 1981, 1983; Klyashtorin, 1982; Parry, 1983; Yarzhombek *et al.*, 1983). Assuming that the average value of feeding metabolism is 30% of the initial energy metabolism, its share of the total energy capacity of a population will be 23% of Q_Σ (total metabolism of the population).

Equation 25 showed that $Q_\Sigma = Q_b + Q_a$, so $Q_a = Q_\Sigma - Q_b$. When calculated by difference Q_a is not, strictly speaking, the actual active metabolism because of the feeding metabolism. Denoting the latter as Q_c and 'proper' active metabolism as $Q_{[a]}$, we get the following equation:

$$Q_a = Q_{[a]} + Q_c \qquad (40)$$

and

$$Q_\Sigma = Q_b + Q_{[a]} + Q_c \qquad (41)$$

Having compared the various types of metabolism (Q_b, $Q_{[a]}$ and Q_c) with the energy equivalent of assimilated food 'A', one can define corresponding coefficients of food utilized for basal (K_2^Q), active ($K_2^{[a]}$) and feeding (K_2^c) metabolism. These coefficients are characterized by a certain vagueness which they share with (K_2^Q). However, in each of the derived coefficients, this element of uncertainty is reduced three-fold on average when compared with the total coefficient. In the efficiency of food assimilation for metabolism and in values of (K_2^p) the fish clearly divide into groups according to their innate mobility rate. In the highly mobile group, 'proper' active metabolism accounts for 60% of the food assimilated (about 50% of the food consumed).

In fish of moderate activity, the corresponding estimates are 40% and 30%, respectively, and in sluggish fish the energy utilized for active metabolism approximates to that for basal metabolism.

The consumption of substance and energy by each age group in populations of Black Sea fish corresponds with their contribution to production (Table 14).

5.5. SUBSTANCE AND ENERGY BUDGET

The balance of substance and energy in fish populations was discussed earlier. Here it is intended to combine all components constituting the balance, so that the totality of substance and energy used by populations of each species can be represented as an integrated characteristic. The flows of substance and energy, the efficiency of their use and the stages of transformation will be presented, together with the quantitative significance of each species in the trophic dynamics of Black Sea ecosystems.

The complete balance equation emerges as follows:

$$C = P_s + P_g + Q_b + Q_{[a]} + Q_c + U + F \qquad (42)$$

This equation, which underlies studies of specific functions intrinsic to a population, has been taken from general physiology where it is even more extended. However, some factors, when applied to population budgets, are of minor significance and may be omitted. Simplified balance equations help to solve problems of ecological and production physiology (Brody, 1945).

Among works of special value to research in population budgets are those conducted with centrarchids (Gerking, 1952, 1966, 1972), white perch from the North American lakes (Wissing, 1974), three-spined sticklebacks from Arctic lakes of Canada (Cameron et al., 1973; Wootton et al., 1980), northern pike (Diana and MacKay, 1979; Diana, 1983), fish occurring in the Thames (Mann, 1965, 1969), in the Vistula and other rivers of Poland (Backiel, 1973; Penczak et al., 1976), goby from estuaries of Scottish rivers (Healy, 1972), Pacific sardine (Lasker, 1970), flounder (Steele, 1965; MacKinnon, 1973), cod (Daan, 1975; Eliassen and Vahl, 1982; Kerr, 1982), frogfish from the Sargasso Sea (Smith, 1973), gobies from waters around Hokkaido (Ochiai and Fuji, 1980), medaka (Hirshfield, 1980), various Antarctic notothenia (Kozlov, 1975; Chekunova and Naumov, 1982; Chekunova, 1983; Clarke, 1985) and sockeye salmon (Brett, 1973) and some meso- and bathy-pelagic fish off Southern California (Childress et al., 1980). Most of these studies were carried out to acquire better understanding of the energy budget of a population, while only a few included protein balance.

In recent decades, substance and energy balance are usually presented as diagrams (Zaika, 1983; Sorokin, 1982), especially impressive in the publications of Niimi and Beamish (1974), Daan (1975), Calow (1977), Brett and Groves (1979), Shatunovsky (1980), Parry (1983), Turetsky (1983) and Amineva and Yarzhombek (1984). Such diagrams make it a simple matter to assess both the total budget of the system and flows and cascades of substance and energy in it.

Figures 69 and 70 show the balance of dry substance and energy for populations of Black Sea fish observed over annual cycles. The total production, metabolic expenditures, and food assimilated and consumed are illustrated. Very similar patterns can be obtained on the basis of wet material and protein. The figures demonstrate that the energy balance of the populations of Black Sea fish alters considerably over the annual cycle, ranging from 0–10 to 30–270 cal g^{-1} day^{-1} for the total production and food consumed, respectively. The balance estimates vary most in anchovy, then,

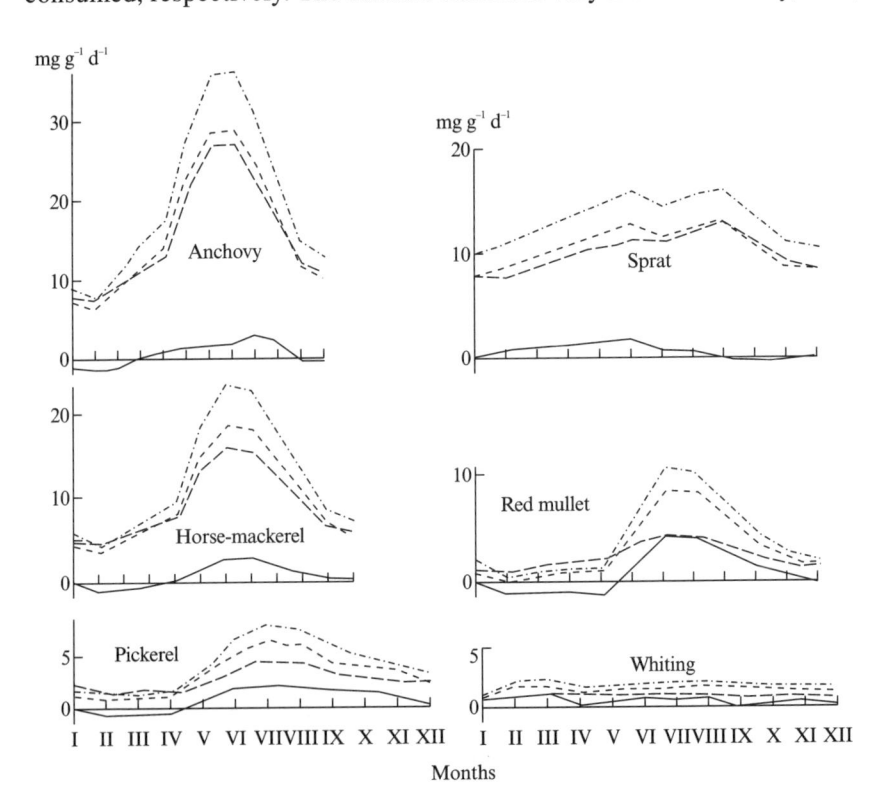

Figure 69   The annual cycle of feeding, metabolism and production in different species: dotted and broken line, food consumed; broken line, food assimilated; solid line, total specific production; broken line with long dashes, expenditure on total metabolism (dry matter).

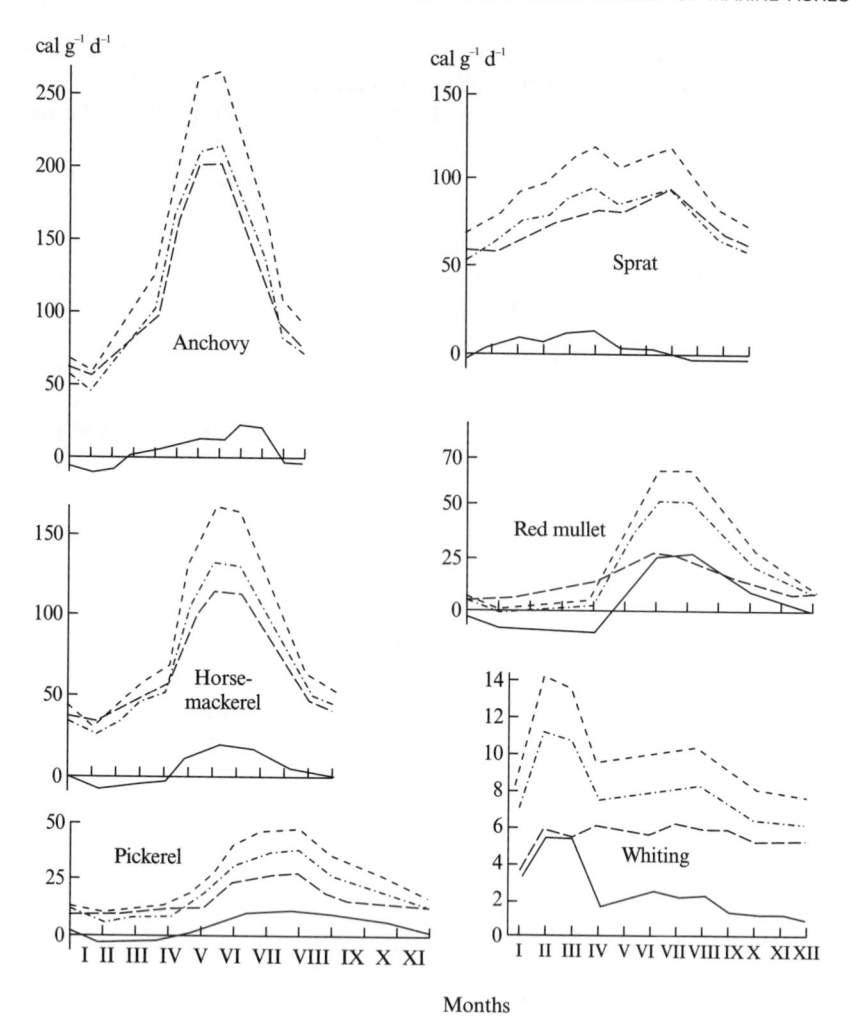

Figure 70   As in Figure 69, calories. Note different key: broken line, food consumed; broken and dotted line, food assimilated; solid line, total specific production; broken line, long dashes, expenditure on total metabolism.

in reducing order, horse-mackerel, sprat, red mullet, pickerel and whiting. This order is the same in the food consumption of the different species, and is directly related to natural swimming activity.

Flows and cascades of substance and energy in any month can be calculated for populations of any species. More significant are diagrams that represent the total annual flows and cascades of substance and energy over a whole year. Figure 71 illustrates such flows of energy in the two extremes of Black Sea species – anchovy and whiting. Corresponding flows of substance have also been published by Shulman and Urdenko (1989). Schemes like these are specific 'fingerprints' for a population of each species studied.

The metabolized portion of assimilated energy is not distinguished here, unlike Brett and Groves (1979), who concluded that metabolized food makes up 81% of the food assimilated. Figure 71 has been drawn from data averaged over many years.

5.6. TROPHIC SIGNIFICANCE OF POPULATIONS

Anchovy, sprat and horse-mackerel eat plankton, so it is possible to define the role they play in the food web of the Black Sea. It is most important to find out the extent to which they use food reserves, and contribute to the transformation and cycle of matter and energy in the ecosystem.

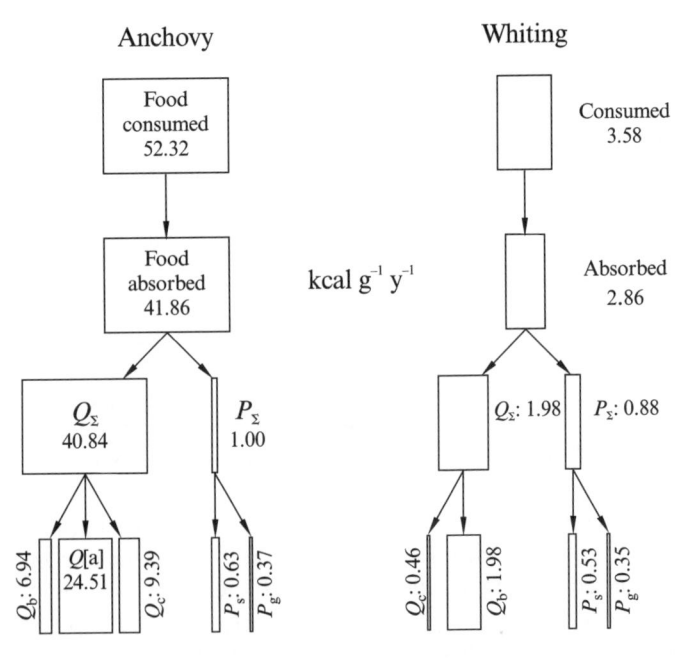

Figure 71 Annual flows of energy (kcal g⁻¹ fresh weight of the population) in anchovy and whiting. Q_Σ, total energy used for metabolism; P_Σ, total energy used for production; Q_b, energy used for basal metabolism; $Q_{[a]}$, energy used in active metabolism without feeding metabolism; Q_c, energy used in feeding metabolism; P_s, energy used for somatic production; P_g, energy used for generative production (gonads).

According to Greze *et al.* (1973) and Greze (1979), the average annual biomass of food crustaceans and chaetognaths over the Black Sea is 3.5 mt and their average annual production is 120 mt, copepods being the basis of the zooplankton. We found that the annual consumption of zooplankton by anchovy was 5 mt, 2.5 mt in sprat and 250 kt in horse-mackerel. Thus over the years the anchovy population consumes about 4% of the average annual production of Black Sea zooplankton, sprat 2% and horse-mackerel 0.2%. Because of their small numbers, the population of pickerel consumes much less of the zooplankton compared with horse-mackerel. Red mullet and whiting do not eat plankton. Thus the main species of planktonivorous fish consume no more than 6–6.5% of the total production. This value would be slightly greater if we allow for food consumed by larvae and fry white were eliminated from the population.

However, other plankton eaters exist in the Black Sea. Among formidable competitors to the fish are the medusa *Aurelia aurita*, other jellyfish and chaetognaths. It was reported that *Aurelia* ate 60% of the total annual production of zooplankton (over an 80-year period), that is, ten times as much as all the planktonivorous fish (Anninsky, 1990). The other jellyfish and the chaetognaths usually consume as much zooplankton as *Aurelia*. In the late 1980s to early 1990s, the stock of Black Sea zooplankton was seriously endangered by the intrusion of yet another plankton-eating species, the ctenophore *Mnemiopsis leidyi*.

The close relationship between the level of lipid deposition in planktonivorous fish and the condition of their nutritive base provides indirect evidence of severe food competition, as noted in Chapter 6.

Values for the relative transformation of substance and energy in populations of fish that live on plankton are close to those describing feeding because a major part of the nutrients consumed is metabolized and excreted. Taking medusae, ctenophores and chaetognaths into account, almost the entire substance and energy provided by the lower trophic links of the pelagic ecosystem are used in transformation processes proceeding in the higher links (Greze, 1979; Vinogradov and Shushkina, 1980; Sorokin, 1982). Despite this, in general, the role of planktonivorous fish in the trophic dynamics is not of great importance.

A useful concept with which to describe the interrelations between trophic levels in an ecosystem is the efficiency of food utilization. Until now we have mostly used the coefficient K_2, but now the coefficient K_1 is needed to examine the energy equivalent of weight. Populations of anchovy, sprat and horse-mackerel (fast swimmers) usually use just 1.9–4.5% of ingested energy for production, while in the relatively sluggish red mullet, pickerel and whiting, this percentage amounts to 16.6–24.6%. Corresponding figures for expenditure on metabolism are 75.5–78.1% and 55.4–63.4 in the two groups. Food consumed for somatic production makes up 1.2–1.6% and 8.6–14.8%, and that for generative production 0.7–2.9% and 5.1–11.5%, respectively. These figures are within the range of values published in the literature.

Of especial interest is the proportion of energy used by populations for different kinds of metabolism. Active fish consume about 50% of the 'metabolic' energy for swimming, 17–18% for feeding (food consumption and assimilation) and 13% for basal metabolism. All in all, the sum of the energy used for these different purposes exceeds considerably the energy used for construction.

The energy budget is distributed somewhat differently in less active species. In red mullet and pickerel, which can be defined as moderately active fish, the energy from food is divided nearly equally between basal and active metabolism (30–35%) while feeding metabolism demands only 13–15%. However, the metabolic energy expenditure still exceeds that used for synthesis, although the latter is several times as great as in the most active fish.

Most of the energy used by whiting is directed towards basal metabolism; active metabolism is very small. We have repeatedly stressed that the proportion of energy expended on active metabolism in active fish species is substantial, but recent data modify this concept in anchovy, horse-mackerel and sprat. Belokopytin (1993) has found the daily metabolic expenditure of these species to be only half what was formerly believed, because in the hours of darkness their physical activity decreases almost to zero. Accordingly, the actual daily rations of Black Sea active fish must be something like 60% of the values cited earlier in this section (Belokopytin and Shulman, 1995). The trophic significance of fish inhabiting the pelagic ecosystem is correspondingly reduced; the values of coefficients K_1 and K_2 should be 1.5 times greater.

In red mullet and pickerel such corrections are less pronounced, daily rations being estimated as 85% of the previous estimates, and K_1 and K_2 being 1.2 times higher. The final estimate of the efficiency of food used for production is, however, still lower in highly mobile fish as compared with moderately mobile and sluggish species.

5.7. ECOLOGICAL METABOLISM

This phrase is understood as a process which maintains a population or community as an entity (Khailov, 1971). Particular metabolites act as 'signals', either attracting or repelling other fish, while others have trophic significance. The former have already been given close consideration in a number of books on behaviour and ethology, so need no further comment here. The latter captured the attention of researchers relatively recently. It was found that the uptake of metabolic products is an important factor that cements aquatic ecosystems, from plants to lower animals (protozoans, coelenterates, molluscs and some others). Metabolites that play a trophic role are inorganic salts and high- and low-molecular organic substances, either dissolved or suspended in the water. Some amino acids in the water (glycine and L-alanine, for example) increase the searching initiative of fish (Bondarenko, 1986; Kasumyan and Taurik, 1993).

Experiments using radioactive tracers such as the isotopes of carbon, phosphorus, sulphur and calcium have demonstrated that fish are also capable of taking up dissolved and suspended substances, not through the gastrointestinal tract exclusively, but also immediately through the skin and gills (Karzinkin, 1962; Fedorova, 1974; Romanenko et al., 1980; Yarzhombek, 1996). These substances may participate in the synthesis of the most complex organic elements of the body – carbohydrates, adenyl nucleotides, lipids and even proteins. It might therefore appear that the theory of Pütter (1909) about epidermal uptake of dissolved organic matter, which was disproved by Krogh (1939) for fish, could be valid after all, but this is not really the case. However impressive the experiments with tracers might be, they do not destroy the basic observation that 90% or more of the requirements of the fish, in contrast to those of some soft-bodied aquatic invertebrates, enter the fish through the mouth. The trophic importance of dissolved compounds is very much a minor factor in fish inhabiting the oceans.

Large concentrations of fish do, however, influence the environment, particularly when they crowd together in a small area, oxygen concentration decreasing and ammonia nitrogen increasing (Nikolsky, 1974; Bray et al., 1986; Konstantinov and Yakovchuk, 1993). Lungfish aerate their brood of young by supplying oxygen to them from their own blood through filaments on their abdominal fins (Cunningham and Reid, 1932). The mass mortality of adult Pacific salmon, followed by decay, enriches the spawning grounds with organic matter which becomes the basis of the subsequent nutrition of the hatched larvae. It therefore regulates the abundance of the young population.

In section of this chapter it was pointed out that the part played by fish in the total food web of the pelagic portion of the Black Sea is relatively small, and that their feeding reduces the plankton by 6–10% or less. A similar picture emerges from the production and transformation of substances by fish, and the effect cannot be compared with that attributable to the medusa Aurelia aurita and the ctenophore Mnemiopsis leidyi, which compete fiercely with the fish for food. The competition provokes drastic interannual changes in the lipid contents of Black Sea pelagic fish. The effect on demersal fish has been little studied.

The relatively modest importance of fish in trophic dynamics is greatly exceeded by their economic importance, and so by the research effort put in by workers from many countries. Further knowledge of the biology of fish is still required. Fish are also suitable subjects for the study of adaptation and evolution. Being a top link in the trophic chain, they can also serve as an indicator of the condition of the whole ecosystem.

It should always be borne in mind that published figures for the ecological efficiency of food assimilated for production are averages from studies made over many years. The picture may be very different in years of shortage or plenty, when efficiency, especially as estimated in the period of growth and accumulation of lipid, differs a great deal from the average.

6. Indicators of Fish Condition

6.1. GENERAL CONCEPTS

Indications of the condition of living systems in ecology are similar to those in medicine, veterinary science and animal husbandry. The aim of research in this area is to be able to identify characteristics which can define the best possible state of organisms and populations, and can also signal and quantify deviations from it. Such indications should have high resolving power, so as to reveal the situation in deeply hidden processes within the biological systems.

Workers studying the ecology of animals first met this problem in the 1950s, since when it has developed along two lines, morpho-physiological and physiological–biochemical. Both approaches must: (1) characterize inherent functional features of the organism or population; (2) cover the whole range of variability of the process examined; (3) be representative of the population; and (4) (ideally) be easily determined under field conditions (Schwartz *et al.*, 1965; Badenko, 1966; Shulman, 1967, 1972a; Shulman and Shatunovsky, 1975; Love, 1980, 1988; Shulman and Urdenko, 1989; Chikhachev, 1991; Sidorov *et al.*, 1993; Yurovitsky and Sidorov, 1993).

The physiological–biochemical approach has been found to be the more satisfactory of the two. Among its advantages are: (1) the greater resolving capacity, grounded in actual features of physiology and metabolism; (2) the reliability of results using much smaller numbers of fish (morpho-physiological characteristics are much more variable); and (3) the frequent possibility of adaptation to field conditions.

Morpho-physiological indices consist of such measurements as the ratio of length to weight of the body or the ratios between the weights of some organs – liver, gonads, heart, spleen, kidneys, brain, etc. – and the weight of the body. A significant drawback of any such indicators is that they are based on the fresh weights of the organ and the whole body, which can vary independently of each other: 60–80% of the fresh weight consists of water, variations in which account for most of the variations in weight. It is much more important to know about the changes occurring in the total content of organic and inorganic compounds.

A simple example illustrates one of the fundamental weaknesses in morpho-physiological measurements. When non-fatty fish such as cod become severely depleted late in the spawning cycle, much protein is removed from the musculature. The contractile material of the cells diminishes but the outline of each cell is relatively unchanged, because the space between the remaining contractile material and the surrounding connective tissue matrix is now filled with fluid (Love and Lavéty, 1977). The fish definitely look thinner in extreme depletion (Love, 1988) but, because of the substitution by liquid, the visible emaciation is not proportional to the loss of protein, and the weight/length ratio therefore underestimates the loss of nutritional 'condition'.

6.2. LIPIDS

Lipids, especially the triacyl-glycerols, display wide variability during certain periods of the annual cycle and changed living conditions, and thus possess great potential as 'markers' of condition. The variations emerge anew each year, defining the starting-point of biochemical change. The fatness of the majority of species is lowest during spawning, so, taking this point as 'zero', one can easily assess the rate of accumulation of lipid reserve in the post-spawning foraging period. Data on lipid accumulation have an advantage over those on increase in protein accumulation, since the starting point of the latter is variable and difficult to determine.

Table 21 illustrates the advantage of triacyl-glycerol data over the 'Condition Factor', defined as $100 \times W/L^3$, where W is the weight of the body and L its length. This table includes data from studies of the effects of the larvae of parasitic nematodes on anchovy (Shulman and Shchepkina, 1983). It shows that, during the entire annual cycle, the parasites provoked a 2–3-fold decrease in the triacyl-glycerol in the body of the fish, while the condition factor remained unchanged. As in the previous example, the low fat content here entailed a substantial increase in the water content. Since the specific gravity of lipid is less than that of water, losses of essential lipid reserve can result in an actual increase in body weight, such has been reported in the Azov anchovy (Shulman, 1960b, 1972a). Other authors who have also pointed out defects in 'condition factor' are Love (1980), Gershanovich et al. (1984) and Shcherbina (1989).

Table 21   Effect of infection by parasites (nematode larvae) on Black Sea anchovy. (After Shulman and Shchepkina, 1983.)

	n	Strength of infection	Condition factor	Triacyl-glycerol level in muscle (%)
January	12	Weak	0.70	2.654 ± 0.75
	12	Strong	0.78	1.144 ± 0.31
March	12	Weak	0.87	1.823 ± 0.25
	10	Strong	0.88	0.322 ± 0.20
May	14	Weak	0.83	0.363 ± 0.07
	14	Strong	0.83	0.111 ± 0.03
October	10	Weak	0.86	3.867 ± 0.40
	10	Strong	0.89	2.554 ± 0.35

Lipid accumulation depends directly upon the availability of food, and is of the utmost importance to fish in the post-spawning period. Replenishing the energy and plastic reserves used during spawning, it allows some species to prepare for wintering migration and the hazardous period in winter when the food supplies give out. 'Food supply' implies not only the amount of food in a unit of water area, but also the number of consumers – both the species under examination and competitors – and specific feeding conditions. These include among others the temperature, upon which digestion rate depends, transparency of the water, which influences the rate of hunting, and the availability of the food actually present.

Clearly, it is not possible to determine all of these factors when studying a population in nature, and we must be content with the definition of food supply as the ratio between food consumed and food required. There is, however, an indirect approach based on the result of feeding, seen as the content of lipid accumulated by the population at the end of the feeding period. Lipid in the form of triacyl-glycerols is not, of course, the only possible index of feeding. Other parameters which describe the success of the process are the level of glycogen accumulation in the liver of 'lean' species (e.g. Gadidae, Gobiidae), glycogen in the mantle cavity of bivalve molluscs (Goromosova and Shapiro, 1984; Shchepkina, 1990) and the level of protein accumulated in the hepatopancreas of squid (Shulman *et al.*, 1992). It is up to the individual research worker to decide which parameter best describes the principal features of the process, by being sufficiently variable and easily evaluated.

It should be borne in mind that the lipid index cannot characterize the food supply over the whole annual cycle, but only that in the period when lipids are accumulated rather than consumed. For warm-water fish of the southern European seas, this is the summer and autumn, while for cold-water fish such as the Black Sea and Mediteranean sprat, it is spring and summer. The food supply for sprat is therefore best estimated from the lipid content in June–July and for anchovy in October–November. Such studies were made over many years on Azov (Shulman, 1974; Luts and Rogov, 1978) and Black Sea (Shulman and Dobrovolov, 1979) anchovy, Azov kilka (Luts, 1986), and Black Sea sprat (Shulman, 1996). The studies on sprat extended over more than three decades (1960–1994).

The index of food supply (I_{fs}) may be interpreted as the ratio between the fatness attained by sprat at the end of feeding in a certain year (F_i) to average fatness ($\overline{F}$) deduced from data of the same period obtained over many years:

$$I_{(fs)i} = F_i / \overline{F} \qquad (43)$$

At $I_{(fs)i} > 1$, the food supply of the population is adequate; at $I_{(fs)i} < 1$, it is inadequate.

Given that the average fatness of sprat calculated over many years is 11.49% ± 0.29%, the years of sufficient food supply were 1960, 1962, 1964, 1976, 1977, 1980, 1981, 1982, 1984, 1988, 1991, 1992 and 1994, while the years 1965, 1968, 1969, 1970, 1971, 1972, 1973, 1983, 1986, 1987, 1989 and 1990 were unfavourable. In 1961, 1963, 1967, 1973, 1974, 1978 and 1985 the supply was about average.

Analysis of these data provides an understanding of processes taking place in the Black Sea. The first period comprises the time from 1960 to 1977. The average fatness of the sprat population was 8–13% at the end of the feeding period, not varying greatly between years (coefficient of variation 11.8%). The lipid content decreased from the early 1960s to the early 1970s, then increased in the mid-1970s. Counting techniques were far from precise at that time, but it appears that the abundance of the stock was fairly low.

The second period ranged from 1978 to 1985. In the first half, the fatness of the sprat showed the highest rise recorded, from comparatively low values to 15.5%, which level persisted for 3 years, followed by a pronounced drop to 9% and later a rise to 12%, the coefficient of variation being 15.2%. The abundance of the stock of sprats changed with the changes in their lipid contents, the stock of 1980 being three times that of any recorded in the period 1960–1977.

The third period (1986–1994) was marked by a return to the range of 9–13.5%, similar to that in the first period but with a higher interannual variation (coefficient of variation 14.7%). Unlike the situation in the second period, the early years of the third showed a negative correlation between fatness and stock density. It is worth mentioning that sprat from around the island of Zmeiny were almost always fatter than their counterparts in waters around the Cape of Tarkhankut, presumably because of the input of nutrients from the Danube.

Certain patterns can be distinguished in all these changes. In the 1960s and most of the 1970s, the nutritive base for the plankton-eating fish was typical of the

Black Sea, but, in the late 1970s, the load of nutrients in the Dnieper, Dniester and Danube rivers resulting from human activity (mostly nitrogen and phosphorus) increased markedly and strongly affected the north-western part. On the shelf, the effect was ruinous because of heavy eutrophication (Vinogradov *et al.*, 1992a), but the pelagic area incorporated the nutrient influx into its ecosystem and increased its biomass of phyto- and zooplankton (Samyshev, 1992). The latter author showed that in 1979–1980 the biomass of phytoplankton and in 1980 that of the zooplankton rose high in the northwestern part of the sea as compared with the preceding years. As we have seen, the fatness of the sprat population increased accordingly, and anchovy, another planktonivorous fish, also increased in numbers at this time (Shlyakhov *et al.*, 1990).

The size of the stock and their fatness declined dramatically in 1983 because of the outbreak of the medusa *Aurelia aurita*, which prospered on the enhanced level of zooplankton and competed with the sprat for food. Surprisingly, the sprat stock overcame the pressure from *Aurelia* the following year, and its fatness increased. Two years after that the medusae declined in numbers and the sprat stock increased dramatically. This reduced the food supply – and so the fatness of the stock declined. The sprats grew fat again only when their numbers dropped in 1988. The food factor therefore influenced both the accumulation of energy and reproduction in the stock.

A further 'attack' on the food supply and abundance of sprat was made by the ctenophore *Mnemiopsis leidyi*, which was introduced into the Black Sea and multiplied greatly in 1989–1990. The anchovy in the Black and Azov Seas were endangered even more than the sprat. However, as soon as the ctenophore decreased in biomass, the sprat revived once again. In 1991–1992 their lipid content increased to a high level which was a record for fish found near Zmeiny island. At the same time the stock was being rapidly restored – sprats seem to be more resilient than anchovies. The main outbreak of *Mnemiopsis* takes place in the second half of the year, while mass feeding of sprat occurs in the first. Living in waters under the thermocline, sprats do not face the same competition for food from the ctenophore as the anchovies. The latter inhabit the surface layer together with *Mnemiopsis*, feeding most strongly in the second half of the year when competition from the jellyfish is strongest.

The broad picture is that gradual long-term variations in the fatness of sprat, found in the 1960s and 1970s, changed to a pattern of frequent drastic changes in the 1980s. The pelagic ecosystem of the Black Sea appears to have lost its stability. The enhanced input of nutrients has made it vulnerable to the expansion of organisms which compete fiercely against plankton-feeding fish, and we should not have to wait long to discover whether these fish can adapt to the new conditions or enter a state of permanent recession.

The extended survey has enabled us to appreciate that data on fatness can contribute to knowledge of the long-term dynamics of food supply in fish and the degree of trophic tension in the water body at large. Through the data we can monitor the population, and it is important that the three-decade survey should continue.

Processed data on the lipid content of Black Sea anchovy are not as complete, covering the period from the late 1950s to the early 1970s, but even so it is clear that the average fatness of anchovy at the end of the feeding period (October–November) correlates well with the average annual temperature regime ('thermal background') of the Black Sea (Volovik, 1975). The thermal background of the sea is calculated using the following equation:

$$T^0 = \frac{\overline{\Sigma\, t_m}}{\sigma} \qquad (44)$$

where t_m is the average monthly temperature obtained from March to August of a certain year in the sea surface near Batumi, the Caucasian coast of the Black Sea (Bryantsev et al., 1987). The correlation is free from complications: the high heat capacity of the sea allows rich production of phyto- and zooplankton late in the year, and variations in it were originally the sole factor governing fish food supply in the period before extra nutrients from human activity upset the balance.

The study of the dynamics of fatness links the particular stock to the basic characteristics that influence the productivity of the ecosystem. It is much more difficult to define the relationship between fish fatness and the biomass of plankton. This is because current techniques for sampling deal solely with residual biomass, but not with the production generated by the most important components of the ecosystem. In addition, these techniques do not take into account factors such as elimination occurring at different trophic levels, so the lipid contents and numbers of fish – the final link in the chain – are the most reliable integrated characteristics of the condition of the ecosystem. Fatness of the fish also identifies rich feeding grounds, such as the north-western part of the Black Sea adjoining the mouth of the Danube.

An additional indicator of food availability is the age-dependent content of the lipid stored in the fish. As noted in Chapter 4, the fatness depends directly upon the food supply, and with age the metabolism shifts towards lipid accu-mulation. When feeding conditions are poor, however, the relationship with age becomes inverted (Shulman, 1974). The phenomenon was first defined in Azov anchovy (Shulman, 1960b, 1964a) and later in Black Sea sprat (Minyuk, 1991) (Figure 72). In years when food is plentiful, the lipid content is greater and the relationship between fatness and age is more pronounced. A similar picture has been found in a variety of fatty fish from northern seas (Shatunovsky, 1980) and in Caspian kilka (Shatunovsky and Rychagova, 1990).

There is a change with age in the ability of Atlantic cod to mobilize protein. This appears to be an attempt to compensate for the fact that, as the fish increase in length with age, the size of the gonads as a proportion of the body weight increases until the fish die through 'over-reproduction' (Love, 1970).

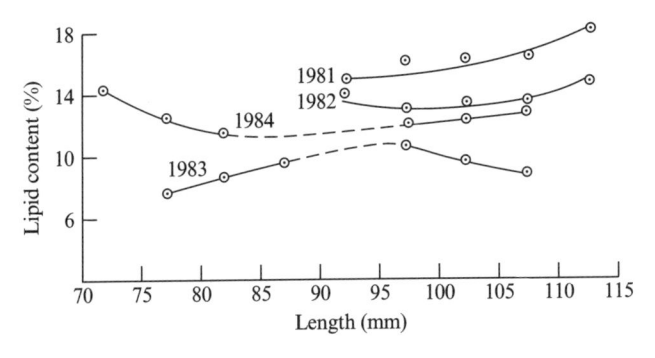

Figure 72　Annual variations in the lipid content of sprat of different sizes. Fish were caught in June or July. (After Minyuk, 1991.)

Laboratory experiments showed that the water content of the muscle of small, immature cod subjected to starvation rose (reflecting protein depletion) from 80% to around 86%, beyond which level the fish died. Larger fish, however, which had spawned several times, could be depleted until the water content of the musculature was over 95% – a remarkable adaptation to the more severe depletion imposed under natural conditions (Love, unpublished). While this phenomenon differs from that described above in fatty fish, it again illustrates a change in the metabolism of fish in response to growth. Likewise, Borisov and Shatunovsky (1973) studied the possibility of using the water content to estimate the natural mortality rate of Barents Sea cod.

As mentioned earlier, the critical level of fatness will trigger migration by the Azov anchovy at a particular range of water temperatures. Knowing the weather forecast and the fatness of the main stocks of anchovy in the Azov Sea in September, it is therefore not difficult to predict the time and character of the migration run through the Kerch Strait (Shulman, 1957, 1960b, 1972a; Taranenko, 1964; Luts and Rogov, 1978). Lean fish (lipid content less than 14%) do not migrate at all and die. Surveys made in September will have high economic value, because the prognosis of the order in which various shoals will pass through the strait enables fishing boats to gather at the right time and place. In former times, the fishermen were often idle for considerable periods. The migration potential of Black Sea anchovy can also be predicted in the same way (Danilevsky, 1964; Chashchin and Akselev, 1990), as can that of Caspian kilka (Rychagova *et al.*, 1987; Rychagova, 1989), Pacific sardine (Shvydky, 1986) and some other species (Filatov and Shvydky, 1988; Shvydky and Vdovin, 1991).

Sprats differ from anchovy in that they make only short trips from the depth towards the shore, but the fatness may still serve to locate the best fishing sites (Shulman *et al.*, 1985). If artificial sources of light are used for catching sprats, they respond much better to the light if their fatness is low (Gusar *et al.*, 1987) – the sated fish are no longer interested.

Low levels of neutral lipids in the muscle and liver characterize a poor swimming performance in horse-mackerel (Figure 22 page 81). Recent experiments with a flume showed that the percentage of weak or 'recessive' fish in populations has been increasing of late, indicating a deterioration of living conditions, perhaps a measure of the increase in pollution. The individual variability of fatness has been suggested as an indicator of condition, since it increases in unfavourable environments (Nikolsky, 1974; Polyakov, 1975; Shatunovsky, 1980). The levels of triacyl-glycerols accumulated in the body and liver also indicate the time and rate of spawning migration (Krivobok and Tarkovskaya, 1960; Shatunovsky, 1980), while the lipid content of the gonads appears to indicate the fecundity and the diameters of the oocytes (Figure 73).

The rate of smoltification of young salmon can also be assessed from their fatness (Malikova, 1962; Sidorov, 1983; Varnovsky, 1990a), which also indicates ability of other species to survive the winter (Mukhina, 1958; Polyakov,

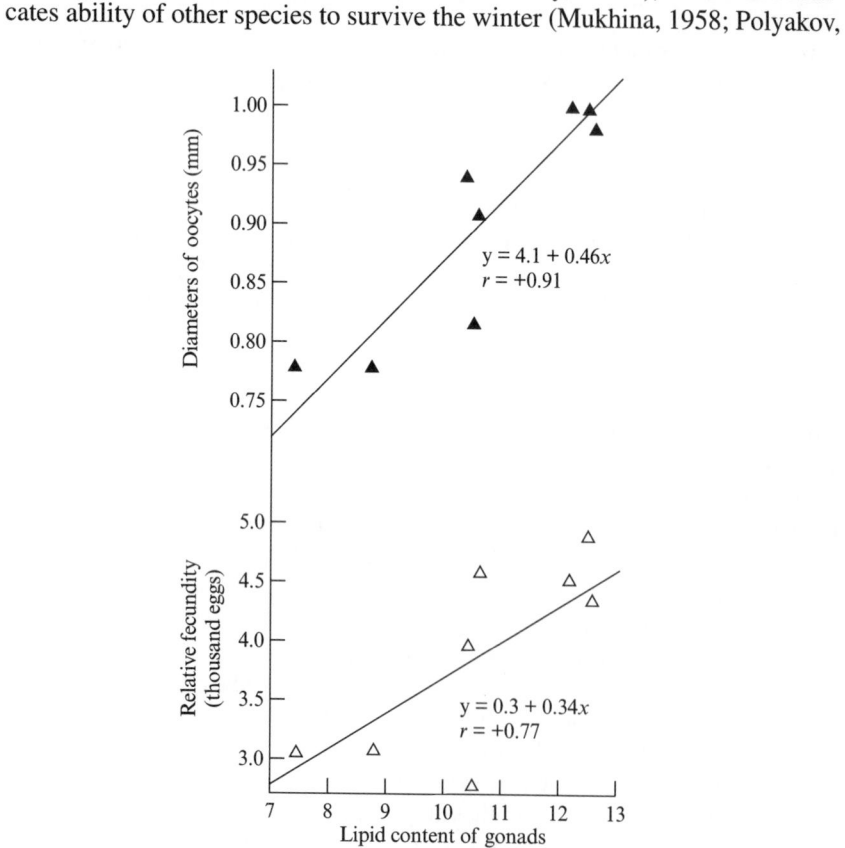

Figure 73   Relationship between the fat content of female gonads of Black Sea horse-mackerel, and the size and number of eggs produced.

1958). On the other hand, obesity in fish, especially fatty degeneration of the liver, is a sign of disturbed metabolism (Faktorovich, 1967; Sidorov, 1983).

All the above examples link the level of neutral lipids to the energy potential of the fish. The fatty acid composition of the same lipids indicates the nature and composition of the diet. Phospholipids also characterize the state of the organism, particularly of their cell membranes. The polyenoic acids, in which phospholipids are rich, follow the degree of metabolic activity. A low level in horse-mackerel swimming in an exercise tunnel has been shown to coincide with poor swimming performance, while vigorous swimmers were found to develop decreased levels as fatigue progressed. An increased level of lysophosphatidyl ethanolamine and lysophosphatidyl choline has been said to indicate pathology and decay of the tissues (Sidorov, 1983; Bolgova et al., 1985a,b; Lizenko et al., 1991). The content of docosohexaenoic in the eye tissues of Pacific saury was found to correlate with the intensity of their reaction to light (Table 8; Yuneva et al., unpublished).

The level of docosohexaenoic acid in the red muscle of male humpback salmon indicates their condition during the reproductive period. The content is higher in leaders and lower in satellites. The reproductive capacity of the females can be deduced from the docosohexaenoic acid content of their triacyl-glycerols (Figure 24, page 85) that supply the major part of the plastic material for the creation of oocytes and, later, eggs. Much of this fatty acid originates in the living food from the marine environment, but it is also essential for the proper development of the initial stages of freshwater fish – an important point to remember in fish farming. Further information on the relationship between the condition of fish and their contents of neutral and polar lipids may be obtained from Lühmann (1953), Morawa (1955), Krivobok (1964), Love (1970, 1980), Kozlov (1972), Shatunovsky (1980), Lapin and Shatunovsky (1981), Sidorov (1983), Sautin (1989), Gershanovich et al. (1991), and Yurovitsky and Sidorov (1993).

6.3. OTHER SUBSTANCES

Any biochemical link in a metabolic chain is potentially an indicator of the condition of an organism (Shulman and Shatunovsky, 1975; Sidorov et al., 1990; Yurovitsky and Sidorov, 1993). The concentration of haemoglobin in the blood is widely used for assessing condition (Puchkov, 1954; Stroganov, 1962; Smirnova, 1967; Kititsina and Kurovskaya, 1991), and is especially important for finding the early stages of pathology developing from faulty nourishment or maintenance in aquaculture. It is also useful in the study of toxicology (Lukyanenko, 1987) and has been extensively used in studies of numerous species of fish from the Azov, Black and Mediterranean Seas

(Kotov, 1976; Tochilina, 1990; Soldatov, 1993). It can also be used to study the progress of fatigue in fish, since the level declines in the blood (Belokopytin and Rakitskaya, 1981). The concentration of myoglobin in the red muscle indicates the customary vigour of swimming of individual fish (Love *et al.*, 1977: Atlantic cod), the colour of the tissue increasing during migration or artificial stimulation of swimming and fading during subsequent rest.

Other useful indicators of condition are serum albumins, α- and β-globulins, and the total concentrations of serum proteins and lipoproteins (Kulikova, 1967; Ipatov, 1970; Kondratyeva, 1977). Used together with the haemoglobin values, they give a good estimate of the quality of brood fish (Badenko *et al.*, 1984; Chikhachev, 1991; Lukyanenko *et al.*, 1991: Azov and Caspian sturgeons). A series of indicators, rather than just one, gives much more information. For example, the protein content of cod muscle decreases during starvation only after the level of liver lipid has dropped below a certain critical value (Black and Love, 1986), so the extent of depletion can be realistically judged only by measuring both. The determination of muscle protein alone fails to detect the early stages of depletion, while the liver lipids do not change further during a long range of subsequent stages. Further, in assessing the state of a fishing ground, it is useful to know whether a given starving fish is beginning to recover or is still deteriorating. Protein and lipid determinations alone do not tell us. However, the bile of a fish feeding actively is small in volume and yellow in colour, while if it has not fed for more than 3 days the colour is deep blue–green and the volume large (Love, 1980). Cellular lysis and the mobilization of proteinaceous structures in muscle are closely linked with the activity of lysosomal enzymes, and the activity of acid phosphatase has often been employed as an index of lysosomal activity (De Duve, 1963). Black (1983) has shown that there is a linear correlation between the acid phosphatase activity of cod muscle tissue and the degree of protein removal during starvation. Further, Beardall and Johnston (1985) have demonstrated that, while a number of other lysosomal enzymes also increase during the starvation of saithe, their levels return to non-starved values in as little as 10 days after the resumption of feeding. Love (1997) has therefore suggested that all these observations could be combined to assess the current status of the starvation of a fish such as cod:

1. Large volume of blue bile, high values of acid phosphatase – severe depletion in progress.
2. Small volume of yellow bile, high values of acid phosphatase – recovery from starvation has begun within the last week or so.
3. Small volume of yellow bile, low value of acid phosphatase, quantity of muscle protein still reduced – fish has been severely starved but has been recovering for some time.
4. Small volume of yellow bile, low value of acid phosphatase, muscle protein normal, water content down to 'fed' values, liver lipid reduced – recovery almost complete.

Since the liver changes markedly in size according to its lipid content, the most useful measure is the quantity of lipid in the whole liver, rather than the concentration in a piece of the organ. Where much lipid is stored, the liver is large and creamy in appearance, while the liver of a starved fish is small and red. It must be emphasized that this particular scheme applies only to 'non-fatty' fish, which store virtually all their lipid reserves in their livers. Fatty fish will behave somewhat differently, although the general principles are the same.

In ecological literature there is still controversy as to whether one parameter or several should be measured to assess the condition of fish (Khailov et al., 1986). It is clear from the above scheme that a set of indicators gives a fuller picture than is possible with a single determination, but sometimes only one parameter can be measured under field conditions. Much information has been gained about the usefulness of serum protein determinations for assessing biological condition (Lukyanenko, 1987). Knowledge of the total protein content of tissues is not useful where it is relatively stable, but Geraskin and Lukyanenko (1972) used deviations of protein content from the norm to assess the effects of toxicants. Knowledge can also be gained about the condition of fish eggs, the content of protein being greater in mature eggs than in over-mature (Konovalov, 1984). The initiation of the reproductive processes in female fish is indicated by the colloid resistance of the serum proteins (Kychanov, 1984; Kychanov and Volodina, 1985).

Indicators related to nitrogenous metabolism include the concentrations of free amino acids (Shcherbina, 1973; Ostroumova, 1983). The ammonia coefficient (O/N), described in Chapter 2, is also useful, values less than 8.67 indicating anaerobic metabolism which can be caused by, for example, increased concentrations of heavy metals in the water, which inhibit the normal supply of oxygen to the tissues (Shulman, 1996). The rate of oxygen uptake is by itself a good indicator of condition, responding as it does to every disturbance of the norm (Stroganov, 1968).

Next in line is the ratio between RNA and DNA, or the content of RNA alone, especially transport RNA rather than total. These show the vigour of protein synthesis associated with growth at the moment that the fish was killed for sampling (Pirsky et al., 1969; Bulow, 1970; Haines, 1973, 1980; Ipatov, 1976; Thorpe et al., 1980, 1982; Bulow et al., 1981; Wilder and Stanley, 1983; Nakano et al., 1985; Varnavsky et al., 1991).

A very important indicator of the rates of various processes is the enzymatic activity of the tissues, although there are problems in determining enzyme activity under conditions approximating those of the (previously) living fish. A promising approach is the determination of alkaline phosphatase activity in the scales. This enzyme is associated with calcification of bones, hence with the rate of 'mineral growth' of the fish. From studies conducted on Black Sea red mullet and Azov round goby (Senkevich, 1967; Shulman, 1974), common goby (Fouda and Miller, 1979) and some freshwater fish (Goncharova, 1978) it was concluded that alkaline phosphatase in the scales correlates positively with growth rate (Figure 74). Similar results are known from other workers (Roche *et al.*, 1940; Motais, 1959). Cytochrome oxidase and ATPase determinations can give information about the quality of eggs and sperm. Their activity is much higher in the ripe sex products than in immature (Gosh, 1985). Enzymatic activity also yields information about the peroxidizing and antioxidizing systems of the fish. In work performed by Rudneva-Titova (1994) on Black Sea fish, much attention has been paid to measurements of lipoxidase, superoxide dismutase, catalase, peroxidase and glutathione reductase activity. The values change considerably from the effects of oil products, pesticides, heavy metals, toxicants and other xenobiotic agents.

The activity of aryl hydrolase, a detoxicating enzyme, can be a useful indicator of chemical pollution (Vetvitskaya *et al.*, 1992). The thiamine content of various organs also gives a good indication of the condition of fish exposed to toxicants (Malyarevskaya, 1979; Malyarevskaya and Karasina, 1991). Signs of internal destructive processes resulting from pollution are detachment of muscles and the occurrence of abnormally thin egg capsules (Sidorov *et al.*, 1991; Nemova *et al.*, 1992; Yurovitsky and Sidorov, 1993). Pathology of this sort engenders a pronounced increase in the activities of the proteinases cathepsin, calpaine and hyaluronidase, and a decrease of hydroxyproline in the connective tissues.

The content of glycogen, glucose, lactate, adenyl nucleotides and creatine phosphate, together with the activities of the enzymes which regulate them, also provide good indications of condition (Morozova *et al.*, 1978a; Trusevich, 1978). Indices of fatigue in fish are a sudden drop in the level of blood sugar, an accumulation of lactate in the blood and a decline of ATP in the muscle, bearing in mind that, in at least some species of fish, the lactate is not released into the blood from the muscle if the fish is exercised stressfully (Wardle,

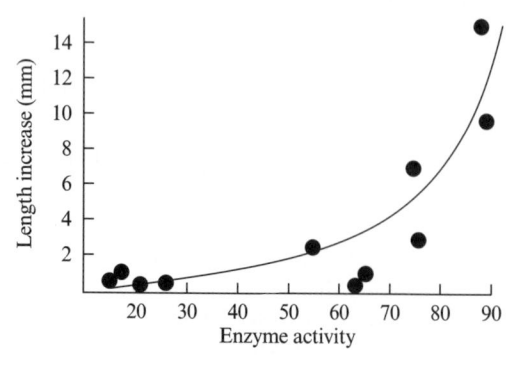

Figure 74   Relationship between the activity of alkaline phosphatase (in milligrams of inorganic phosphate released from 100 g of scales per hour) in the scales of red mullet and the rates of growth of the body. (After Senkevich, 1967.)

1978). Hochachka (1962) showed that the low content of glycogen occasionally found in salmon indicated their inability to go upstream through man-made rapids. We state again that the purpose of glycogen mobilization is to provide bursts of short duration that fish have to make at times. The content of glucose in the blood of fish subjected to oil pollution drops dramatically (Kotov, 1976). The concentration of lactate is increased in the overmature eggs of fish (Gosh, 1985). Developing eggs of Atlantic salmon infested with the fungus of *Saprolegnia* have a reduced glycogen content (Timeyko, 1992). A shift in the osmoregulatory characteristics of the blood is a classic indicator of migratory readiness of diadromous fish (Fontaine, 1948; Zaks and Sokolova, 1961; Natochin, 1974; Krayushkina, 1983; Smirnov and Klyashtorin, 1989; Klyashtorin and Smirnov, 1990; Varnavsky, 1990a; Chernitsky, 1993).

The number and size of mitochondria indicate the level of energy metabolism and vary according to the condition of the fish (Ozernyuk, 1985; Savina, 1992), as does the level of nucleolus–protein biosynthesis (Arkhipchuk and Makarova, 1992).

Indices of condition like those mentioned above have also been used in studies of other water organisms. The food available to the medusa *Aurelia aurita* and the ctenophore *Mnemiopsis leidyi* can be judged from their level of carbohydrates (Anninsky, 1990), and that available to copepods such as *Calanus helgolandicus* can be judged from their content of neutral lipids, especially wax esters (Yuneva and Svetlichny, 1996).

Since biochemical and physiological techniques for monitoring fish populations give values on numerical scales, it is essential to define the optimum values and the range of deviations in each species. This has been done in the case of lipid content of Azov anchovy as a measure of its preparedness to migrate (Shulman, 1957, 1960b, 1972a; Taranenko, 1964; Luts and Rogov, 1978). The water content of cod muscle is a comparable index, values of 80.0–80.9% indicating no mobilization of protein and 81% or more indicating various degrees of plundering the protein resources (Love, 1960).

Future studies must include a search for modifications to laboratory techniques that would enable them to be used at sea, and for the correlations between the basic biochemical characteristics of the fish and the values obtained by such modified techniques. Some examples of simple techniques are listed below. Some are more accurate than others, but all give an idea of some aspect of nutritional condition.

1. In fatty fish, the total lipid content correlates negatively with the content of dry matter and water. The 'fat-water line' has been recognized in dozens of species of fish (Kizevetter, 1942; Levanidov, 1950; Lühmann, 1953, 1955; Brandes and Dietrich, 1958; Krivobok and Tarkovskaya, 1960; Shulman, 1961). The correlation is a good one ($r = -0.8$ to -0.9), but each species has its own equation. In Black Sea sprat, for example, the relationship is: $Y = -13.28 + 0.84X$, where Y is the total lipid content and X the content of dry matter

(Shulman *et al.*, 1989). The weight of the fresh sample and the dry weight will therefore give a good estimate of the total lipid, the determination of which in the field would be difficult.

2. Determinations of protein are relatively slow, but a 'protein-water line' has been described (Love, 1970, Figure 85) in the muscle of Atlantic cod, a non-fatty species ($r = -0.63$). The water content (derived from the dry weight) therefore gives a reasonable estimate of the protein content of this and other non-fatty species.

3. The flesh of non-fatty species is normally translucent, but during the spawning cycle is seen to become progressively opaque as the fish deplete their protein reserves, perhaps because of a change in the balance of protein, water and minerals. This observation was quantified by Love (1962), who measured the absorption of light of small pieces of cod muscle, after removing the connective tissue, squeezed to constant thickness between two glass plates. The close correlation between the opacity and the increase in water content (the 'negative' measure of the protein content) is shown in Figures 75 and 76. Although spawning takes place in early March

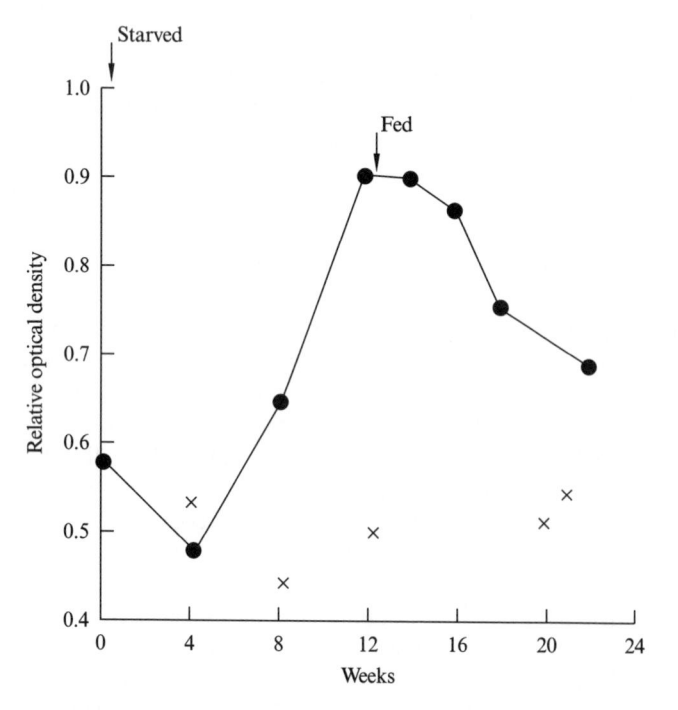

Figure 75   The effect of starvation and subsequent feeding on the optical density of the muscle of summer-caught cod in an aquarium. The optical density is relative to an arbitrary fixed standard. ●, starved fish; ×, controls caught from the same fishing ground as the starved fish at various times during the experiment. (After Love, 1962.)

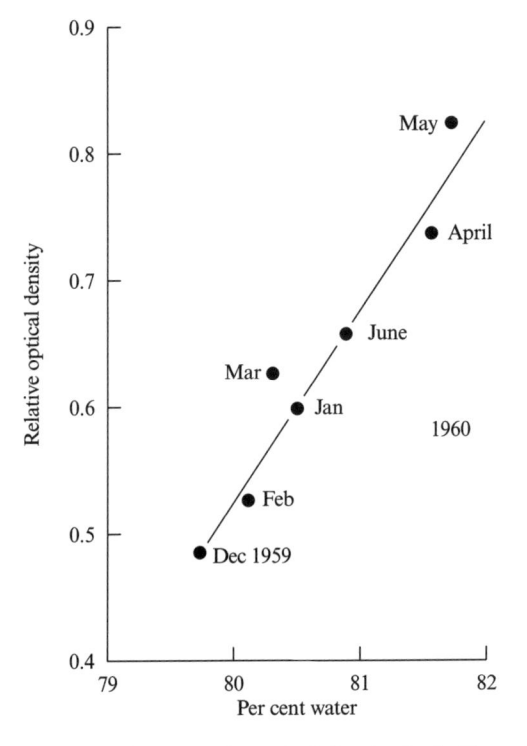

Figure 76   Relationship between the optical density (as in Figure 75) and the water content of cod white muscle caught at different times of the year. (After Love, 1962.)

in this stock of cod, it will be seen that they experienced maximum depletion in May in the year that the work was carried out. This equipment can be used on a moving boat, but the technique appears not to have been adopted by others.

4. Examination of the gall bladder for colour and size is readily done in the field. Using this technique, it was sometimes observed that all the fish caught on one ground could be starving, while those from a nearby ground caught within a day or two were all actively feeding. Extra information of this kind is useful in ecological studies.

5. The pH of the muscle 24 h after death gives a good measure of the carbohydrate resources of the fish at the point of death (Black and Love, 1988). This is now easy to measure at sea.

6. There is a close correlation between the docosohexaenoic acid content of fish lipids and the iodine number. Shulman and Yuneva (1990a, b) showed that the relationship in humpback salmon is: $Y = -25.69 + 0.185X$, where Y is the C22:6ω3 content and X the iodine number of the lipids. Again, iodine numbers can readily be obtained at sea, eliminating the need for gas–liquid

chromatography. Other quick methods are described by Zhukinsky and Gosh (1988), Varnavsky (1990b) and Shvydky and Vdovin (1994).

There is no space here to consider fully the application of condition measures to the larval stages of fish. The biochemical and other indices that regulate survival are reviewed by Ferron and Legett (1994).

The study of biological condition has come a long way from the weight/length ratio, and the information now available gives a much deeper insight into the workings of the organism. It is also of great importance to those working in a number of applied fields – ecology, conservation, natural process modelling (simulation), fishery and fish culture.

7. Intraspecific and Interspecific Differentiation of Fish

Genotypic studies involve large-scale examination of the structure within a population. Originally, they were based on haemagglutination, i.e. defining the blood type (Sindermann and Mairs, 1961; Altukhov, 1969, 1974; Limansky, 1970) and later involved the study of polymorphism of proteins, including haemoglobin, albumin, transferrin, α-, β- and γ-globulins and enzymes (Tsuyuki et al., 1965; Salmenkova, 1973; Altukhov, 1974, 1983; Kirpichnikov, 1978, 1987; Dobrovolov, 1980; Chikhachev, 1984; Jamieson, 1974; Jamieson and Birley, 1989; Lukyanenko et al., 1991). Study of protein polymorphism can reveal the genetic structure of a population, interpopulation invasion and the nature of local fish stocks and shoals, and has made it possible to evaluate their degree of displacement. For example, the separation of the Baltic cod from the Atlantic stock has been confirmed (Jamieson, 1974). The genetic results also provide a better understanding of the distinctiveness of allied species, and allow their taxonomic status to be defined more exactly from associated morphological characters (Dobrovolov, 1980; Smith et al., 1990; Lukyanenko et al., 1991). Additional progress in this field has been achieved from studies of DNA (Mednikov et al., 1977; Ginatulina et al., 1988; Cherkov and Borchsenius, 1989; Smith et al., 1989; Bentzen et al., 1996). Other recent work in fish genetics is contained in a volume edited by Matthews and Thorpe (1995).

The genotypic aspect of research into intra- and interspecific differentiation should be regarded as an independent field, with a less direct bearing on ecology. Ecology emerges only when the adaptive significance of the various characteristics is explored and related to the environmental variables. In the initial stages of research, the adaptive importance of genetic characters such as allozymes is not easy to establish with certainty.

Much progress has also been achieved in the phenotypic approach to studies of intra- and interspecific differentiation of fish. It is this approach in the context of a 'species within its area' that has been vigorously explored by ecologists (Volskis, 1973). Each species needs to be regarded as a discrete ecological unit, not an abstract speculation but a distinct component of the biosphere in which it plays a special and versatile part.

7.1. INTRAPOPULATION VARIABILITY

Variabilities within and between species represent adaptation evolved to provide a dynamic balance of a population, race or species inhabiting a changeable environment. This concept has been recognized and developed by many authors (Nikolsky, 1965; Polyakov, 1975; Shatunovsky, 1980). Following from it, any growing instability of the environment leads to an increase of intra- and interpopulation variability.

Shatunovsky (1980) reported that the growth rate, oxygen uptake and mortality rate of steelhead trout and Baltic cod changed during the lifetime of one generation (Figure 77), the fish composing the population becoming heterogeneous. We have already referred to heterogeneity manifesting itself as variations in the fatness of various species from the Azov and Black Seas (Shulman, 1967, 1972a). Another manifestation is the varying content of lipid fractions and their unsaturation in populations of Black Sea horse-mackerel and Pacific humpback salmon (Chapters 2 and 3). The physiological and biochemical characteristics become more variable during periods of enhanced functional activity, for example, in the periods before spawning and during spawning, when the coefficient of variation in the lipid content may amount to 80%. This phenomenon is widespread, for example, the replenishment of glycogen in the liver or the large increase in non-esterified fatty acids in the blood of cod during refeeding after starvation. The variation between individual fish is greatest where the curve is steepest and least when values have settled to a steady state (Black and Love, 1986). Indeed, a plot of the variations in the results can be as revealing as a plot of the results themselves, showing, as it does, the metabolic activity at each stage of the experiment.

Such heterogeneity fosters optimum functioning of the population, and guarantees long-lasting and efficient spawning in those fish that exhibit intermittent maturation and batch reproduction. It results in balanced productivity and recruitment of the stock and in the more efficient use of the nutritive base. If living conditions become hard, individual variability will rise within the population (Nikolsky, 1974; Polyakov, 1975), offering a better chance for the population to survive. Shatunovsky (1980) recorded

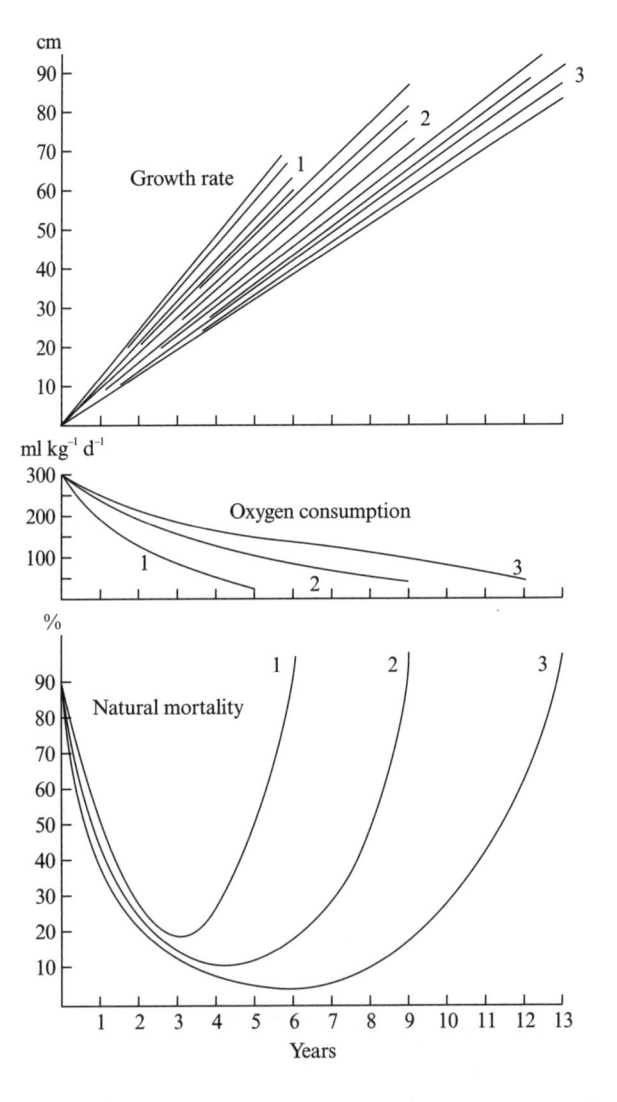

Figure 77   Effect of different rates of growth of individual fish on other properties (schematic): 1, fast growth rate; 2, middle rate; 3, slow rate. (After Shatunovsky, 1980.)

heterogeneity in growth rate between different generations of White Sea flounder. Individual variability depends significantly upon the age structure of a population. Of especial importance is the divergence between individuals in protein accumulation, which accounts for the productivity, quality of reproductive products and feeding mode followed by individuals (Shatunovsky, 1980). It was convincingly shown that narrowing the nutritive base may lead, through depressed protein growth, to emergence of monocyclic dwarf forms (as in smelt), altering the character of the originally polycyclic population (Kriksunov and Shatunovsky, 1979).

7.2. INTERPOPULATION VARIABILITY

Whereas individual variability promotes stability of the population, the interpopulation or intraspecific variability performs a similar function for species within a particular biosphere. An increase in intraspecific variation is a sign of biological progress.

In northern seas, fish such as herring, cod, haddock and flounder show intraspecies population differences in life span, structure, age at maturity and various physiological and biochemical characteristics (Shatunovsky, 1963, 1970; Love, 1970, 1980; Storozhuk, 1971; Lapin, 1973). In contrast, fish from the Black, Azov and Mediterranean Seas have shorter life spans and display less distinct intraspecifc differences. Moreover, within one sea population, fish of one species do not differ much in age structure, fecundity or spawning character. The most pronounced difference in the latter group of species is the rate of lipid accumulation in sprat from different sites in the Black Sea and anchovy in the Azov Sea (Shulman, 1972b). Comparison of anchovy from the Azov, Black and Mediterranean Seas shows more marked differences in the levels of accumulated lipid and the growth rate (Chapter 2).

Anchovies arranged in the following order – Mediterranean (*Engraulis encrasicholus mediterraneus*), Black (*E.e. ponticus*) and Azov (*E.e. maeoticus*) Seas – show a decrease in linear growth and somatic production rate (Figure 78). The most likely explanation is the shortening of the growth period during the annual cycle resulting from reduced time of intensive feeding at the optimum ambient temperature. In the Mediterranean Sea, the water is warmest and so favourable for growth over most of the year. The Black Sea is not as warm, and the favourable conditions last only from May to October, while in the Azov Sea the favourable period is even shorter.

Lipid accumulation differs from somatic growth, since food plankton increases in abundance from the Mediterranean, via the Black to the Azov Sea, a situation that promotes the greatest accumulation of lipids in Azov fish.

Species that live in northern seas may be structurally differentiated by environmental conditions, which vary considerably from the centre to the periphery of the area. Crucial factors are the temperature and salinity of the water and the character of the nutritive base. Populations of cod, herring and flounder exist in two forms, oceanic and coastal (Shatunovsky, 1980).

The food supply of oceanic populations of cod, herring and haddock in the north-east Atlantic varies considerably from year to year. In the North Sea the

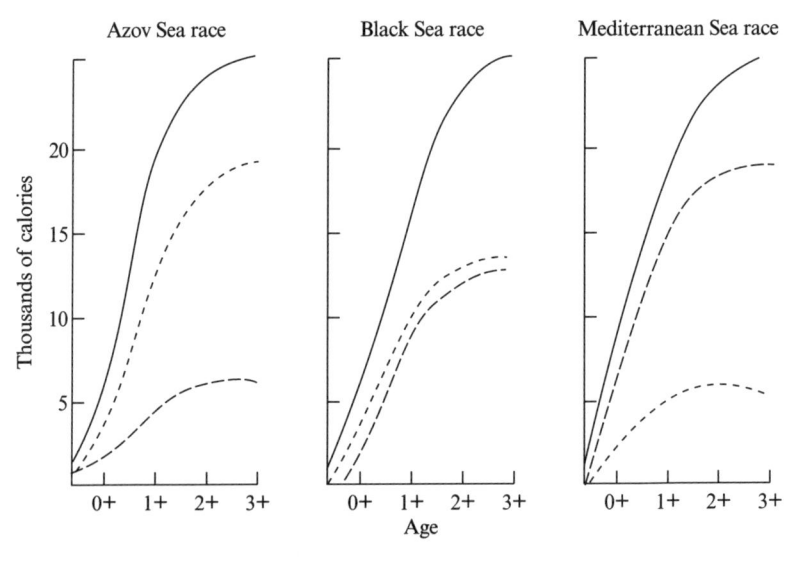

Figure 78   Production (calories) in different races of European anchovy: solid line, total calories; dotted line, protein calories; broken line, lipid calories. (After Shulman and Urdenko, 1989.)

food supply remains more or less stable, so there are smaller year-to-year variations in fatness of fish at the end of their feeding period. The best-nourished cod are those confined to the small area of the Faroe Bank. There they are very corpulent compared with other stocks, possessing the highest concentration of protein in their musculature of any cod examined and carrying relatively enormous, creamy livers. The ground appears to provide a good food supply as a result of upwelling of oceanic currents. In this stock, even the small percentage of lipids found in the musculature of cod is larger. The carbohydrate content of the musculature is high, leading to a lower post-mortem pH. This engenders a tougher texture after cooking and the ready breakdown of connective tissue, which causes the fillets to 'gape'. This unique stock of cod was illustrated by Love (1970, Figure 62), and its properties were described fully by Love *et al.* (1974, 1975) and Love (1986).

The colour of the skin of many fish species changes according to that of the sea bottom. In the black volcanic larva off north Iceland, cod have a very dark skin, while on the white shell bottom at the Faroe Bank, all species of fish are almost devoid of pigmentation. However, Love (1974) has shown that, although live Faroe Bank cod and the darker Aberdeen Bank cod, when placed together in an aquarium of intermediate colour, tended to adopt that colour, they were still easily distinguished after 8 months of living together. The range of adaptation to environmental colour appears to be limited within each population. Conditions on fishing grounds are critical in the nutrition of the fish and can vary greatly. As noted previously, studies on the bile of cod have shown that all the fish from one ground can be actively feeding, while those from a nearby ground are all starving. The growth rate of cod populations decreases consistently from high in the North Sea to lesser in the Baltic to least in the White and Barents Seas. However, North Sea cod do not usually live longer than 8 years, while Arctic cod exist that are more than 25 years old. A similar difference in life span is found in herring from the North and Baltic Seas (Shatunovsky, 1980). The level of liver lipid is greater in cod from the Baltic Sea and the Arctic sector of the Norwegian Sea, when compared with those from the North Sea. However, those from the White Sea have lower levels coupled with a slower growth rate, so the relationship of low growth-high lipid is not universal.

Characteristics of the protein and lipid metabolism of a population are directly related to those of the reproductive system – the period of maturation, absolute and relative fecundity and the weight and content of organic matter in gonads and their products (Shatunovsky, 1980). Different groups of North Sea herring differ somewhat in content of dry matter in individual eggs (Blaxter and Hempel, 1966). There are comparable differences in Baltic and White Sea flounder, and the spring brood of Baltic and Atlantic–Scandinavian herring (Shatunovsky, 1963). The weights of mature ovaries are 20–50% more in Baltic populations of cod, herring and flounder than those recorded in oceanic and North Sea stocks (Shatunovsky, 1980). More rapid growth and accelerated maturation are characteristic of populations that have stable and sufficient food supplies, such as North Sea cod and herring from the central region of the area. Slow growth but accelerated maturation both occur in marginal populations with a plentiful food supply; among such are White Sea cod and Black Sea flounder. In Mediterranean sprat, a good food supply exists at the boundary of the population area, giving high somatic and generative production.

The duration and scope of generative production are longer in southern forms that have multiple spawning (Koshelev, 1984). Their generative production is 20–30% of their body weight, compared with 4–10% from northern forms (Shulman and Urdenko, 1989). The southern forms are younger at maturity than the northern, and their abundant yield of sexual products reduces their somatic growth. Living at lower temperatures, the northern forms create more somatic material and have the longer life span.

Geography and climate strongly influence many species of fish. For example, the northern (Baltic) race of sprat spawn in the summer and attain maximum lipid content in winter (Mankowski et al., 1961; Biryukov, 1980). In contrast, lipid is maximal in summer in sprat of the Black and Mediterranean Seas. A similar discrepancy in the timing of peak fatness is found in different races of herring (Marti, 1956; Wood, 1958; Krivobok and Tarkovskaya, 1960; Shatunovsky, 1980), cod (Dambergs, 1963; Love, 1970; Shatunovsky, 1980); flounder (Dambergs, 1963; Shatunovsky, 1963, 1970); and North Atlantic horse-mackerel (Podsevalov and Perova, 1973; Chuksin et al., 1977).

Marked interpopulation distinctions in characteristics such as fatness have been observed in a number of freshwater fish – silver carp, bighead, bleak, wild goldfish and mongolian grayling (Marti, 1988; Konovalov, 1989; Lapin and Basaanzhov, 1989; Lapina, 1991).

The phenotypic distinctions between fish of different populations discussed in this section are often more profound than interspecific ones.

7.3. INTERSPECIFIC VARIABILITY

This factor reveals physiological and biochemical distinctions between phenotypes of related species. Differences between forms that belong to different families, orders and even classes are probably not relevant. Distinctions should be named 'specific' only between allied species of the same genus.

Species specificity is relevant to ecology, as it involves the identification of characters peculiar to a given species but different from others, however closely related. Such knowledge points to the understanding of the adaptive significance of the characters, and the nature of the microevolutionary processes involved in the emergence of the particular form. Precise analysis of the species specificity of such characters will require comprehensive examination of their variability within the species, accurate determination of the structural and functional polymorphism and the extent of any interspecific transgression. An example is the thermal resistance of isolated tissues and proteins (including enzymic proteins), studied by Ushakov (1963), detailed in Chapter 2. The study revealed a distinct relationship between the thermal stability of the tissues and proteins, and the temperature conditions of the environment where the organisms lived. Andreeva (1971) found small differences in the shrinkage temperatures of the collagens of cod and whiting which depended on the place of capture, warmer habitats leading to slightly higher shrinkage temperatures, i.e. greater stability. However, Lavéty et al. (1988) could find no shrinkage difference between the collagens of two groups of turbot maintained for a long period at temperatures 6–10°C apart. Possibly the adaptation reported by Andreeva had been established at an early larval stage of the fish.

The studies reported in Chapter 2, on five related species of Azov goby, brought to light a close relationship between the degree of unsaturation of lipids and the concentration of ambient oxygen.

Species specificity of phenotypes has been little studied through the physiological or biochemical approach because of insufficient knowledge of their adaptive role. While the many papers on protein polymorphism are important in the context of population genetics, they cannot yet contribute to the progress of the present subject. An exception is the encouraging interpretation of the haemoglobin and serum protein polymorphism observed in related species of sturgeon (Lukyanenko *et al.*, 1991).

Looking into the variability shown at levels higher than species, one immediately encounters difficulties. Many published accounts are mere descriptions, but what is required is knowledge of the nature of the evolutionary processes and the environmental factors that engender them. The task is not easy, because the divergence and convergence of physiological and biochemical characteristics within a large taxon are so great and the forces that form them so varied that a clear picture can hardly be drawn yet. We warn against adopting too superficial an approach when tackling this problem, since environmental aspects have usually been treated as of secondary importance.

8. Conclusions

8.1. ECOLOGICAL PRINCIPLES

We have shown many examples of the way in which ecological science can be enriched through knowledge of the effects of the environment on living systems. The interactive units are populations and species, rather than individual organisms. Data obtained from individuals need to be extrapolated to higher organizational levels.

8.1.1. The Principle of Multidimension

Physiological and biochemical characteristics are dynamic, not fixed. A feature or process in a single organism should not be regarded as a point. Even the activity of an enzyme should be regarded only as a quasi-constant, which includes kinetic elements such as temperature, K_m, etc. Enzymatic activities undergo considerable changes during ontogeny (Parina and Kaliman, 1978). Metabolic parameters such as tissue respiration change similarly (Zotin, 1974; Ozernyuk, 1985), so the quasi-constant point becomes an ontogenic vector.

Along with the age factor, there are at least two environmental factors which must be taken into account when studying specific adaptations: the seasonal rhythm and geography. The quasi-constant point therefore transforms into a rather complicated three-dimensional figure. There are, however, more than three dimensions, because of other contributory factors. These include variability of the environment (temperature, salinity) at a certain instant, and varying actions such as eating, defence and attack, and reactions to food supply and infestation with parasites. Significant metabolic shifts in the organism are caused by changes in the optimum temperature of enzymatic reactions, concentration/dilution of enzymes and substrates, enzyme–substrate specificity and, finally, altered enzymatic activity.

We are not looking here at a change in one enzymatic reaction, but a whole interactive collection of them that greatly alter the rate and character of metabolic activity. In fish, the phenomenon may emerge as a changed ratio between proteolysis, lipolysis and glycogenolysis, or between glycolysis and the pentose phosphate shunt or a changed rate of ATP resynthesis (Romanenko *et al.*, 1991). Just as in higher animals, fish display a distinct circadian rhythm of metabolism, taking up glycogen during the day and triacyl-glycerols at night (Kondrashova and Mayersky, 1978). Most marked are the age, seasonal and distributional changes found in the lipid reserves (Shulman, 1974). In multi-dimensional analysis, the so-called star diagram – analogous to a phenomenon in meteorology – has been applied as a means of describing the integral variation of relevant processes. Observing the principle of multidimensions contributes to progress in ecological biochemistry.

8.1.2. Principle of Taxonomic Specificity

These studies require a thorough knowledge of the taxonomic status of the animal. There are, of course, features common to all living creatures: all have genetic coding, transcription, translation, enzyme–substrate interaction, energy transfer through the redox chain, the tricarboxylic acid cycle, transamination, oxidative phosphorylation and so on. However, superimposed on these are features of biochemical processes peculiar to a species, a race, a population or even a single organism. It is these features that allow the development of the amazing diversity of small taxa.

In different species or races, or even breeds developed by deliberate selection, whether they be fish, other animal or plant, the metabolism may assume a special character and rate depending on the specific ensemble of enzymatic reactions. For instance, the three races of anchovy described in this account differ in the ratio between their rates of protein biosynthesis and neutral lipid accumulation (Figure 78, page 225). What is most remarkable is that each of the processes is based on similar biochemical reactions, but the environments are different, as are the responses of the fish to them.

Differences in the character and rate of protein, lipid and carbohydrate metabolism, as well as in the contents of proteins and lipid fractions in the serum and muscle, have been found in a variety of cultured fish (Shcherbina, 1973; Cowey and Sargent, 1979). It is most important to identify the biochemical specificity of a taxon and also to recognize that such specificity does exist. Otherwise, there will be difficulty in comparing data from congeneric species, or races of the same species taken from different geographical zones, or from similar fish differing in breeding conditions and so on. Genetic studies on the transferrins have indicated the presence of up to four distinct races of haddock in the area extending from the North Sea to the Faeroes and Rockall (Jamieson and Birley, 1989). Nevertheless, large geographical distances do not necessarily lead to large differences between oceanic stocks. Studies on more than 20 species of teleosts, using enzyme electrophoresis, indicate little divergence between stocks separated by wide areas of the ocean (Smith *et al.*, 1990). The latter authors concluded that most of the genetic variations were found within, rather than between stocks. The cod stocks in the North Sea and the Grand Banks off Newfoundland are isolated, but their mitochondrial DNA shows no major difference; the two cannot be separated by this character alone (Smith *et al.*, 1989). In contrast, mitochondrial DNA analysis has successfully separated the British Isles population of whitefish into three geographically separated groups in Scotland, Wales and northern England (Hartley, 1995). Moreover, recent investigations of DNA by the minisatellite and microsatellite techniques (Galvin *et al.*, 1995; Bentzen *et al.* 1996; see Creasey and Rogers, 1999, for methods) show that the northern stock of cod is not panmictic but is composed of genetically distinguishable subunits, each of which is connected with a particular spawning ground. Further work is needed on separation of fish stocks by these promising molecular methods, but there is a need to bear in mind ecological factors.

8.1.3. Principle of Environmental Specificity

Ignoring environmental characteristics when studying an organism results in flagrant misinterpretation of the physiological and biochemical situation. It is first essential to be aware of the innate mobility rate, temperature preferences and other properties of the creature being studied. Originally, fish biochemistry adopted the classic concept that the source of immediate energy in ATP resynthesis is glucose, the supplementary source being non-esterifed fatty acids. The main energy reserve for the same purpose was supposed to be glycogen and the supplementary one triacyl-glycerol.

These ideas have arisen from studies based on mammals – rat, mouse, guinea-pig and man. In addition, the domestic hen and frog have been used, with fish represented by carp and goldfish, molluscs by *Anodonta*, and crustaceans by *Daphnia*. Although all these creatures are widely diverse, they possess one feature in common: they possess rather low innate mobility, which probably made them convenient for use in experiments in the first place. Replacing these animals with highly mobile ones gives different results. Here are the pairs of opposites: rabbit–hare; laboratory rat and mouse–house rat and mouse; domestic hen–wild duck; carp and goldfish–mackerel and tuna; *Anodonta*–squid; and *Daphnia–Calanus*. Using active animals, we find that the basic sources of energy for ATP resynthesis are non-esterified fatty acids and triacyl-glycerols, supplementary ones being glucose and glycogen. Even this assertion is probably not fully complete (Kutty, 1972; Hochachka and Somero, 1973, 1984; Waarde, 1983; Shulman *et al.*, 1983). Other distinctions in the mode of life may well impact on the character of energy substrates.

Another case in point is the temperature preference displayed by fish and other poikilothermic animals. It is a tradition of biochemical studies to examine enzymatic activity at 37°C when dealing with homothermic animals and at 25°C with poikilotherms. However, the temperature optima of enzymes from the latter group may well differ from this traditional value. Hochachka and Somero (1973), Klyachko and Ozernyuk (1991) and Klyachko *et al.* (1992) found that the enzyme activity/temperature curves were closely linked to the degree to which the animal required cold or warm water (Figure 79). One should therefore always determine the temperature kinetics of the process being investigated. This thesis has been further illustrated in the work of Ugolev and Kuzmina (1993), who studied the digestive enzymes of fish.

As noted in Chapter 2, Emeretli (1994a) described the temperature kinetics of LDH and SDH in three fish with different temperature preferences, illustrating the 'actual' rate of enzymatic reactions occurring in these fish over the whole annual cycle (Figure 1, page 11). Such data correspond far more appropriately to the ecology of the fish than those routinely obtained at 25°C.

It has also become evident that organisms dwelling permanently in hypoxic waters have a peculiar metabolism in which protein is the main energy substrate, used anaerobically (see Chapter 2). This knowledge must not be ignored when performing experiments on organisms inhabiting hypoxic areas of the ocean.

8.1.4. Principle of Functional Rhythmicity ('Macroscale Time')

When discussing the multidimensional character of biochemical features, it is stressed again that they may display clear seasonal, circadian and other variations, such being inherent in all living creatures. The annual cycle divides into a number of periods, each differing in its rate of biochemical activity. For example, incorporation of ^{14}C-acetate into the liver proteins of Black Sea scorpion fish differs markedly between the pre-spawning and pre-wintering seasons (spring and autumn, respectively), reflecting differences in the anabolism and catabolism of proteins.

The enzyme Na–K-ATPase regulates ion transport; when salmon migrate from river to sea, it acts by excreting excess minerals across the gill membranes, while on the return from sea to river it enables the fish to take up minerals from the river water (Hochachka and Somero, 1973, 1977; Natochin,

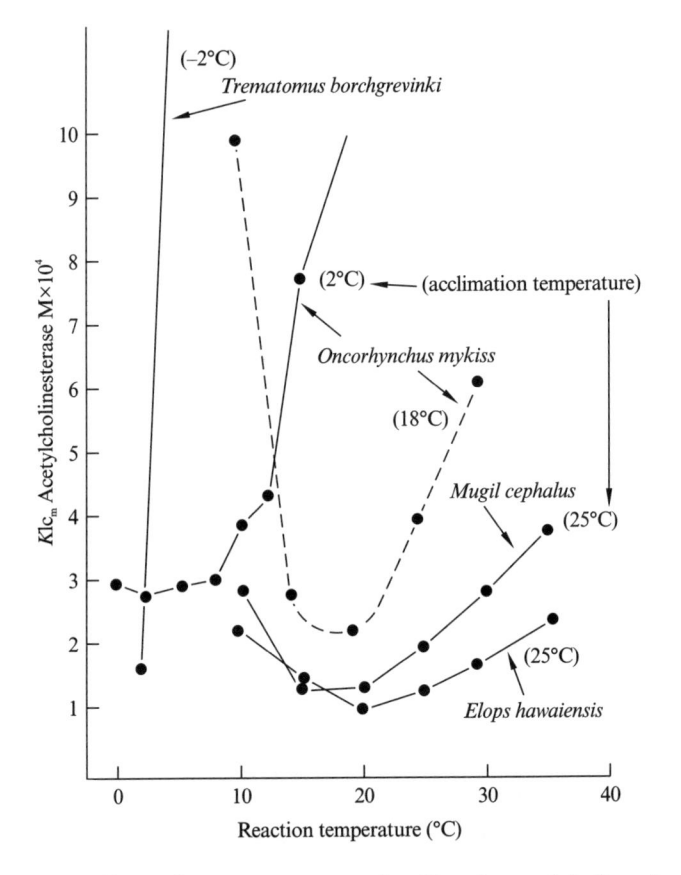

Figure 79   Effect of temperature on the K_m of acetylcholine for acetyl-cholinesterase of four species of fish. (After Hochachka and Somero, 1973.) *Trematomus* is an Antarctic species, *Oncorhynchus* has wide tolerance, and the other two species are warmwater fish.

1976). The enzyme responds sensitively to changes in the concentration of the substrate that occur in different periods of the annual cycle. The same applies to the activity of many isozymes.

8.1.5.   Principle of Functional State ('Microscale Time')

Organisms may differ in their functional state during one period of their seasonal or circadian rhythm – resting, various degrees or forms of swimming or feeding. Their biochemical characteristics will differ correspondingly. Different rates of swimming result in the use of different energy substrates: triacyl-glycerols, protein or glycogen (Black, 1958; Johnston and Goldspink,

1973; Morozova *et al.*, 1978a,b; Waarde, 1983). Under laboratory conditions, fish swimming slowly ('cruising') reduce the level of glucose during the first few minutes (Figure 18), after which it stabilizes and remains stable for long periods, depending on the species. Lipid is now the main source of energy. When fatigue sets in, glycogen is mobilized again. Figure 20 (page 74) shows the variety of resources that support the resynthesis of ATP.

8.1.6. Acclimation Principle

Handling fish causes large changes in their energy resources. Trusevich (1978) reported that the glycogen content of both muscle and liver of red mullet dropped to zero immediately after catching, and did not recover for several hours. However, it seems unlikely that the level regained in the laboratory would be the same as in the wild. It can take a week or two for the fish to recover from capture in the wild, but fish transferred to aquaria after capture rapidly lose much of the stamina possessed by wild fish. Love (1970), reviewing the differences between fish in the wild and those kept in captivity, concluded that 'The greatest caution must be used in interpreting analytical results obtained from fish that have been kept under restraint of any kind. Sometimes what appears to be negligible restraint has a far-reaching effect.' As regards the level of glycogen pertaining in fish just *before* capture, however, the measurement of muscle pH one day after death appears to give a very satisfactory assessment, so that it is not necessary to rest the captured fish under conditions which probably result in unrealistic levels (Black and Love, 1988).

There are special difficulties in comparing enzymic activities of different species of fish that live at different temperatures. Because of the adaptation of the enzymes, their activity measured in the laboratory may appear to be greatest in the fish adapted to the lowest temperatures, merely because the temperature of measurement is higher than that in nature. The same result may be shown if the species has been examined at different periods of the annual cycle, which differ markedly in temperature conditions. To avoid adaptation artefacts, the research worker should be aware of the kinetics of the enzyme under study and try to establish the 'usual' temperature of the environment, adjusting the temperature of measurement accordingly.

A similar caution applies to pH and the concentration of substrate at which activity is measured, the aim being to match these also to natural conditions. In nature, the enzyme and substrate interact under minimum concentration of substrate, and the activity changes rapidly as the amount of substrate alters (Hochachka and Somero, 1973). In experimental work, however, excess substrate is the usual condition and is surely unrealistic.

As in the situation above concerning carbohydrate concentrations, much enzymic work is carried out on well-acclimated organisms, but this means acclimation to the specific conditions of the laboratory experiment, not to natural living conditions. Acclimation takes a relatively long time, and develops in an irregular way. Only when the curve flattens can the process be regarded as stable, and the pattern should always be kept in mind when adaptations of metabolic processes are being investigated.

It is difficult to establish a 'normal' point at which to start an experiment. Animals brought to the laboratory are stressed to a greater or lesser degree, as observed from behaviour, breathing and heart rate, and from responses of endocrines such as catecholamines and from the sugar content of the blood. According to Selye (1973), three stages can be observed in stress: initial (excitation), stabilization and finally depression. At first sight, it appears reasonable to delay experiments until obvious traces of stress have disappeared (Meyerson, 1981) but, as stated above, the act of resting an animal will itself engender a situation different from the 'natural'. The essential pre-conditions for biochemical study are therefore an environment as close as possible to the natural, and knowing the elusive normal condition of the animal itself. Further information on the acclimation principle is given by Khlebovich (1986).

8.1.7. Integration Principle

The specific methodology of biochemistry involves studies at the biomolecular, subcellular and cellular levels in organs and tissues, and only rarely involves the 'whole animal' approach. Is it valid to extrapolate such data to organisms, populations or species?

It has been found that the respiration rate decreases steadily in the series: mitochondrion, cell, tissue, organism and population (Alekseeva, 1959; Ivlev, 1959; Ozernyuk, 1985; Savina, 1992). This means that the total respiration of the isolated cells of a tissue surpasses the respiration of the tissue itself and the total respiration of all the isolated tissues is greater than that of the whole organism. The total respiration of all the organisms of a population which are not in contact with each other exceeds the consumption of oxygen by the population (the 'team effect'). Thus, integration results in inhibition of metabolism, a phenomenon that does not occur in the sponges. In that case, the respiration of the total isolated tissue is equal to that of the whole organism (Ivlev, 1959), evidence of the primitive organization of the sponge. This primitiveness can also be deduced by the behaviour of the sponge *Grantia compressa*: when it reproduces, the whole organism disappears and is replaced by a solid mass of young sponges (Orton, 1929).

The inhibition of metabolic activity by integration of the systems is also seen in enzymatic activity. Ilyin and Titova (1963), working with rats, showed that enzymatic reactions proceeded at higher rates in denervated tissue than in innervated. Many hormones also inhibit enzymatic activity, while others such as thyroxine, somatotrophin and insulin stimulate it. The final situation is therefore the result of the action of interactive systems.

The integration principle applies to both homeothermic and poikilothermic animals. Below 0°C, purified enzymes *in vitro* crystallize and lose their native properties, while above 40–50°C most of them become denatured. In the organism, however, these enzymes may be more stable. For example, the enzymes in Antarctic icefish, which live at around –1.8°C, have temperature optima below zero (Hochachka and Somero, 1973; Schmidt-Nielsen, 1979). The enzymatic activity of plants in areas where the temperature can fall to as low as –60°C is very greatly reduced, but not destroyed (Wiser, 1970). Some organisms can live at temperatures as high as +70°C, such as chironomids in hot springs (Galkovskaya and Suschenya, 1978), or even above the boiling point of water, like certain bacteria from hot springs and hydrothermal vents (Miroshnichenko *et al.*, 1990; Blöchl *et al.*, 1997; Jannasch, 1997). In all these organisms that live in extreme environments, the enzymes examined in the whole organism usually display characteristics quite different from those found after isolation and purification. Within the organism, many enzymes exist in functional complexes with coenzymes. Hochachka and Somero (1973) were right to warn against the study of coenzyme-free enzymes, but to pay very close attention to the enzyme–coenzyme complexes that were actually present.

Biochemical adaptation is not always the complete answer to unfavourable ecological conditions; sometimes a fish can solve its problems by adapting its behaviour, as the following example shows. In the Sea of Azov, two allied species of goby are found, the round goby and the syrman goby. Their main biological difference is their response to anoxic conditions. The syrman goby tolerates anoxia better, having better skin respiration than the other species. For efficient operation, skin respiration depends less on the partial pressure of oxygen than does branchial respiration. In addition, the haemoglobin of the syrman binds oxygen more efficiently, and the tissues are better at mobilizing energy anaerobically (Shulman *et al.*, 1957; Soldatov, 1993). In the Azov Sea, especially in the summer time, the dissolved oxygen content may drop dramatically, reaching zero at times. This is the result of increasing eutrophication, which leads to a summer kill of fish. The syrman goby would be expected to deal with the situation better than the round goby, but in fact does not. Unlike the 'unprotected' round goby, the syrmans tend to remain in the hazardous area, while the round gobies leave the area immediately and congregate near the shore, where the water is better aerated. When the summer kill persists for several days, the majority of the syrman gobies die, while all the round gobies survive. This explains why the biomass of round goby is considerably higher in the Azov Sea than that of the 'better-equipped' syrman goby.

This example shows how population behaviour strategy works. Hochachka and Somero (1973) considered it to be superior to biochemical strategy and, in the case of gobies, it largely accounts for the outcome of the unintended competition between the two species.

8.1.8. Species Centricity Principle

As already stated, studies on individual organisms may not reveal the overall pattern of a population. For example, when fish and birds are preparing to migrate, only a relatively small proportion of them can accumulate the maximum level of lipid. These are known as leaders (Dolnik, 1965; Shulman, 1972a), and they determine the time at which migration should take place, carrying with them their less well-prepared counterparts. Mass mortality of the latter is frequently observed during the migration. Clearly, examination of the lipid accumulation of even a considerable number of individuals may not help in identifying the stage of readiness to migrate.

Many species sacrifice some individuals, or even whole generations, in order to survive. Pacific salmon all die after spawning, even though their life may be prolonged for some weeks by feeding them (McBride *et al.*, 1965). The mass mortality appears to be essential for the survival of the race, organic material from the corpses supplying nutrients to the growing fry (Nikolsky, 1963). Only the fish best prepared by accumulating maximum lipid reserves are directly involved in reproduction; the remainder do not reach the spawning grounds or provide sufficient generative tissue. On completing spawning, male gobies and some other species die of starvation. Another way in which individuals are eliminated is by the mass mortality common during wintering. These mechanisms regulate the numbers in a population, but are not readily perceived through the study of a few individuals. A classic example of self-regulation in a species is the mass suicide known in lemmings, which is thought to be a response (based on catecholamines) to an inadequate food supply following a population upsurge.

The strategy of individuals is therefore subordinated to that of the species. Leaders are the main providers of species strategy.

8.2. THE FORCES OF PHYSIOLOGICAL AND BIOCHEMICAL ADAPTATIONS AND EVOLUTION

8.2.1. Environmental Factors

Accepting Darwin's thesis that variability within species makes selection possible, one gains a clear understanding of the importance of environmental factors in relation to adaptation and evolution. In aquatic animals it is such factors as temperature, salinity, dissolved oxygen and food supply that account for variations among species and adaptation strategy. The underlying biochemical and physiological principles have been discussed by Fry (1958), Love (1970, 1980), Shulman (1972), Hochachka and Somero (1973, 1977), Brett (1979), Brett and Groves (1979), Prosser (1979), Schmidt-Nielsen (1979), Shatunovsky (1980) and Shulman and Urdenko (1989).

The relationship between substance and energy balance on the one hand and temperature on the other is a case in point, temperature being the most important environmental factor. According to Fry (1958) and Brett and Groves (1979), temperature regulates the metabolic rate, the rate of substance and energy uptake and the rate at which they can be used for production. It has the most pronounced effect on the rhythms of processes affected by the annual cycles, especially when it varies greatly during the year. Beyond food consumption and the generation of production in fish populations, its influence extends to the activities of tissue enzymes, structural and functional features of lipids, and the energy capacity of the transport system of the organism.

The rate and orientation of rhythmic processes are influenced not only by temperature but also by the temperature requirements of the species. In Black Sea fish which require warm water, such as anchovy, horse-mackerel and red mullet, the basic processes which make up vital activity are limited to the narrow temperature range at which they operate during the warm season. Changes in rhythmic processes are therefore sharply defined. During the warm season (May–October), these populations create body and gonad tissue at rates that depend on feeding intensity. During the cold season, a marked reduction in metabolic processes and food intake occurs, so that tissue production ceases. What is more, the food intake fails to support the energy reserves and structural substances at a steady level, so endogenous feeding becomes operative. Stored lipids and proteins are then actively used to provide both energy and the early stages of vitellogenesis.

In fish which require cold water (sprat and whiting), the annual rhythms are not nearly so well defined. Their preferred temperature zone is in deep water, where the temperature is relatively stable throughout the year. The temperature of the regions of the Black Sea inhabited by warm-water fish may vary by as much as 10–15° over the year, but the habitat of cold-water fish varies by only 3–5°. The latter fish have not only a smaller metabolic variation throughout the year, but also a higher consumption of food in winter (sprats consume 7.3% of their body weight daily at that time) than that of the species requiring warm water.

The less-marked annual rhythm of the cold-water species compensates for their lower metabolic rate. It is in the cold season that the sprat and whiting create reproductive products through enhanced exogenous and endogenous feeding, although their endogenous feeding is less than that of the warm-water species. There is a species difference between whiting and sprat: whiting are able to produce both somatic and generative tissue at the same time, while sprat have to separate them, producing somatic growth in the summer and generative development in the winter.

In the Black Sea, the production of protein and accumulation of lipid occur simultaneously in cold-water fish, while in the warm-water fish the activities are separated. Pickerel, which are intermediate as regards temperature preference, display similarities to both other forms. Like the warm-water fish, the major part of production occurs during the warm season, although the amplitude of the biological rhythms is not as pronounced. Like the cold-water whiting, the pickerel use protein, not lipid, as a main reserve.

It is probably the temperature rhythms that provoke the proper succession of basic metabolic processes throughout the year. Firstly, the energy and plastic reserves are to be laid down and, secondly, mobilized for generative production followed by, or even together with, somatic production. There is a direct relationship between the extent of the discrepancy between somatic and generative production and the degree to which a fish requires cold water.

Temperature may influence the organism or the population both directly and indirectly. An indirect factor linked to temperature is the duration of daylight, which influences the behaviour pattern followed by a population, governing the 'biological clock' – setting the time of pre-spawning, spawning and pre-wintering seasons. The seasonal rhythms of fish in high and temperate latitudes depend greatly on temperature, but day-length marks the arrival of varying temperature conditions and is a recognized 'trigger' of the spawning cycle. Another factor governed by temperature is the food supply, which is the most important factor according to Fry (1958) and Brett (1979). As the temperature rises, plankton increases, providing favourable feeding conditions for planktonivorous fish.

However, although responsible for production rate, temperature does not necessarily influence the efficiency of food utilization for production or growth. It has been suggested that, in cold-loving aquatic organisms, K_2 is usually higher than in fish from warmer waters (I.V. Ivleva, 1981). Following this assumption, the anabolic rate of low-temperature poikilotherms decreases less than the rate of catabolism and food consumption. This concept has been disputed (Vinberg, 1956, 1986; Brett et al., 1969; Brett and Groves, 1979). Data obtained by the latter authors indicates that K_2 is not inversely proportional to the degree to which a species requires lower temperatures. Studies of annual cycles demonstrated that the maximum values of K_2 arise in the area of temperature optimum for each species. It is concluded that all elements of energy balance are directly proportional to ambient temperature, and no trace of a compensating enhancement of production rate has been found in cold-water fish compared with warm-water fish.

Salinity is a regulating factor (Fry, 1971; Brett, 1979), of greatest importance to diadromous and semi-anadromous fish. In order to keep in balance, fish in sea water must drink copiously and continually excrete salt, while in fresh water they must acquire salts and excrete water. Fish which live in both media at different

stages of their lives undergo a 'pre-adaptation' in which the content and composition of the blood changes, altering the osmotic pressure and signalling them to
migrate to the other medium. For solely marine and solely freshwater fish, salinity
is the barrier serving to separate species areas by means of tolerance zones. For
euryhaline fish, these zones are more extensive than for stenohaline species.

Concentration of oxygen is a limiting factor which has a profound effect on
the distribution and evolution of a species (Fry, 1959; Brett, 1979). Low values
lead to a marked drop in the rate of energy metabolism, hampering functional
activity, production and abundance. Even so, the diversity and abundance of
creatures dwelling in vast hypoxic zones of the oceans (planktonic crustacea,
squid, lantern fish and many other forms) suggest that conversion of metabolism to provide more energy anaerobically, and reduction of physical
activity, may stimulate production to a high level. For instance, in well-aerated
surface waters of the Arabian Sea, squid grow to 30 cm in length and
0.7–0.9 kg in weight (Chesalin, 1993), while at 350 m depth where the oxygen
concentration is low, these values increase respectively to 65 cm and 8–9 kg.
Factors other than oxygen concentration may be involved.

In this context it is natural to consider the potential adaptability of species, as
investigated by Karpevich (1976). Knowledge gained in this field is, however,
far from satisfactory and we continue to underestimate the ability of species to
adapt to extreme conditions. This aspect acquires special significance because
of the increasing pollution of water bodies by human activities, so biochemical
mechanisms of adaptation must become the object of much future study.

8.2.2. Motor Activity

In the course of evolution, fish, like other members of the animal kingdom, have
developed features that enable them to use their nutritive base with the greatest
efficiency. They have a much greater production rate than, for example, birds or
mammals. The reasons for this high production are several. For example, fish do
not have to use energy for thermoregulation; the body-temperature of some
species does rise during physical activity, but only because the heat cannot be
conducted away rapidly enough (Zharov, 1965). Although the density of water
causes resistance to swimming, a fish is more or less weightless, so needs little
or no energy for support. The food of many species of fish is rich in protein,
which promotes a greater rate of protein biosynthesis in their bodies.

It has been mentioned earlier that feeding is the greatest 'channel of
communication' between the organism and its environment; it is swimming
activity that ensures the successful exploration and exploitation of the
nutritive base.

For a species to break into an area and become a significant presence, it
can either expand widely, populating ecological sites densely, or develop

specialized vital functions which enable it to occupy narrow ecological niches uncongenial to competitors. Developing new areas requires high mobility (Orbeli, 1958; Arshavsky, 1966; Shmalgausen, 1969; Leibson, 1972; Shulman, 1974). Among functionally active terrestrial forms are hoofed and preying mammals and migratory birds, while among fish are the sharks and tunas (predatory fish with high mobility) and planktonivorous fish such as herrings and flying fish. Highly mobile fish carry the largest weight of muscular tissue, which can be as much as 58% of their total body wet weight.

The narrow specialization needed in order to fill special habitats does not necessarily require high mobility. Total muscle performance in these specialized forms is therefore not large, but the fish develop a capacity for sudden motor reaction to defend themselves against predators or to ambush prey.

In terrestrial animals, this sort of specialization can be found in ground squirrels, some other rodents and beasts of prey. Among fish, there are bottom and near-bottom forms like morays, scorpion fish, pike, cod and gobies. Whether high or low, functional activity fosters the biological progress of different species, and, like temperature, it can be regarded as a controlling factor of metabolism (Fry, 1959). However, neither Fry nor Brett (1979) ever actually took innate mobility rate as an adaptation factor, although Brett regarded activity as being responsible for a certain level of energy metabolism in fish.

Given the same temperature conditions, the natural mobility rate is the determinant of specific oxygen uptake, energy metabolism, the ratio between protein and lipid used in metabolism, the level of tissue enzyme activity, the rate of self-replenishment of proteins, the energy capacity of tissues, the concentration of erythrocytes and haemoglobin in the blood, the lipid fraction and fatty acid composition of lipids in the liver and other tissues, and the pattern of mobilization in endogenous feeding.

It is self-evident that a high natural mobility rate reduces the proportion of food available for production. Coefficient K_2 in highly mobile fish is several times lower than it is in sluggish fish. However, mobile fish are capable of high production through an intensified food consumption, while sluggish fish, also capable of high production, achieve it by channelling most of their food into body-building. Differences between highly mobile and sluggish fish can be summarized as follows:

Active forms	Sluggish forms
High level of energy uptake	Low level of energy uptake
High level of active metabolism	Low level of active metabolism
Occupation of vast areas	Occupation of narrow niches
Highly efficient use of food for metabolism	Highly efficient use of food for production
Highly extensive production	Highly intensive production

It is not easy to decide which of the two strategies is of greater worth, since both promote biological progress. However, on reflection, being active seems the more advantageous, since greater numbers and production result, despite the less efficient use of food.

Fish have also developed alternative tactics of metabolism. Active fish have been found to increase their metabolic rate by increasing the aerobic phase during periods of enhanced activity (summer, spawning season). At less active times (winter), the anaerobic phase predominates, as it does during any period when the swimming rate changes from fast to slow. The functional state therefore also affects the efficiency of food utilization for growth and metabolism (Brett, 1973). Sex differences arise because of (1) greater quantities of gonads generated by females and (2) greater physical activity by males during spawning (Beamish, 1968).

The swimming activity and temperature act independently on the metabolism of fish, sometimes cancelling, sometimes reinforcing each other, but activity appears to have the greater influence. Increases in both parameters produce very similar effects, accelerating the metabolism and shifting it towards the uptake of predominantly aerobic substrates. It is therefore reasonable to consider them as a pair. For example, anchovy and sprat are both active, but anchovy inhabits the warm upper layers of the Black Sea, while the sprat lives below the thermocline at lower temperatures. The food plankton thrive in warm water and become very abundant in the warm season above the thermocline, so the anchovy enjoy more favourable feeding conditions than the sprat, and their distribution area, numbers and production are all greater. It appears that the specific production (production/biomass) is inversely proportional to the species range, being greater for sprat than for anchovy. Similarly, in red mullet, pickerel and whiting, the values are higher than in horse-mackerel. Absolute production and numbers are greater and the range is more extensive in anchovy and horse-mackerel than in the others. Thus, high absolute production is attained through the extensive mode of production and high specific production through an intensive mode. It is not certain as yet whether this phenomenon is general or simply a case special to the Black Sea.

It appears that the concept of 'ecological efficiency', seen by Slobodkin (1962) as the ratio between production and consumed food (K_1) does not indicate the efficiency of vital activity. It should in fact be deduced from the biomass, absolute production and the extent of the range of the species. This concept integrates all the elements implied by 'biological progress' (Severtsev, 1934) apart from intraspecific differentiation. Ecological efficiency must not be isolated from biological progress. The determining factor maintaining actual efficiency is motor activity. High expenditure of energy and the corresponding 'cost' of production are indeed central to ecological efficiency and biological progress. In summary, temperature, food and motor activity are the most significant factors underlying adaptation strategy in fish.

8.2.3. Energy Optimum and Structural and Functional Homeostasis

We have seen how high and apparently unnecessary expenditure of energy may result in a major biological effect. The energy principle is often taken as a core, not only in ecology but biological evolution in general (Mezhgherin, 1970, 1990; Dolnik, 1978; Lapkin, 1979; Zotin and Krivolutsky, 1982; Ozernyuk, 1985). Both idioadaptation, which leads to speciation, and apomorphosis, which promotes a qualitative leap in the evolution of large taxa (e.g. order, class and type) are regarded as factors which allow a certain level of active energy metabolism to be reached. The same thesis applies to stages of ontogenesis (Vinberg, 1956; Zotin, 1974; Ozernyuk, 1985, 1988; Zotin *et al.*, 1990; Ozernyuk *et al.*, 1993) and metabolic rhythms and population behaviour over the annual cycle (Kalabukhov, 1950). The energy principle also works successfully in the morpho-physiological progress of individual organisms (Severtsev, 1934), but it is simply a tool which sustains biological progress, i.e. high production and abundance, wide distribution and differentiation between and within species. Positive biological progress may be found in the forms that differ greatly in their levels of energy metabolism – compare, for example, migratory birds and the parasites living in them. Having said that, however, we should point out that biological progress does not invariably follow accelerated energy metabolism, otherwise a shrew would by now be superior to the primates.

It seems that the core of progress on all fronts is the level of structural and functional homeostasis, i.e. the dynamic stability and adjustment of internal systems, rather than the rate of metabolism. As Slobodkin (1964) stated, the strategy of evolution aims at keeping the internal environment sustainable at a steady level. The steady state is maintained via the stream of information received from the environment – physical and chemical impulses together with nutrients and the oxygen to oxidize them. Homeothermic organisms are superior to poikilotherms, and primates possessing a cerebral cortex are superior to other mammals. Pavlov (1932) called this cortex the controller of the balance between the organism and the external environment; the energetics are just the cost of homeostasis (attributed to Ozernyuk).

The energy optimum of ontogenic development, annual cycle and loco-motion is supposed to coincide with minimum expenditure of energy (Schmidt-Nielsen, 1975; Shilov, 1985). However, average fish in a population of anchovy or horse-mackerel swim 23–28 km day^{-1}, while red mullet, pickerel or whiting only cover 5–8 km (Belokopytin and Shulman, 1987). The energy taken up by the first group of fish for swimming is several times higher than that by the second, yet it is just high energy uptake that ensures that all the fish thrive.

Nature does not always follow the thrifty mode of 'management', as the levels of energy expenditure prove. The term energy optimum implies the rate of energy metabolism adequately fostering maximum biological progress possible for a living system, be it a population, species, community or other. Shugaev (1989) concluded, for example, that from the viewpoint of bioenergetics, the very existence of predators is sheer squander! Yarzhombek (1975) pointed out that the swimming velocities of fish that save the most energy are not fit either for chasing the prey or escaping the predator. Fish require surplus energy (Khlebovich, 1986; Kulinsky and Olkhovsky, 1992). Bernatchez and Dodson (1987) claim that the efficient use of energy would not ensure optimum conditions for anadromous migration.

It is, perhaps, better to assume that living systems thrive on a combination of 'maximum strategy', which ensures biological progress of the system (implemented in geological macro-time) and 'minimum strategy' which makes the system sustainable (implemented in ecological micro-time). This concept is in harmony with the views of Slonim (1979) and Kulinsky and Olkhovsky (1992), who considered the combined effects of maximum and minimum strategies in metabolic adaptations, and concluded that functional excess was a key factor in metabolic processes.

Epstein (1992) considers that reliability and complexity are as important as sustainability in the basic characteristics of living systems. Supporting the three requires the organism to receive a flow of information as strong as the energy flow.

Concepts of metabolic processes may be distorted if they are based solely on energy. We have already shown that protein production and the accumulation of lipids differ between the three races of European anchovy, but the energy equivalents of the body weights alter identically in all three (Shulman and Urdenko, 1989), even though the metabolic patterns differ. No fish farmer will consider only the caloric content of the feed, but will concentrate on a proper balance of constituents. Only considerations of both matter and energy will give a true understanding of all the processes involved.

Adaptation to any environmental factor causes changes in the functional activity of a biological system, even when it does not entail altering motor activity. For instance, changes in salinity induce increases in the rate of ion transport through the gill lamellae. Changes in temperature affect the composition of membrane lipids, which, in turn, alter the rate of intracellular metabolism. Adaptations to changes in dissolved oxygen level affect the oxygen transport in tissues.

Adaptation to varying rates of locomotory activity is not as simple as it might first appear. One aspect is the higher metabolic rate sometimes displayed by less-active fish that enables them to compete with more-active species. Does the rate of energy metabolism provide the best index of functional activity (the traditional view), or is the phenomenon more complicated than that? The latter view apears to be better but authors persist in treating the rate of oxygen uptake as the best 'description' of functional activity in an organism. The entire web of problems associated with functional activity has been given serious consideration by Ginetsinsky (1963), Leibson (1972), Natochin (1976, 1988), Polenov (1983) and Natochin and Menshutkin (1993).

It is not our intention to ignore the diversity of ecological approaches in which the starting point is energy. Problems in linking energy and substance in studies of fish have been expounded by Shulman and Urdenko (1989). Brett (1979) has also used data from both energy and plastic metabolism. What modern ecological physiology and biochemistry need is an approach which would integrate the energy- and substance-based methods of study.

8.3. APPLIED ASPECTS

Basic physiological and biochemical studies hold great potential for solving applied problems. The analogy with medical science is pertinent: enriched by advanced methods, medicine has been revolutionized in diagnostics, understanding the mechanisms of pathologies and curing diseases. Ichthyology has been improved in this way only recently. It has not been unknown for workers in the fish industry either to oppose new approaches or to remain indifferent to scientific progress. To non-scientists, basic science seems both irrelevant and remote, and research methods seem much too complicated for them to use. This is why the fishing industry progresses so slowly compared with, for example, animal husbandry or plant cultivation.

It is important to stress the value of scientific methods in the following areas: making fisheries more rational and efficient; mariculture; fish processing technology, including the determination of nutritional value and of biologically active substances; toxicology; and protection of commercial stocks. We shall paint broad outlines only, since a comprehensive survey would require a separate monograph. Data on the nature of metabolic processes going on in an organism have already found wide use in mathematical modelling. Such knowledge should be disseminated further.

There is continual change in the problems requiring solution. This stems from ever-increasing human activities, including overfishing and pollution, which dangerously disturb the factors sustaining natural resources. It is therefore important to restore the balance and guard fish stocks. Thus the demand for fish protein will have to be satisfied more and more through aquaculture.

In the past, the fish used as food for man or farm animals was evaluated on the basis of protein and total lipid content, but nowadays the emphasis is on the biologically active substances such as polyenoic fatty acids, vitamins and antioxidants. Recent problems have arisen because of contamination with sewage, mineral oil and its by-products, heavy metals and radioactive pollutants. The bizarre finding of certain Tasmanian oysters, bred downstream from a zinc smelting plant, which contained 10% of zinc on a dry weight basis is an extreme case of what can happen. The discharge of hot water into the sea is another form of pollution. Interference with the natural flow of rivers sometimes results in estuary water becoming more salty, sometimes less salty, the desalination sometimes spreading to adjacent areas including the sea shelf. Such changes can be ruinous for native species if there is also eutrophication.

8.3.1. Protecting Fish Stocks

Protection of fish stocks is not only a matter of controlling fishing, but implies taking steps to sustain their optimum biomass. It also includes the protection of other constituents of the food chain, as in the case of overfishing of capelin in the Barents Sea, which resulted in a dramatic drop in the cod stock which feed on them (Orlova et al., 1990; Matishev and Pavlova, 1994). The methods, discussed in Chapter 6, for monitoring the 'well-being' of fish stocks, are able to give early warning of trouble, those designed or modified for use at sea being particularly valuable. From the colour and quantity of bile, one can easily identify grounds where the fish are not feeding. The effect of the outbreak of the medusa Aurelia aurita in the Black Sea in 1981 was to reduce greatly the lipid content of the sprats. An even greater deterioration in the nutritive base of the sprats resulted from the introduction of the ctenophore Mnemiopsis leidyi, which unbalanced the whole pelagic ecosystem; this was again identified from the reduction in lipid stores. A unique approach was adopted by Cowie and Little (1966), who classified Atlantic cod on the Aberdeen market as to suitability for processing solely on the basis of the pH of the muscle. A pH of 7 or more, from a depleted fish, indicated a sloppy textured fish which was unsuitable for selling 'fresh', while those with a pH below 6.6 were too tough for freezing and the fillets 'gaped' (broke up in pieces). Using a probe electrode on the pH meter, they did not damage the fish, which could still be sold.

Physiological and biochemical indicators are widely applied for the assessment of commercial fish stocks in the Azov and Black Seas (Shulman, 1960a,b, 1967, 1972a; Danilevsky, 1964; Dobrovolov, 1970; Lutz and Rogov, 1978; Chashchin and Akselev, 1990; Studenikina et al., 1991; Shulman et al., 1993a), the Caspian (Geraskin and Lukyanenko, 1972; Rychagova, 1989), Baltic (Ipatov, 1970; Krivobok and Storozhuk, 1970; Shatunovsky, 1980) and Barents Seas (Shatunovsky, 1980), and in seas of the Far East (Kalyuznaya, 1982; Shvydky, 1986).

Shatunovsky (1980) suggested a system of biochemical approaches for taking the maximum number of fish from commercial stocks in northern seas without exhausting the populations. He consulted data about somatic and generative increase in a number of age groups, the sex ratio and the natural death rate. A mathematical model of fish stock exploitation was created by

Bulgakova *et al.* (1973), using biochemical techniques. The studies quoted in Chapter 5 may also be used to plan the rational use of resources.

Lukyanenko (1971, 1989) defined specific and population heterogeneity in Caspian sturgeons. He distinguished between Persian and Russian sturgeon areas, the Ural and Volga stocks, the life cycles of winter and spring races, and identified local stocks. It was as a result of his work that the fisheries were reorganized so that the populations were damaged as little as possible.

At present, the basin of the Caspian Sea has been seriously endangered because of growing pollution, reduced river inflow from the overdammed Volga and uncontrolled fishery. These factors, together with the chaos resulting from the recent major political change, threaten the existence of the commercial sturgeon stock. Lukyanenko (1990, 1992) has suggested a programme to save these fish.

Fish hatcheries may inadvertently alter genetic structure of the stock as, for example, with Pacific salmon (Altukhov, 1983). Maintaining the original genetic diversity ensures a healthy stock. The correct rearing and transplanting of the fry of Caspian and Azov sturgeons is of great importance (Gerbilsky, 1941; Barannikova, 1984). These fish have long been the most valuable fishery resource of the former USSR.

The notorious project of the Soviet Union, to 'make the rivers flow upstream', was successfully stopped by strenuous objections from physiologists. Convincing proofs were furnished that the waters of the Black Sea had quite a different mineral content from that of the Caspian Sea, so that their transfer would have destroyed the stock of Caspian sturgeons (Natochin *et al.*, 1975). The study of energy balance in Azov anchovy during their spring migration from the Black to the Azov Sea (Belokopytin, 1993) demonstrated that the proposed dam across the Kerch Strait would have stopped migration completely, since the powerful counter-stream down the proposed fishways would have been too swift for the fish.

It is essential to know the threshold environmental parameters when considering the 'well-being' of a population. Such knowledge makes it possible to sound the alarm whenever a sudden environmental change occurs (Klyashtorin, 1982; Varnavsky, 1990b).

8.3.2. The Rationalization of Fisheries

A physiological and biochemical approach permits forecasting, which saves money and time. Data on the lipids stored in the anchovy enables the prediction of the date and character of the winter migration through the Kerch Strait, so that fishermen can start essential preparations a month beforehand and arrive at the Strait just as the migration commences. In earlier times they had to wait, sometimes for weeks, for the anchovy to come. Proper fisheries strategy saves millions of dollars on the anchovy alone, and is based on work on the Black Sea anchovy (Danilevsky, 1964; Chashchin and Akselev, 1990), Caspian kilka (Rychagova et al., 1987; Rychagova, 1989), Pacific ivasi (Shvydky, 1986) and pollock (Shvydky and Vdovin, 1991).

Forecasting the behaviour and distribution patterns of wintering fish is also possible. Depending on the lipid content stored, Azov and Black Sea anchovy form wintering stocks of different density, occupying different depths and developing different mobility rates, knowledge of which again helps the tactics of the fishermen. Black Sea sprats assemble in either coastal or offshore waters depending on the fatness they happen to have acquired during the summer feeding. When they are in the inshore waters, the sprats gather into dense shoals, but offshore they are more diffuse.

The lipid content of fish also indicates whether the fisherman should use lights to attract the fish (Gusar and Getmantsev, 1985), since this method of capture works only on fish with low reserves. It has in addition been used to predict migration time, routes, behaviour and distribution in fish from northern seas and deep-water oceanic areas, but the forecasts have been sporadic; regular annual monitoring has been more common in fish from southern seas and those of the Far East.

Shatunovsky (1980) has reviewed work involving long-term forecasting. This involved lipid analyses of different age-groups and elements of mathematical modelling, and revealed the degree to which fish were prepared for reproduction and wintering, also the proportion of the stock which was likely to be eliminated. Long-range forecasts have occasionally been used for the fisheries in the Black and Azov Seas. Low lipid content in anchovies at the end of the feeding season suggested the probability of considerable elimination (winter death) and hence a low yield for the spring fishery. Lipid analyses also allow rough estimates of the numbers of fish in a stock, since there is a positive correlation between fatness and numbers. The dramatic decline in sprats and anchovies following invasion of the Black Sea by *Aurelia*, which competed for food, was preceded by a sudden loss of fatness. The lipid data were the only source of knowledge to explain what had happened.

8.3.3. Rearing Marine Fish

Pisciculture is of special importance, since it provides not only protein for mankind, but maintenance and restoration of damaged populations. Although reared in freshwater reservoirs, salmon and sturgeons contribute to the replenishment of marine stocks. Almost no other field of applied science has benefitted so much from the use of physiological and biochemical methods as fish farming.

Gerbilsky (1941) suggested the injection of pituitary extract as a means of accelerating the maturation of female sturgeons, and the method is still in use (Kazansky, 1963; Barannikova, 1984).

Work designed to make salmon immune to their own reproductive tissue, which they destroy internally, ensures that all the food, which is costly, is directed towards somatic growth and not to gonads (Secombes et al., 1985). A similar benefit has been achieved by Rowe and Thorpe (1990) and Rowe et al. (1991), who suppressed the maturation of cultured male Atlantic salmon parr by fasting them on alternate weeks in the spring.The mesenteric lipids of these fish then failed (by May) to reach the level needed for maturation, so again food was not wasted on gonad development. The importance of lipid stores in the maturation of this species has been further demonstrated by another observation. In contrast to the Oncorhynchus species of the Pacific, a proportion of Atlantic salmon recover after spawning and return to the sea, spawning again in the river the following year. The second maturation depends on their starting to feed again in January and February; later resumption of feeding fails to remature the spawned fish (Johnston et al., 1987).

Selection of high-quality brood fish from their biochemical characteristics (Badenko, 1966; Lukyanenko, 1971; Geraskin and Lukyanenko, 1972; Chikhachev, 1991) has helped in supplying sufficient numbers of eggs for hatching. The quality of the eggs themselves can be gauged from a number of characteristics. The hatching rate is slightly better in eggs with a higher content of riboflavin, and strikingly better with higher content of iron (Hirao et al., 1955). In the latter case, the critical factor may be the ability to form a sufficient quantity of blood pigment to support life. Movement of the larva within the egg consumes energy, and the egg uses increasing quantities of oxygen from fertilization to hatching (Hayes et al., 1951). Both lipids and proteins are consumed (Suyama and Ogino, 1958) but the water content increases, keeping the weight more or less constant (Phillips et al., 1958). It follows, then, that high-quality eggs should have adequate protein and lipid resources to allow for this. Vitamin A is needed by the developing eyes and is not synthesized in the egg, so an adequate level of this substance is also an indicator of quality. During development of the egg, it changes from the ester form to the alcohol (retinol) (Yamamura and Muto, 1961). A low concentration of carotenoids in rainbow trout eggs results in high mortality rate (Georgiev, 1971), although it is possible to rear the eggs of this species when they are completely devoid of carotenoid (M. Hata, personal communication). Many authors have observed that heavily pigmented eggs are more likely to be fertilized than those with little pigment (Love, 1988). The exact role of carotenoids in eggs has still to be defined, but there seems little doubt that they are beneficial.

The condition of the parent fish is a vital factor in the production of high-quality eggs. Shelbourne (1974, 1975) found that more abnormalities occur in the progeny of unacclimated, overcrowded, injured or starving females, so the trouble is probably transmitted to the eggs while they are still in the ovary. Where parent rainbow trout receive supplements of vitamin A and vitamin E, higher concentrations show up later in the eggs (Kinumaki et al., 1972). Similarly, Watanabe et al. (1984) showed that the fatty acids in the diet supplied to the brood stock of red sea bream is transferred to the eggs. It is therefore possible to adjust the level of 22:6ω3 in the eggs if it is too low; Shimma et al. (1977) found that carp eggs which contained less than 10% of this fatty acid in their total lipids showed a poor rate of hatching. A supplement of ascorbic acid fed to rainbow trout brood stock was shown by Sandnes et al. (1984) to increase the hatching rate of the eggs. Zhukinsky and Gosh (1973) suggested that the intensity of oxidative phosphorylation or of ATPase activity might be used for the measurement of egg quality. Maybe the activity of any enzyme in the major metabolic cycles would indicate intrinsic vigour (Love, 1980).

The metamorphosis of salmon parr to smolt can be monitored (Smirnov et al., 1986; Klyashtorin and Smirnov, 1990; Varnavsky, 1990b; Varnavsky et al., 1991; Zaporozhets, 1991), using such measurements as swimming performance, electroconductivity of blood plasma and the RNA/DNA ratio.

Stress is one of the greatest dangers to cultured fish. Almost any disturbance or environmental change affects or inhibits the heartbeat (Roberts, 1973), and beyond that the fish stop eating. I. Lester (personal communication) was compelled to feed his salmon during the evenings, because the almost inaudible sound of construction work a mile away during the day put the fish off their feed. If the stress is not removed, the fish will die sooner or later, and once actual deaths begin it may be too late to save the remaining stock. Identifying stress in the stock by measuring blood glucose or corticosteroids, which rise, is not practicable because of the stress caused by handling, but finding large, blue, gall bladders in the dead fish would at least show that the fish had already stopped eating. The next step would then be to identify a cause of stress. The *effects* of stress can be moderated by lowering the salinity of the water. This relieves the fish of some osmotic effort so that it can deal better with the situation.

It is hardly possible to make fish fry grow unless their nutrition satisfies basic biochemical criteria. Diets balanced with the correct proportions of proteins, lipids, carbohydrates, essential fatty and amino acids, minerals, vitamins and antioxidants have been devised by Malikova (1967), Shcherbina (1973), Cowey and Sargent, 1979; Sorvachev (1982), Ostroumova (1983) and many others. Fish that have starved for a considerable time cannot be fed normal food, which kills them. Bouche et al. (1971) devised a synthetic diet, consisting of casein and vitamins set in gelatin, which saved the lives of carp and has been successfully used in work with starving cod.

Enriching the water with mineral additives greatly promotes the growth of fish fry (Karzinkin, 1962; Romanenko et al., 1980, 1982). Being incorporated into organic compounds, these substances stimulate plastic metabolism.

Detailed studies on food digestion and assimilation (Pegel and Remouev, 1967; Shcherbina, 1973; Cowey and Sargent, 1979; Sorvachev, 1982; Ostroumova, 1983; Ugolev and Kusmina, 1993) and on substance and energy balance in growing fish fry (Ivlev, 1939; Gerking, 1952; Karzinkin, 1952; Krivobok, 1953; Ryzkov, 1976; Brett and Groves, 1979; Cho et al., 1982; Gershanovich et al., 1987) have led to definitions of the nutritional requirements of species used in aquaculture, and in the development of brood stocks. The general strategies for applying biochemical methods to marine fish culture were devised by Brett and Groves (1979) and Shatunovsky (1980), who used mathematical modelling in their work.

8.3.4. The Technology for Using the Fish Resources of the Ocean

The need to assess the quality of the fish used as raw material for industry has been recognized for a long time. The procedures for measuring moisture, protein, lipid and mineral substances in whole bodies, organs and tissues of fish were devised a century ago (Atwater, 1892; Paton, 1898). Fish can be grouped according to their protein and lipid contents (Levanidov, 1968), those possessing more than 20% crude protein (N × 6.25) being called 'protein fish'. Those with more than 15% of lipid are termed fat fish, those with around 15% lipid 'medium fat' and less than 1.5% 'lean fish'. The proportions of these two constituents and their seasonal changes have been the subject of long-term studies, for example, at the Torry Research Station in Aberdeen and the Pacific Institute of Marine Fisheries (TINRO) in Vladivostok. One of the main objectives of this work was to identify the seasons when a wide range of marine products would be of the greatest nutritive value, for example, the quality and shelf life of fresh or frozen fish delivered to the consumer, suitability for salting, smoking, drying, canning and converting to fishmeal. Studies carried out with commercial interests in mind nevertheless yielded much physiological and biochemical information, while academic studies provided technologists with much practical information.

Studies of this kind are not always of a routine nature, and surprising facts can emerge. For example, cod from the Faroe Bank are the most richly nourished members of this species, with the most thick-set bodies (illustrated in Love, 1970), but make unsatisfactory subjects for frozen storage. Their firm texture becomes unacceptably tough after even a short term in the cold store, while their higher content of polyunsaturated lipids rapidly leads to a strong rancid flavour, again after only brief cold-storage (Love, 1975). The low post-mortem pH, the cause of the very firm texture in the first place, also makes the frozen fish gape badly after thawing, so that it cannot then be hung for smoking. The latter author concluded that the post-mortem fall in pH (arising from the breakdown of tissue glycogen) was the most important factor influencing the suitability of fish for processing. In fatty fish it is probably the lipid content, since starving fatty species taste dry and fibrous, rather than succulent.

Technological advances have been made by studying substances known to be biologically active in the nutrition of man, animal husbandry, poultry breeding and fish farming, especially the polyunsaturated fatty acids, enzyme modulators, antioxidants, leucotriene and prostaglandin precursors. The results of fatty acid analyses reported by Lovern (1942, 1964), Ackman (1964, 1967) and Rzhavskaya (1976) show the essential lipid fractions present in commercially valuable species. The data illustrate how fish can enrich the diets of man and domestic animals, and can be used in medicine (Borisochkina, 1991). Medicines based on polyunsaturated fatty acids are effective in the treatment of cardiovascular disorders, stimulating vitality and helping to prolong life. Squalene, a hydrocarbon from the liver of deep-water sharks, is widely used in perfumery and pharmacology. The most valuable source of trace elements and vitamins has been the liver of the cod and the halibut, although many other marine fish are useful in this regard.

8.3.5. Toxicology

Toxic compunds are routinely disposed of in the seas and fresh water. Such disposal is against common sense, and the effects can be long-lasting. Even previously 'pure' water bodies such as Lake Baikal have been affected. In addition to the input of toxicants, such as oil and oil by-products and the hydrocarbons resulting from their decomposition, pesticides, heavy metals and a variety of sewage effluents, the seas and oceans receive excess crop nutrients from land drainage, causing eutrophication. Among the severe consequences of this are growths of toxic algae and the production of an oxygen deficiency when the blooms die, which inevitably causes fish kills.

Understanding the physiological and biochemical mechanisms of the impact and the underlying metabolic mechanisms holds the key to solving this problem. Stroganov (1979) determined the critical toxic limits for many aquatic organisms, and Malyarevskaya (1979) and Malyarevskaya and Karasin (1991) pointed out the role of nicotinamide coenzymes, which occur naturally in aquatic organisms, as protectants against toxic substances. Complex mechanisms within the nitrogen metabolism are involved in the detoxication of injurious products formed by pollutants (Nemova and Sidorov, 1990; Kurant and Arsan, 1991; Nemova, 1991; Nemova et al., 1992; Konovets et al., 1993; Zhidenko et al., 1994). Lukyanenko (1967, 1987, 1990) presented a wide variety of evidence which suggested that toxicants may affect the organisms both directly and indirectly, the latter relating to changes which they induce in the temperature, gas exchange, ion content, salinity, pH and so on. The total effect is therefore much increased, as evidenced by the falling resistance to toxicants recently manifested by fish.

The combined impact of toxicants and deteriorated environments result in effects which comply with the laws governing Selye's non-specific physiological reactions (Lukyanenko, 1967, 1990; Masmanidi, 1974; Kotov, 1979).

Toxicants and their associated effects also disturb the balance in the peroxic–antioxidant system which results in the accumulation of free radicals. These generate changes in the structure and function of cell membranes (Telitchenko, 1974). As a result, the phospholipid bilayer of the membrane is destroyed, with fatal consequences (Sidorov, 1983). This phenomenon may be the cause of destruction of muscular tissue of fish. The reproductive system of fish is the part most susceptible to the influence of toxicants. Generative metabolism is the first to be affected by the adverse changes in the environment, which throws the entire reproductive cycle of a population into disorder.

Stroganov (1979) and Sidorov (1983) shed new light on viral and parasitic diseases in fish. In particular, it was found that many pathological infections affected the lipid metabolism. In some cases they led to a sharp fall in the reserve lipid content, and in others to excessively high values – the so-called fatty degeneration of the liver. The proportions and quantities of structural lipids are inevitably changed in one or another direction as soon as any pathology appears.

Acknowledgements

The first author (GES) would like to thank his associates and students from the Department of Animal Physiology at the Institute of Biology of Southern Seas, National Academy of Sciences of Ukraine (Sevastopol), the Laboratory of Biochemistry of the Karadag branch of IBSS, and colleagues from the Laboratory of Fish Physiology of the Azov and Black Seas (formerly the Southern Institute for Marine Fishery and Oceanography in Kerch), who generously contributed material and comments, showing much interest in the step-by-step progress of the manuscript. It gives special pleasure to name G.I. Abolmasova, K.D. Alekseyeva, L.P. Astakhova, Y.S. Belokopytin, E.V. Ivleva, T.P. Kondratyeva, N.I. Kulikova, G.S. Minyuk, A.L. Morozova, Z.A. Muravskaya, L.V. Rakitskaya, L.S. Svetlichny, N.K. Senkevich, E.N. Silkina, E.N. Stavitskaya, A.Y. Stolbov, V.V. Trusevich, V.Y. Shchepkin, A.M. Shchepkina, I.V. Emeretly, T.V. Yuneva and K.K. Yakovleva. Thanks are also due to Olga Klimentova, who translated the first draft into English.

The second author (RML) would like to record his indebtedness to Professor K.J. Whittle for access to computer services at The Torry Research Station, Aberdeen, in its last days, and to Professor A.D. Hawkins for use of the library facilities at the Marine Laboratory, Aberdeen. Without such help this work could not have been completed.

Both authors would like to thank Professor J.H.S. Blaxter for the original invitation to write this review and Professor A.J. Southward for his assistance through many revisions. Mr C. Silver generously helped in conversion of the early computer files to IBM format, and redrafted many of the diagrams. Dr Eve Southward helped correct the references and taxonomic details. Last but not least, both authors thank their wives, Svetlana and Muriel, for their support during this arduous undertaking.

REFERENCES

Abramova, N.B. and Vasilyeva, N.M. (1973). Some properties of the embryonal mitochondria of loach (In Russian). *Ontogenesis* **4**, 288–293.

Ackman, R.G. (1964). Structural homogeneity in unsaturated fatty acids of marine lipids. *Journal of the Fisheries Research Board of Canada* **21**, 247–254.

Ackman, R.G. (1967). Characteristics of the fatty acid composition and biochemistry of some freshwater fish oils and lipids in comparison with marine oils and lipids. *Comparative Biochemistry and Physiology* **22**, 907–922.

Ackman, R.G. (1980). Fish lipids, Part 1. *In* "Advances in Fish Science and Technology" (J.J. Connell, ed.), pp. 86–103. Fishing News Books, Oxford.

Ackman, R.G. (1983). Marine lipids. Fats for the future. *In* "Proceedings of the International Conference on Oils, Fats and Waxes", pp.1–15. Duromark Publishing, Auckland.

Ackman, R.G. and Eaton, C.A. (1966). Some commercial Atlantic herring oils. Fatty acid composition. *Journal of the Fisheries Research Board of Canada* **23**, 991–1006.

Ackman, R.G. and Eaton, C.A. (1976). Variations in fillet lipid content and some percent lipid–iodine value relationships for large winter Atlantic herring from southeastern Newfoundland. *Journal of the Fisheries Research Board of Canada* **33**, 1634–1638.

Addison, R.F., Ackman, R.G. and Hingley, J. (1968). Distribution of fatty acids in cod flesh lipids. *Journal of the Fisheries Research Board of Canada* **25**, 2083–2090.

Akulin, V.N. and Pervuninskaya, T.A. (1978). Some features of fatty acid composition of lipids in tropical fish (In Russian). *Nauchnye Soobschenie Instituta Biologii Morya, Vladivostok* 1978, (3), 5–8.

Akulin, V.N., Chebotareva, M.A. and Kreps, E.M. (1969). Fatty acids of phospholipids in the brain, muscle and liver of diadromous sockeye salmon in the freshwater and marine environment (In Russian). *Zhurnal Evolutsionnoy Biokhimii i Physiologii* **5**, 446–456.

Aleev, Yu.G. (1957). Trachuridae of the seas of the USSR (In Russian). *Trudy Sevastopolskoi Biologicheskoi Stantsii* **9**, 167–242.

Aleev, Yu.G. (1963). "Functional Foundations of the External Structure of Fish" (In Russian). Academii Nauk SSSR, Moscow, 247 pp.

Alekin, O.A. (1966). "Chemistry of the Ocean" (In Russian). Gidrometeoizdat, Leningrad, 343 pp.

Alekseeva, K.D. (1959). Metabolic rates of some marine fish kept in groups and as individuals (In Russian). *Trudy Sevastopolskoi Biologicheskoi Stantsii* **12**, 379–295.

Alekseeva, K.D. (1978). The levels of energy metabolism in fish fry (In Russian). *In* "Elements of Physiology and Biochemistry in Total and Active Metabolism of Fish" (G.E. Shulman, ed.), pp.64–68. Naukova Dumka, Kiev.

Alexandrov, V.Ya. (1975). "Cells, Macromolecules and Temperature" (In Russian). Nauka, Leningrad, 329 pp.

Alikin, Yu.S. (1975). On some regularities of carbon dioxide excretion in Baikal fish during swimming (In Russian). *Vestnik Sibirskogo Otdeleniya AN SSSR, Biologicheskie Nauk* **15**, 63–69.

Altufiev, Yu.V., Romanov, A.A. and Sheveleva, N.N (1992). Histopathology of cross-striated muscular tissue in livers of Caspian sturgeons (In Russian). *Voprosy Ikhtiologii* **32**, 157–171.

Altukhov, Yu.P. (1962). Study of thermoresistance of isolated muscle and serological analysis of Black Sea 'large' and 'small' horse-mackerel (In Russian). *Trudy Karadahs'koyi Biolohichnoyi Stantsyi* **18**, 3–16.

Altukhov, Yu.P. (1969). On the immunogenetic approach to the problem of intraspecific differentiation of fish (In Russian). *In* "Advances in Modern Genetics", Vol. 2, pp.161–195. Nauka, Moscow.

Altukhov, Yu.P. (1974). "Population Genetics of Fish" (In Russian). Pishchevaya Promyshlennost, Moscow, 247pp.

Altukhov, Yu.P. (1983). "Genetic Processes in Populations" (In Russian). Nauka, Moscow, 279pp.

Amineva, V.A. and Yarzhombek, A.A. (1984). "Fish Physiology" (In Russian). Moscow, Legkaya i pishchevaya promyshlennost, 200pp.

Ananyev, V.J. (1965). Changes in the qualitative composition of reserve fat in carp fingerlings in relation to breeding and feeding conditions (In Russian). *Doklady Timiryazevskoy Selskokhozyaistvennoy Akademii* **110**, 281–287.

Anderson, T.R. (1970). Temperature adaptation and the phospholipids of membranes in goldfish. *Comparative Biochemistry and Physiology* **33**, 663–687.

Ando, K. (1968). Biochemical studies on the lipids of cultured fish. *Journal of the Tokyo University of Fisheries* **54**, 61–98.

Ando, S. (1986). Studies on the food biochemical aspects of changes in chum salmon during spawning migration. Mechanisms of muscle deterioration and nuptial colouration. *Memoirs of the Faculty of Fisheries, Hokkaido University* **33**, 1–95.

Andreeva, A.P. (1971). The collagen thermostability of some species and subspecies of the gadoid fish (In Russian). *Tsitologiya* **13**, 1004–1008.

Anninsky, B.E. (1990). Energy balance of the medusa *Aurelia aurita* in the Black Sea (In Russian). *In* "Bioenergetics of Aquatic Organisms" (G.E. Shulman and G.A. Finenko, eds), pp.11–32. Naukova Dumka, Kiev.

Antsupova, L.V., Stepanyuk, J.A., Golovenko, V.K. and Petkevich, T.A. (1989). Functioning of biochemical systems in Black Sea aquatic organisms of different taxonomic groups (In Russian). *Gidrobiologicheskii Zhurnal* **25**, 49–54.

Ardashev, A.A., Korotaev, A.A., Isakova, P.V. and Glushchenko, A.N. (1975). Content of 11-oxycorticosteroids in the blood plasma of spawning keta and gorbuscha (In Russian). *Zhurnal Evolutsionnoy Biokhimii i Physiologii* **11**, 308–309.

Aristarkhov, V.M., Arkhipova, T.V. and Pashkova, G.K. (1988). Changes in biochemical parameters of mussels under the combined effect of oxygen deficiency, temperature and invariable magnetic field (In Russian). *Izvestiya Akademii Nauk SSSR, Seriya Biologiya* 1988 (2), 238–245.

Arkhipchuk, V.V. and Makarova, T.A. (1992). The rhythmicity of nucleolar activity in cells of Cyprinidae (In Russian). *Gydrobiologicheskii Zhurnal* **28**(6), 81–85.

Arsan, O.M. (1986). The role of water temperature in regulation of the processes of glycolysis and the tricarboxylic cycle in fish (In Russian). *Gydrobiologicheskii Zhurnal* **22**(5), 71–74.

Arsan, O.M., Solomatina, V.D. and Romanenko, V.D. (1984). The role of environmental phosphorus in regulating bioenergetic processes of fish (In Russian). *Gydrobiologicheskii Zhurnal* **20**, 53–57.

Arshavsky, I.A. (1966). On the physiological mechanisms causing aromorphoses and adaptations according to A.N. Severtsev (In Russian). *Zoologicheskii Zhurnal* **45**, 1308–1322.

Assaf, S.A. and Graves, J.D. (1969). Structural and catalytic properties of lobster muscle glycogen phosphorylase. *Journal of Biological Chemistry* **244**, 5544–5555.

Astakhova, L.P. (1983). Dependence of heart and brain indices in Black Sea fish on their natural mobility (In Russian). *Zhurnal Evolutsionnoy Physiologii i Biokhimii* **19**, 594–596.

Atchison, G.J. (1975). Fatty acid levels in developing brook trout eggs and fry. *Journal of the Fisheries Research Board of Canada* **32**, 2513–2515.

Atwater, W.O. (1892). The chemical composition and nutritive values of food fishes and aquatic invertebrates. *Report of the United States Commission on Fish and Fisheries for 1888* Part 16, 679–868.

Aveldano, M.I. and Sprecher, H. (1983). Synthesis of hydroxy fatty acids from 4,7,10,13, 16, 19- (1-C-14) docosohexaenoic acid by human platelets. *Journal of Biological Chemistry* **585**, 9339–9343.

Avella, M., Schreck, C.B. and Prunet, P. (1991). Plasma prolactin and cortisol concentrations of stressed coho salmon *Oncorhynchus kisutch* in fresh or salt water. *General and Comparative Endocrinology* **81**, 21–27.

Backiel, T. (1973). Production and food consumption of predatory fish in the Vistula River. *Journal of Fish Biology* **3**, 369–405.

Badenko, L.V. (1966). Estimation of the physiological condition of sturgeon and sevryuga (stellate sturgeon) fry from natural spawning and from fish farms on the Don (based on characteristics of the blood) (In Russian). *Trudy Azovski Morei Nauchno-issledovatelskogo Instituta Rybnogo Khozyaistva* **8**, 61–78.

Badenko, L.V., Dorosheva, N.G., Kornienko, G.G. and Chikhacheva, V.P. (1984). The ecologo-physiological base for increasing the efficiency of commercial rearing of Azov sturgeons (In Russian). *In* "Replenishment of Fish Stocks in the Caspian and Other Seas" (I.B. Bukhanovich, ed.), pp.88–101, Moscow.

Bailey, T.G. and Robison, B.H. (1986). Food availability as a selective factor on the chemical composition of midwater fishes in the eastern North Pacific. *Marine Biology* **91**, 131–141.

Bal, N.E., Geraskin, P.P. and Mishin, E.A. (1989). The composition of water soluble muscle proteins in Russian sturgeon and changes during tissue destruction of varying degree (In Russian). In "Sturgeon Culture in Water Bodies of the USSR" (V.I. Lukyanenko, ed.) Vol.1, pp.16–18, Astrakhan.

Baldwin, E. (1948). "An Introduction to Comparative Biochemistry". 3rd edn. Cambridge University Press.

Baldwin, J. (1971). Adaptation of enzymes to temperature: acetylcholine esterases in the central nervous system of fishes. *Comparative Biochemistry and Physiology* **40 B**, 181–187.

Barannikova, I.A. (1975). "Functional Aspects Underlying Migration of Fish" (In Russian). 210 pp. Nauka, Leningrad.

Barannikova, I.A. (1984). Hormonal regulation of reproductive function in fish of different ecology (In Russian). *In* "Biological Foundations of Fish Culture. Actual problems of ecological physiology and biochemistry of fish" (M.I. Shatunovsky, ed.), pp.178–217, Nauka, Moscow.

Batty, R.S. (1984) Development of swimming movements and musculature of larval herring (*Clupea harengus*). *Journal of Experimental Biology* **110**, 217–229.

Bayne, B.L. (1975). Aspects of physiological condition in *Mytilus edulis* with special reference to the effect of oxygen tension and salinity. *In* "Ninth European Marine Biology Symposium, Oban, Argyll" (H. Barnes, ed.), pp.213–238. Aberdeen University Press, Aberdeen.

Bayne, B.L., Bayne, C.J., Carefoot, T.C. and Thompson, R.J. (1976). The physiological ecology of *Mytilus californianus*. 2. Adaptation to low oxygen tension and air exposure. *Oecologia, Berlin* **22**, 229–250.

Beamish, F.W.H. (1968). Glycogen and lactic acid concentrations in Atlantic cod in relation to exercise. *Journal of the Fisheries Research Board of Canada* **25**, 837–851.

Beamish, F.W.H. (1974). Apparent specific dynamic action of largemouth bass. *Journal of the Fisheries Research Board of Canada* **31**, 1763–1769.

Beamish, F.W.H. (1978). Swimming capacity. *In* "Fish Physiology: 7. Locomotion" (W.S. Hoar and D.J. Randall, eds), pp. 101–189, Academic Press, London,

Beamish, F.W.H. (1981). Swimming performance and metabolic rate of three tropical fishes in relation to temperature. *Hydrobiologia* **83**, 245–254.

Beamish, F.W.H. and Legrow, M. (1983). Bioenergetics of the southern brook lamprey. *Journal of Animal Ecology* **52**, 575–590.

Beamish, F.W.H., Potter, J.C. and Thomas, E. (1979). Proximate composition of the adult anadromous sea lamprey in relation to feeding migration and reproduction. *Journal of Animal Ecology* **48**, 1–19.

Beardall, C.H. and Johnston, I.A. (1985). Lysosomal enzyme activities in muscle following starvation and refeeding in the saithe (*Pollachius virens* L.). *European Journal of Cell Biology* **39**, 112–117.

Bell, M.V., Henderson, R.J. and Sargent, J.R. (1986). The role of polyunsaturated fatty acids in fish. *Comparative Biochemistry and Physiology* **83B**, 711–719.

Bellamy, D. and Chester Jones, I. (1961). Studies on *Myxine glutinosa*. I. The chemical composition of the tissues. *Comparative Biochemistry and Physiology* **3**, 175–183.

Belokopytin, Yu.S. (1968). Level of basal metabolism in some Black Sea fishes (In Russian). *Voprosy Ikhtiologii* **8**, 382–385.

Belokopytin, Yu.S. (1978). Levels of energy metabolism in fully grown fish (In Russian). *In* "Elements of Physiology and Biochemistry in Total and Active Metabolism of Fish" (G.E. Shulman, ed.), pp.46–63, Naukova Dumka, Kiev.

Belokopytin, Yu.S. (1990). Bioenergetics and daily rhythms of locomotory activity in marine fish under experimental conditions and in nature (In Russian). *In* "Bioenergetics of Aquatic Organisms" (G.E. Shulman and G.A. Finenko, eds), pp.149–160, Naukova Dumka, Kiev.

Belokopytin, Yu.S. (1993). "Energy Metabolism in Marine Fishes" (In Russian). Naukova Dumka, Kiev, 128pp.

Belokopytin, Yu.S. and Rakitskaya, L.V. (1981). Haematological characteristics of marine fishes of different ecology (horse mackerel and red mullet) at rest and under muscle load (In Russian). *Voprosy Ikhtiologii* **21**, 504–511.

Belokopytin, Yu.S. and Shulman, G.E. (1987). On the temperature dependence of energy metabolism in fish of the Black and Azov Seas (In Russian). *Gidrobiologicheskii Zhurnal* **23**, 61–67.

Belokopytin, Yu.S. and Shulman, G.E. (1995). Evaluation of energy expenditure and production efficiency in Black Sea fish (In Russian). *Doklady Akademii Nauk* **341**, 281–283.

Belyaev, V.I., Nikolaev, V.M., Shulman, G.E. and Yuneva, T.V. (1983). "Tissue metabolism in fish" (In Russian). Naukova Dumka, Kiev, 142 pp.

Belyanina, T.N. (1966). Seasonal variation of fatness in White Sea sparling in relation to gonad maturation (In Russian). *In* "Regularities of Population Dynamics of White Sea Fish" (G.V. Nikolsky, ed.), pp.156–180, Nauka, Moscow.

Bentzen, P., Taggart, C.T., Ruzzante, D.E. and Cook, D. (1996). Microsatellite polymorphism and the population structure of Atlantic cod (*Gadus morhua*) in the northwest Atlantic. *Canadian Journal of Fisheries and Aquatic Sciences* **53**, 2706–2721.

Berdichevsky, L.S. (1964). "The Biological Basis of Efficient use of Fish Stocks" (In Russian), Nauka, Moscow, 36 pp.

Berdyshev, G.D. (1968). "Ecologo-genetic Factors of Ageing and Longevity" (In Russian). Nauka, Leningrad, 202 pp.

Berezhnaya, G.A., Verbitskaya, V.B., Pushchina, L.I. and Myagkova, G.N (1981). Impact of some environmental factors on metabolic processes of carp fingerlings (In Russian). *In* "Proceedings of IV Congress of All-Union Hydrobiological Society" (G.G. Vinberg, ed.), pp.105–106. Naukova Dumka, Kiev.

Berman, A.L., Svetashev, V.I., Rychkova, M.N. and Shnyrev, V.L. (1979). Lipid composition of the photoreceptor membranes of the retina and the temperature stability of rhodopsins in marine fish (In Russian). *In* "Physiology and Biochemistry of Marine and Freshwater Animals" (E.M. Kreps, ed.), pp.172–180. Nauka, Leningrad.

Berman, Sh.A. (1956). On the problem of physiological readiness of carp fingerlings for wintering (In Russian). *Proceedings of the Academy of Science of the Latvian SSR* **106**, 75–82.

Bernatchez, L. and Dodson, J.J. (1987). Relationship between bioenergetics and behaviour in anadromous fish migrations. *Canadian Journal of Fisheries and Aquatic Science* **44**, 399–407.

Bilinsky, E. and Gardner, L.J. (1968). Effect of starvation on free fatty acid level in blood plasma and muscular tissues of rainbow trout. *Journal of the Fisheries Research Board of Canada* **25**, 1555–1560.

Bilinsky, E. and Jonas, R.E. (1970). Effects of coenzyme A and carnitine on fatty acid oxidation by rainbow trout mitochondria. *Journal of the Fisheries Research Board of Canada* **27**, 857–864.

Bilyk, T.I. (1989). Impact of environmental factors on the adenyl nucleotide metabolism in fish muscle and liver (In Russian). *Gidrobiologicheskii Zhurnal* **25**, 58–65.

Bilyk, T.I. (1991). Seasonal variations in the adenyl nucleotide content of fish liver (In Russian). *Gidrobiologicheskii Zhurnal* **27**, 67–71.

Biryukov, N.P. (1980). "Baltic sprat. Biological Condition and Commercial Use" (In Russian). Leningrad University Press, Leningrad, 142 pp.

Bishop, D.G., James, D.G. and Olley, J. (1976). Lipid composition of slender tuna (*Allothunnus fallai*) as related to lipid composition of their feed (*Nyctiphanes australis*). *Journal of the Fisheries Research Board of Canada* **33**, 1156–1161.

Black, D. (1983). The metabolic response to starvation and re-feeding in fish. Ph.D. thesis, University of Aberdeen, Scotland.

Black, D. and Love, R.M. (1986). The sequential mobilisation and restoration of energy reserves in tissues of Atlantic cod during starvation and refeeding. *Journal of Comparative Physiology* **156B**, 469–479.

Black, D. and Love, R.M. (1988). Estimating the carbohydrate reserves in fish. *Journal of Fish Biology* **32**, 335–340.

Black, E.C. (1958). Energy stores and metabolism in relation to muscle activity in fishes. *Journal of the Fisheries Research Board of Canada* **15**, 573–587.

Black, E.C., Robertson, A.C. and Parker, R.R (1961). Some aspects of carbohydrate metabolism in fish. In "Comparative Physiology of Carbohydrate Metabolism in Heterothermic Animals". (A.W. Martin, ed.), pp.89–124. University of Washington Press, Seattle.

Blaxter, J.H.S. (1969.) Development: eggs and larvae. Fish Physiology 3, 177–252.

Blaxter, J.H.S. and Hempel, G. (1966). Yolk utilization by herring larvae. Journal of the Marine Biological Association UK 46, 219–234.

Blaxter, J.H.S. and Hunter, J.R. (1982). The biology of the clupeid fishes. Advances in Marine Biology 20, 3–194.

Blaxter, J.H.S., Wardle, C.S. and Roberts B.L. (1971). Aspects of the circulatory physiology and muscle systems of deep-sea fish. Journal of the Marine Biological Association UK 51, 991–1006.

Blazka, P. (1958). The anaerobic metabolism of fish. Physiological Zoology 31, 117–128.

Blier, P. and Guderley, H. (1988). Metabolic responses to cold acclimation in the swimming musculature of lake whitefish, Coregonus clupeaformis. Journal of Experimental Zoology 246, 244–252.

Blöchl, E., Rachel, R., Burggraf, S., Hafenbradl, D., Jannasch, H.W. and Stetter, K.O. (1997). Pyrolobus fumarii gen. and sp. nov. represents a novel group of Archaea, extending the upper temperature border of life to 113°C. Extremophiles 1, 14–21.

Boëtius, J. and Boëtius, J. (1980). Experimental maturation of female silver eels. Estimates of fecundity and energy reserves for migration and spawning. Dana 1, 1–28.

Boëtius, J. and Boëtius, J. (1985). Lipid and protein content in Anguilla anguilla during growth and starvation. Dana 4, Special Issue, 1–17.

Boeuf, G. (1987). Physiological bases of salmonid culture: the smoltification phenomenon. Pisciculture Français 88, 5–21.

Boeuf, G., Le Bail, G. and Prunet, P. (1989). Growth hormone and thyroid hormones during Atlantic salmon smolting and after transfer to sea water. Aquaculture 82, 257–268.

Bogdan, V.V., Lysenko, P.V. and Yarzhombek, A.A. (1983). Lipid composition of liver and muscle of one-year carp after wintering (In Russian). In "Comparative Biochemistry of Aquatic Animals" (V.S. Siderov and R.V. Vysotskaya, eds), pp. 66–72. Akademii Nauk SSSR, Karelian Branch, Petrozavodsk.

Bogoyavlenskaya, N.P. (1959). Study of calcium metabolism aimed to apply Ca^{45} as a tracer for fish (In Russian). Rybnoye Khozyaistvo 1959, 1–55.

Bokdawala, F.D. and George, J.C. (1967a). A histochemical study of the red and white muscle of the carp, Cirrhina mrigala. Journal of Animal Morphology and Physiology 14, 60–68.

Bokdawala, F.D. and George, J.C. (1967b). A quantitative study of fat, glycogen, lipase and succinic dehydrogenase in fish muscle. Journal of Animal Morphology and Physiology 14, 223–230.

Boldyrev, A.A. (1979). The role of lipids in the function of Na,-K-activated adenosine triphosphatase (In Russian). Biologicheskii Nauki 1979 (3), 5–17.

Boldyrev, A.A. and Prokofieva, V.D. (1985). How is the activity of membrane enzymes regulated? (In Russian). Biologicheskii Nauki 1985 (9), 5–13.

Bolgova, O.M. (1993). On the relation between the level of eicosotrienoic acids with the ecologo-physiological condition of the fish (In Russian). In "Biochemical Methods in Ecological and Toxicological Research" (O.I. Potapova and Y.A. Smirnov, eds), pp.170–174. Petrozavodsk.

Bolgova, O.M. and Shurov, J.L. (1987). Adaptive changes in fatty acid spectra of tissue lipids in wild and farmed fry of Atlantic salmon in the process of smoltification (In Russian). *Zhurnal Evolutsionnoy Biokhimii i Physiologii* **23**, 211–215.

Bolgova, O.M., Sidorov, V.S. and Smirnov, Yu.A. (1976). Fatty acid composition of muscles of salmon fry in natural and cultured populations (In Russian). *In* "Salmonidae of Karelia", pp. 163–167. Petrozavodsk.

Bolgova, O.M., Lizenko, E.I. and Yarzhombek, A.A. (1985a). Qualitative composition of lipids in tissues of one year carp after wintering (In Russian). *Rybnoye Khozyaistvo* **1985** (12), 27–29.

Bolgova, O.M., Lizenko, E.I., Klepkina, M.N. and Lysenko, P.V. (1985b). Fatty acid composition of triacyl glycerols of liver, muscle and gill in 'strong' and 'weak' one-year carp after wintering (In Russian). *In* "Biochemistry of Fry of Freshwater Fishes", pp. 19–23. Petrozavodsk.

Bondarenko, V.F. (1986). The effect of amino acid spectra on nutritive reactions of fry, yearlings and one-year-old carp (In Russian). *Rybnoye Khozyaistvo* 1986 (10), 39–40.

Bone, Q. (1966). On the function of the two types of myotomal muscle fibre in elasmobranch fish. *Journal of the Marine Biological Association UK* **46**, 321–349.

Bone, Q. (1972). Buoyancy and hydrodynamic functions of integument in the castor oil fish, *Ruvettus pretiosus*. *Copeia* 1972 1, 78–87.

Bone, Q., and Roberts, B.L. (1969). The density of elasmobranchs. *Journal of the Marine Biological Association UK* **49**, 913–937.

Borisochkina, L.I. (1991). Modern trends in production and use of fish oils (In Russian). *Rybnoye Khozyaistvo* 1991 (4), 76–79.

Borisov, V.M. and Shatunovsky, M.I. (1973). On the possibility of applying humidity index for estimating the natural mortality rate of Barents Sea cod (In Russian). *Trudy VNIRO* **93**, 311–321.

Borlongan, I.G. and Benitez, L.V. (1992). Lipids and fatty acid composition of milkfish (*Chanos chanos* Forskål) grown in fresh water and sea water. *Aquaculture* **104**, 79–89.

Bouche, G., Murat, J.C. and Parent, J.P. (1971). Etude de l'influence de régimes synthétiques sur la protéosynthèse et les réserves glucidiques et lipidiques dans la foie de la carpe dénutrie. *Comptes Rendus des Séances de la Société de Biologie, Paris* **165**, 2202–2205.

Boukhchache, D. and Lagarde, M. (1982). Interactions between prostaglandin precursors during their oxygenation by human platelets. *Biochimica et Biophysica Acta* **713**, 386–392.

Boulekbache, H. (1981). Energy metabolism in fish development. *American Zoologist* **21**, 377–389.

Boyd, T.A., Cha, C.-J., Forster, R.P. and Goldstein, L. (1977). Free amino acids in tissues of the skate *Raja erinacea* and the sting ray *Dasyatis sabina*: effects of environmental dilution. *Journal of Experimental Zoology* **199**, 435–442.

Brækkan, O.R. (1956). The function of the red muscle in fish. *Nature, Lond.* 178, 747–748.

Brækkan, O.R. and Boge, G. (1962). Vitamin B6 and the reproductive cycle of ovaries in cod. *Nature, Lond.* **193**, 394–395.

Brandes, C.H. and Dietrich, R. (1958). Betrachtungen über die Beziehungen zwischen dem Fett- und Wassergehalt und die Fettverteilung bei Konsumfischen. *Veröffentlichungen des Institut für Meeresforschung in Bremerhaven* **5**, 299–305.

von Brand, T. (1946). "Anaerobiosis in Invertebrates". Biodynamica, Normandy, Missouri, 328 pp.

Brawn, V.M. (1969). Buoyancy of Atlantic and Pacific herring. *Journal of the Fisheries Research Board of Canada* **26**, 2077–2091.

Bray, R.N., Purcell, L.J. and Miller, A.C. (1986). Ammonium excretion in a temperate reef community by a planktivorous fish, *Chromis punctipinnis* (Pomacentridae), and potential uptake by young giant kelp *Macrocystis pyrifera* (Laminariales). *Marine Biology* **90**, 327–334.

Brenner, R.R., Vazza, D.V. and De Tomas, M.E. (1963). Effect of a fat-free diet and of different dietary fatty acids (palmitate, oleate and linoleate) on the fatty composition of freshwater fish lipids. *Journal of Lipid Research* **4**, 341–345.

Brett, J.R. (1970). Fish – the energy cost of living. *In* "Marine Aquiculture" (W. J. McNeil, ed.), pp.37–52. Oregon State University Press, Corvallis.

Brett, J.R. (1973). Energy expenditure of sockeye salmon during sustained performance. *Journal of the Fisheries Research Board of Canada* **30**, 1799–1809.

Brett, J.R. (1979). Environmental factors and growth. *In* "Fish Physiology: 8, Bioenergetics and Growth" (W.S. Hoar and D.J. Randall, eds), pp. 599–677. Academic Press, New York, London.

Brett, J.R. (1983). Life energetics of sockeye salmon *Onchorynchus nerka*. *In* "Behavioral energetics; the cost of survival in vertebrates" (W.P. Aspey and S.I. Lustick, eds), pp. 29–63. Ohio State University Press, Columbus.

Brett, J.R. (1986). Production energetics of a population of sockeye salmon. *Canadian Journal of Zoology* **64**, 555–564.

Brett, J.R. and Glass, N.R. (1973). Metabolic rates and critical swimming speeds of sockeye salmon (*Oncorhynchus nerka*) in relation to size and temperature. *Journal of the Fisheries Research Board of Canada* **30**, 379–387.

Brett, J.R. and Groves, T.D.D. (1979). Physiological energetics. *In* "Fish Physiology: 8, Bioenergetics and Growth" (W.S. Hoar and D.J. Randall, eds), pp. 280–352. Academic Press, New York, London.

Brett, J.R., Shelbourne, J.E. and Shoop, C.T. (1969). Growth rate and body composition of fingerling sockeye salmon in relation to temperature and ration size. *Journal of the Fisheries Research Board of Canada* **29**, 2363–2394.

Brichon, G., Chapelle, S. and Zwingelstein, J. (1980). Phospholipid composition and metabolism in haemolymph of *Carcinas maenas* – effect of temperature. *Comparative Biochemistry and Physiology* **67**, 647–652.

Brill, R.W. and Dizon, A.E. (1979). Red and white muscle fibre activity in swimming skipjack tuna, *Katsuwonus pelamis*. *Journal of Fish Biology* **15**, 679–685.

Brizinova, P.N. (1958). Changes in the fatness during the ontogenesis of carp (In Russian). *In* "Proceedings of 8th Conference on Fish Physiology" (G.S. Razinkin and G.A. Malyukina, eds), pp. 244–250.

Brockerhoff, H., Ackman, R.G. and Hoyle, R.J. (1963). Specific distribution of fatty acids in marine lipids. *Archives of Biochemistry and Biophysics* **100**, 9–12.

Brockerhoff, H., Yurkovski, M., Hoyle, R.J. and Ackman, R.G. (1964). Fatty acid distribution in lipids of marine plankton. *Journal of the Fisheries Research Board of Canada* **21**, 1379–1384.

Brody, S. (1945). "Bioenergetics and Growth". Reinhold, New York, 1023 pp.

Brubakk, A.O., Hemmingsen, B.B. and Sundness, G. (eds) (1989) "Supersaturation and bubble formation in fluids and organisms". The Royal Norwegian Society of Sciences and Letters, Trondheim.

Bryantsev, V.A., Kovalchuk, L.A., Novikov, N.P., Panov, B.N. and Chashchin, A.K. (1987). Formation of the wintering stock of Black Sea anchovy (In Russian). *Rybnoye Khozyaistvo* 1987 (4), 49–52.

Bulgakova, Yu.V. (1993). Daily dynamics of the Black Sea anchovy feeding and the factors responsible (In Russian). *Voprosy Ikhthyologii* **33**, 395–400.

Bulgakova, T.J., Zasosov, A.V. and Shatunovsky, M.I. (1973). On the modelling of commercial fishery and fish culture systems, taking into account ecologo-physiological factors (In Russian). *Trudy VNIRO* **94**, 9–23.

Bullock, T.H. (1955). Compensation for temperature in the metabolism and activity of poikilotherms. *Biological Reviews of the Cambridge Philosophical Society* **30**, 311–342.

Bulow, F.J. (1970). RNA/DNA ratios as indicators of recent growth rates of a fish. *Journal of the Fisheries Research Board of Canada* **27**, 2343–2349.

Bulow, F.J., Zeman, M.E., Winningham, J.R. and Kuddson, W.F. (1981). Seasonal variations in RNA–DNA ratios and in indicators of feeding, reproduction, energy storage and condition in a population of bluegill. *Journal of Fish Biology* **18**, 237–244.

Burd, A.C. (1974). The north-east Atlantic herring and the failure of a fishery. In "Sea Fisheries Research" (F.R. Harden Jones, ed.), pp.167–191. Elek Science, London.

Burlachenko, I.V. (1987). Experience of use of artificial foods for mullet fry (In Russian). *In* "Problems of Physiology and Biochemistry of Fish Feeding" (V.K.Vinogradov and M.A. Shcherbina, eds), pp. 66–75. All-Union Institute of Pond Fish Culture, Moscow.

Burlakova, Ye.B., Storozhuk, N.M. and Khrapova, N.G. (1988). Relationship between the activity of antioxidants and substrate oxidisability in lipids of natural origin. *Biophysics* **33**, 840–846.

Burt, J.R. (1969). The course of glycolysis in fish muscle. *In* "Food Science and Technology, Chemical and Physical Aspects of Food" (J.M. Leitch, ed.). Vol.1, pp.193–198. (Proceedings of the 1st International Congress on Food Science and Technology, London, September 18–21, 1962.) Gordon and Breach, London.

Bustomante, G. and Shatunovsky, M.I. (1981). Seasonal dynamics of some morphophysiological and biochemical characteristics of black jack from the south-west shelf off Cuba (In Russian). *Voprosy Iktiologii* **21**, 1134–1140.

Caldwell, R. (1969). Thermal compensation of respiratory enzymes in tissue of the goldfish. *Comparative Biochemistry and Physiology* **31**, 79–93.

Caldwell, R. and Vernberg, F.J. (1970). The influence of acclimatisation temperature on the lipid composition of fish gill mitochondria. *Comparative Biochemistry and Physiology* **37**, 179–181.

Calow, P. (1977). Conversion efficiencies in heterotrophic organisms. *Biological Reviews of the Cambridge Philosophical Society* **52**, 385–409.

Calvo, J. and Johnston, I.A. (1992). Influence of rearing temperature on the distribution of muscle fibre types in the turbot, *Scophthalmus maximus*, at metamorphosis. *Journal of Experimental Marine Biology and Ecology* **161**, 45–55.

Cameron, J.N., Kostoris, J. and Penhale, P.A. (1973). Preliminary energy budget of the ninespike stickleback in an arctic lake. *Journal of the Fisheries Research Board of Canada* **30**, 1179–1189.

Campbell, C.M. and Davis, P.S. (1978). Temperature acclimation in the teleost, *Blennius pholis*, changes in enzyme activity and cell structure. *Comparative Biochemistry and Physiology* **61**B, 165–167.

Caulton, M.S. (1978a). The effect of temperature and mass on routine metabolism in *Sarotherodon (Tilapia) mossambicus* (Peters). *Journal of Fish Biology* **13**, 195–201.

Caulton, M.S. (1978b). Tissue depletion and energy utilisation during routine metabolism by sub-adult *Tilapia rendalli*. *Journal of Fish Biology* **13**, 1–6.

Cetta, C.M. and Capuzzo, J.M. (1982). Physiological and biochemical aspects of embryonic and larval development of the winter flounder. *Marine Biology* **71**, 327–337.

Chapman, D. (1975). Phase transitions and fluidity characteristics of lipids and cell membranes. *Quarterly Reviews of Biophysics* **8**, 185–235.

Chashchin, A.K. (1985). On change of the population structure of anchovy in the basin of Azov and Black Seas (In Russian). *Voprosy Ikhtiologii* **25**, 583–589.

Chashchin, A.K. and Akselev, O.I. (1990). Migrations of the stock and availability of Black Sea anchovy to the fishery in autumn and winter (In Russian). *In* "Biological Resources of the Black Sea" (V.A. Shlyakhov, ed.), pp. 80–93. VNIRO, Moscow.

Chebanov, N.G., Varnavskaya, N.V. and Varnavsky, R.S. (1983). Assessment of the spawning performance success of male sockeye salmon of different hierarchical status, using genetic–biochemical markers (In Russian). *Voprosy Ikhtiologii* **23**, 774–778.

Chebotareva, M.A. (1983). Comparative biochemical study of cholesterol of the brain of fish (In Russian). *Zhurnal Evolutsionnoy Biochimii i Physiologii* **19**, 420–427.

Chebotareva, M.A. and Dityatev, A.E. (1988). Brain cholesterol in vertebrates of different classes (In Russian). *Zhurnal Evolutsionnoy Biochimii i Physiologii* **24**, 426–432.

Chekunova, V.I. (1983). Ecological groups of marine cold-requiring fish and their energy metabolism (In Russian). *Voprosy Ikhtiologii* **23**, 829–838.

Chekunova, V.I. and Naumov, A.G. (1982). Energy metabolism and food requirements of marbled notothenia (In Russian). *Voprosy Ikhtiologii* **22**, 294–302.

Chepurnov, A.V. and Tkachenko, N.K. (1983). Lipids, glycogen, free amino acids in gonads, embryos and larvae of some Black Sea fishes (In Russian). *Ecologiya Morya* **15**, 40–46.

Cherkov, V.M. and Borchsenius, S.N. (1989). The determination of Phylogenetic relations between Pacific salmon species of genus *Oncorhynchus* by the method of DNA molecular hybridisation (in Russian). *Biologiya morya* 1989 (2), 23–29.

Chernitsky, A.G. (1993). Smoltification of sea trout (In Russian). *Doklady Akademii Nauk SSSR* **333**, 818–819.

Chernitsky, A.G., Gambaryan, S.P., Karpenko, L.A., Lavrova, E.A. and Shkurko, D.S. (1993). The effect of abrupt salinity changes on blood and muscle electrolyte content in the smolts of the Atlantic salmon, *Salmo salar*. *Comparative Biochemistry and Physiology* **104 A**, 551–554.

Cherry, D.S., Dickson, K.L. and Cairns, J. (1975). Temperatures selected and avoided by fish at various acclimation temperatures. *Journal of the Fisheries Research Board of Canada* **32**, 485–491.

Chesalin, M.V. (1993). The character of distribution and biology of the squid *Stenoteuthis analaniensis* in the Arabian Sea (In Russian). *Gydrobiologicheskii Zhurnal* **29**, 16–27.

Chikhachev, A.S. (1984) Genetic control over the population structure and hybridization of valuable fish stocks in artificial breeding (In Russian). *In* "Genetics, Selection, Hybridization of Fish" (V.S. Kirpichnikov, ed.), pp. 16–20. Rostov-on-Don.

Chikhachev, A.S. (1991). Principles of physiologo-biochemical and genetic monitoring of mariculture products (In Russian). *Rybnoye Khozyaistvo* 1991 (12), 57–59.

Childress, J.J. and Nygaard, M.H. (1973). The chemical composition of midwater fishes as a function of depth of occurrence in southern California. *Deep-Sea Research* **20**, 1093–1109.

Childress, J.J., Taylor, S.M., Cailliet, G.M. and Price, M.H. (1980). Patterns of growth, energy utilisation and reproduction in some meso- and bathypelagic fishes of southern California. *Marine Biology* **61**, 27–40.

Childress, J.J., Price, M.H., Favurri, J. and Cowles, D. (1990). Chemical composition of midwater fishes as a function of depth of occurrence off the Hawaiian Islands. Food availability as a selective factor. *Marine Biology* **105**, 235–246.

Chizhevsky, A.L. (1976). "The Earth Echo of the Sun Storms"(In Russian). Mysl, Moscow, 368pp.

Cho, C.Y., Slinger, S.J. and Bayley, H.S. (1982). Bioenergetics of salmonid fishes: energy intake, expenditure and productivity. *Comparative Biochemistry and Physiology* **73B**, 25–41.

Christian, J.S. (1950). The adreno–pituitary system and population cycles in mammals. *Journal of Mammalogy* **31**, 247–259.

Christiansen, J.A. (1984). Changes in phospholipid classes and fatty acid desaturation and incorporation into phospholipids during temperature acclimation of green sunfish, *Lepomis cyanellus*. *Physiological Zoology* **57**, 481–492.

Chuksin, Yu.V., Akhramovich, A.P., Mikhailov, A.Yu. and Arrhipov, A.Yu. (1977). Environmental factors of seasonal changes in the distribution and behaviour of mackerel, scad and blue whiting *poutassou* westward of the British Isles (In Russian). *Proceedings of All-Union Institute of Marine Fisheries and Oceanography* **121**, 11–24.

Clarke, A. (1985). The physiological ecology of polar marine ectotherms: energy budget, resource allocation and low temperature. *Oceanis* **11**, 11–26.

Clarke, W.C., Shelbourne, J.E. and Brett, J.R. (1978). Growth and adaptation to sea water in under-yearling sockeye and coho salmon subjected to regimes of constant or changing temperature and day length. *Canadian Journal of Zoology* **56**, 2413–2421.

Comfort, A. (1964). "Ageing: the Biology of Senescence". Routledge and Kegan Paul, London.

Connell, D.W. and Miller, G.J. (1984) "Chemistry and Exotoxicology of Pollution". John Wiley and Sons, New York, 444pp.

Conover, R.J. and Corner, E.D.I. (1968). Respiration and nitrogen excretion by some marine zooplankton in relation to their life cycles. *Journal of the Marine Biological Association UK* **4**, 49–75.

Conte, F.P. (1969). Salt secretion. *In* "Fish Physiology 1", (W.S. Hoar and D.J. Randall, eds), Vol 1, pp. 241–292. Academic Press, London and New York.

Constantz, G.D. (1980). Energetics of viviparity in the gila topminnow Pisces: Poeciliidae. *Copeia* 1980 (4), 876–878.

Corner, E.D.S., Denton, E.J. and Forster, G.R. (1969). On the buoyancy of some deep-sea sharks. *Proceedings of the Royal Society of London B* **171**, 415–429.

Cossins, A.R. and Prosser, C.L. (1978). Evolutionary adaptation of membranes to temperature. *Proceedings of the National Academy of Sciences USA* **75**, 2040–2043.

Cossins, A.R., Friedlander, M.J. and Prosser, C.L. (1977). Correlations between behavioural temperature adaptations of goldfish and the viscosity and fatty acid composition of their synaptic membranes. *Journal of Comparative Physiology* **120**, 109–121.

Cowey, C.B. (1967). Comparative studies on the activity of D-glyceraldehyde-3-phosphate dehydrogenase from cold- and warm-blooded animals, with reference to temperature. *Comparative Biochemistry and Physiology* **31**, 79–93.

Cowey, C.B. (1976). Use of synthetic diets and biochemical criteria in the assessment of nutrient requirements of fish. *Journal of the Fisheries Research Board of Canada* **33**, 1040–1050.

Cowey, C.B. and Sargent, J.R. (1979). Nutrition. *In* "Fish Physiology" (W.S. Hoar, D.J. Randall and J.R. Brett, eds), Vol. 8 pp.1–70. Academic Press, New York and London.

Cowey, C.B., Daisley, K.W. and Parry, G. (1962). Study of amino acids, free or as components of protein, and of some B vitamins, in the tissues of the Atlantic salmon, *Salmo salar*, during spawning migration. *Comparative Biochemistry and Physiology* **7**, 29–38.

Cowie, W.P. and Little, W.T. (1966). The relationship between the toughness of cod stored at −29°C and its muscle protein solubility and pH. *Journal of Food Technology* **1**, 335–343.

Creasey, S.S. and Rogers, A.D. (1999). Population genetics of bathyal and abyssal organisms. *Advances in Marine Biology* **35**, 1–151.

Crockford, T. and Johnston, I.A. (1993). Developmental changes in the composition of myofibrillar proteins in the swimming muscles of Atlantic herring. *Marine Biology* **115**, 15–22.

Culkin, F. and Morris, R.J. (1970). The fatty acids of some marine teleosts. *Journal of Fish Biology* **2**, 107–112.

Cunningham, J. and Reid, D. (1932). Experimental researches on the emission of oxygen by the pelvic filaments of the male *Lepidosiren* with some experiments on *Symbranchus marmoratus*. *Proceedings of the Royal Society of London B* **110**, 234–248.

Daan, N. (1975). Consumption and production in North Sea cod. An assessment of the ecological status of the stock. *Netherlands Journal of Sea Research* **9**, 24–55.

Dabrowsky, K.R. (1982). Seasonal changes in the chemical composition of the fish body and nutritional value of the muscle of the pollan from Lough Neagh, Northern Ireland. *Hydrobiologia* **87**, 121–141.

Dabrowsky, K.R. (1986). A new type of metabolism chamber for the determination of active and postprandial metabolism of fish, and consideration of results for coregonid and salmon juveniles. *Journal of Fish Biology* **28**, 105–117.

Dabrowsky, K.R., Kaushik, S.I. and Luquet, P. (1984). Metabolic utilisation of body stores during the early life of whitefish. *Journal of Fish Biology* **24**, 721–729.

Daikoku, T. and Sakaguchi, M. (1990). Changes in level of trimethylamine and trimethylamine oxide during adaptation of young eel *Anguilla anguilla* to a seawater environment. *Bulletin of the Japanese Society of Scientific Fisheries* **56**,1895 (only).

Dambergs, N. (1963). Extractives of fish muscle. 3. Amounts, sectional distribution and variations of fat, water-solubles, protein and moisture in cod fillets. *Journal of the Fisheries Research Board of Canada* **20**, 909–918.

Dambergs, N. (1964). Extractives of fish muscle. 4. Seasonal variations of fat, water-solubles, protein and water in cod (*Gadus morhua* L.) fillets. *Journal of the Fisheries Research Board of Canada* **21**, 703–709.

Dando, P.R. (1969). Lactate metabolism in fish. *Journal of the Marine Biological Association UK* **49**, 209–223.

Danilevsky, N.N. (1964). The most important factors determining the term and location of the appearance of Black Sea anchovy fishable stocks (In Russian). *Izvestiya Ribno Resursov* **22**, 115–124.

Danilevsky, N.N. and Mayorova, A.A. (1979). Anchovy (In Russian). *In* "Natural Resources of the Black Sea" (V.N. Greze, ed.), pp. 25–73. Pishchevaya promyshlennost, Moscow.

Danilevsky, N.N., Ivanov, L.S., Kautish, I. and Veriati-Marinescu, F. (1979). Fisheries potential resources (In Russian). *In* "The Principles of Biological Productivity of the Black Sea" (K.S. Tracheva and Yu.K. Benko, eds), pp.291–299. Naukova Dumka, Kiev.

Davison, W. and Goldspink, G. (1977). The effect of prolonged exercise on the lateral musculature of the brown trout. *Journal of Experimental Biology* **70**, 1–12.

Dawson, A.S. and Grimm, A.S. (1980). Quantitative seasonal changes in the protein, lipid and energy content of the carcass, ovaries and liver of adult female plaice. *Journal of Experimental Biology* **76**, 493–504.

De Duve, C. (1963). General properties of lysosomes: the lysosome concept. *In* "Ciba Foundation Symposium: Lysosomes" (A.V.S. De Reuck and M.P. Cameron, eds), pp. 1–35. Little, Brown and Co., Boston, Massachusetts.

Dekhnik, T.V. (1979). Abundance dynamics, survival and elimination of the eggs and larvae of dominant fishes (In Russian). *In* "The Principles of Biological Productivity of the Black Sea" (V.N. Greze, ed.), pp.272–279. Naukova dumka, Kiev.

Demin, V.I., Androsova, I.M. and Ozernyuk, N.D. (1989). Energy metabolism adaptations in fish: the influence of swimming speed and temperature on the cytochrome system of skeletal muscle (In Russian). *Doklady Akademii Nauk SSSR* **308**, 241–246.

Deng, J.O., Orthoefer, F.T., Dennison, R.A. and Watson, M. (1976). Lipid and fatty acids in mullet (*Mugil cephalus*). Seasonal and locational variations. *Journal of Food Science* **41**, 1479–1483.

Denisenko, V.V., Shevchenko, V.A. and Golovenko,V.K. (1971). Biochemical composition of Black Sea plankton (In Russian). *Biologiya morya* **2**, 86–106.

Dergaleva, Zh. and Shatunovsky, M.I. (1977). Notes on the lipid metabolism of striped perch larvae and fry (In Russian). *Voprosy Ikhtiologii* **17**, 947–949.

Derkachev, E.F., Alekseev, V.A. and Konstantinov, M.V. (1976). Regulatory changes of pathways in mitochondria and cytosol (In Russian). *In* "Mitochondria" (S.E. Severin, ed.), pp.135–256. Nauka, Moscow.

Deufel, J. (1975). Physiological effect of carotenoids on Salmonidae. *Hydrologia* **37**, 244–248.

De Vlaming, V., Fitzgerald, R., Delahunty, G., Cech, J., Selman, K. and Barkley, M. (1984). Dynamics of oocyte development and related changes in serum oestradiol-17β, yolk precursor, and lipid levels in the teleostean fish *Leptocottus armatus*. *Comparative Biochemistry and Physiology* **77A**, 599–610.

De Vries, A.L. (1970). Freezing resistance in Anatarctic fishes. *In* "Antarctic Ecology" (M.W. Holdgate. ed.), Vol.1, pp.320–328. Academic Press, London, New York.

De Vries, A.L. (1971). Glycoproteins as biological antifreeze agents in Antarctic fishes. *Science, New York* **172**, 1152–1155.

De Vries, A.L. and Eastman, J.T. (1978). Lipid sacs as a buoyancy adaptation in an Antarctic fish. *Nature, Lond.* **271**, 352–353.

De Vries, A.L. and Eastman, J.T. (1981). Physiology and ecology of notothenoid fishes of the Ross Sea. *Journal of the Royal Society of New Zealand* **11**, 329–340.

De Vries, A.L. and Wohlschlag, D.E. (1969). Freezing resistance in some Antarctic fishes. *Science, New York* **163**, 1073–1075.

De Witt, K.M. (1963). Seasonal variations in cod liver oil. *Journal of the Science of Food and Agriculture* **14**, 92–98.

Diana, J.S. (1983). An energy budget for northern pike. *Canadian Journal of Zoology* **61**, 1968–1975.

Diana, J.S. and Mackay, W.C. (1979). Timing and magnitude of energy deposition and loss in the body, liver and gonads of northern pike. *Journal of the Fisheries Research Board of Canada* **36**, 481–487.

Dietrich, D., Schlatter, C., Blau, N. and Fischer, M. (1989). Aluminium and acid rain: mitigating effects of NaCl on aluminium toxicity to brown trout (*Salmo trutta fario*) in acid water. *Toxicological and Environmental Chemistry* **19**, 17–23.

Dobrovolov, I. (1970). Chemical composition of Black Sea anchovy (*Engraulis encrasicolus ponticus* Alex) near the Bulgarian shore (In Bulgarian). *Izvestiya na Instituta Okeanografiya po Ribno Stopanstvo, Varna* **10**, 115–124

Dobrovolov, I. (1980). Gene markers in fish (In Bulgarian). *Izvestiya Instituta Ribno Resurcov, Varna* **17**, 10–14.

Dobrusin, M.S. (1978). The study of seasonal dynamics of fractional and fatty acid composition of lipids in organs and tissues of horse-mackerel of the north-east Atlantic (In Russian). *Trudy VNIRO* **120**, 44–50.

Dolnik, V.P. (1965). Physiological principles of migration in birds (In Russian). *In* "Biological Significance and Functional Determination of Migratory Behaviour of Animals" (A.D. Slonin, ed.), pp. 12–22.

Dolnik, V.R. (1969). Bioenergetics of flying birds (In Russian). *Zhurnal Obshchey Biologii* **30**, 273–291.

Dolnik, V.R. (1978). Energy metabolism and the size of the animal: the physical rules which underlie this relationship (In Russian). *Zhurnal Obshchey Biologii* **39**, 805–816.

Donaldson, E.M., Fagerlund, U.H.M., Higgs, D.A. and McBride, J.R. (1979). Hormonal enhancement of growth. *In* "Fish Physiology: 8" (W.S. Hoar and D.J. Randall, eds), pp.456–578. Academic Press, New York.

Douglas, E.L., Friede, W.A. and Pickwell, G.V. (1976). Fishes in oxygen minimum zones: blood oxygenation characteristics. *Science, New York* **191**, 957–959.

Driedzic, W.R. and Hochachka, P.W. (1975). The unanswered question of high anaerobic capabilities of carp white muscle. *Canadian Journal of Zoology* **53**, 706–712.

Driedzic, W., Selivonchick, D.P. and Roots, B.J. (1976). Alk-1-enyl ether-containing lipids of goldfish brain and temperature acclimation. *Comparative Biochemistry and Physiology* **53B**, 311–314.

Driedzic, W.R., Gesser, H. and Johansen, K. (1985). Effects of hypoxic adaptation on myocardial performance and metabolism of *Zoarces viviparus*. *Canadian Journal of Zoology* **63**, 821–823.

Drummond, G. (1967). Muscle metabolism. *Fortschritte der Zoologie* **18**, 359–429.

Drury, D.E. and Eales, J.G. (1968). The influence of temperature on histological and radiochemical measurements of thyroid activity in the eastern brook trout. *Canadian Journal of Zoology* **46**, 1–9.

Dryagin, P.A. (1961). The main trends of fish life history studies (In Russian). *Nauchno-Technicheskii Byulletin Gosudarstvennogo Nauchno-issledovatelskogo Instituta Ozernogo i Rechnogo Rybnogo* **14**, 113–117.

Dubrovin, I.Ya., Rogov, S.F. and Prokopenko, E.I. (1973). On the fatness dynamics of Azov anchovy at the wintering grounds (In Russian). *Rybnoye Khozyaistvo* **3**, 506–509.

Duchâteau, G., Florkin, M. and Jeuniaux, C. (1959). Composante aminoacide des tissus, chez les crustaces. I. Composant amino-acide des muscles de *Carcinus maenas* L. lors du passage de l'eau de mer a l'eau saumatre et au cours de la mue. *Archives Internationales de Physiologie et de Biochimie* **67**, 489–500.

Dunn, J.F. (1988). Muscle metabolism in antarctic fish. *Comparative Biochemistry and Physiology* **90B**, 539–545.

Dunn, J.F. and Johnston, I.A. (1986). Metabolic constraints and burst swimming in the Antarctic teleost *Notothenia neglecta*. *Marine Biology* **91**, 433–440.

Duthie, G.G. (1982). The respiratory metabolism of temperature-adapted flatfish at rest and during swimming activity and the use of anaerobic metabolism at moderate swimming speeds. *Journal of Experimental Biology* **9**, 359–373.

Eales, J.G., Chang, J.P. and Krank, G. (1982). Effect of temperature on plasma thyroxine and iodine kinetics in rainbow trout. *General and Comparative Endocrinology* **47**, 295–307.

Eastman, J.T. (1988). Lipid storage system and the biology of two neutrally buoyant Antarctic notothenoid fishes. *Comparative Biochemistry and Physiology* **90B**, 529–537.

Eccles, D.H. (1986). The effect of temperature on the rate of passage of food in the smallmouth yellowfish *Barbus aeneus* Burchall (Teleostei, Cyprinidae). *South African Journal of Zoology* **21**, 68–72.

Eckberg, D.R. (1962). Anaerobic metabolism of carp gills in relation to temperature. *Comparative Biochemistry and Physiology* **5**, 123–128.

Efimova, V.M. (1982). Electromyogram as an indicator of the adaptation to prolonged swimming (In Russian). *In* "Biology of Shelf Waters of the World Ocean" (A.J. Kafanov, ed.), pp.166–167. (Proceedings of the II Conference on Marine Biology, Far East Scientific Centre of Arctic Science of USSR.)

Egorova, M.N. (1968). Ecological characteristics of the blood in some Black Sea fishes (In Russian). *In* "Biological Studies of the Black Sea and its Natural Resources" (V.A. Vodyanitsky, ed.), pp.223–227. Nauka, Moscow.

Eldridge, M.B., Whipple, J. and Eng, D. (1981). Endogenous energy sources as factors affecting mortality and development in striped bass (*Morone saxatilis*) eggs and larvae. *Rapport et Procès-verbaux des Réunions, Conseil Permanent International pour l'Exploration de la Mer* **178**, 568–570.

Eliassen, E., Leivestad, H. and Müller (1960). The effect of low temperature on the freezing point of plasma and on the potassium/sodium ratio in the muscles of some boreal and subarctic fishes. *Bergens Universitet Årbok, Matematisk-Naturvitenskapelig Ser.* 1960 (14), 24pp.

Eliassen, J.-E. and Vahl, O. (1982). Seasonal variations in biochemical composition and energy content of liver, gonad and muscle of mature and immature cod, *Gadus morhua* L, from Balsfjorden, northern Norway. *Journal of Fish Biology* **20**, 707–716.

Eliseeva, E.J., Storozhuk, A.Ya., Pisareva, N.A., Zhakevich, M.L., Dobrusin, M.S. and Guleva, I.B. (1985). Seasonal dynamics of lipid metabolism in the jack mackerel (In Russian). *Rybnoye Khozyaistvo* 1985 (6), 35–38.

Elliott, J.M. (1976). Energy losses in the waste products of brown trout. *Journal of Animal Ecology* **45**, 561–580.

Elliott, J.M. (1982). The effects of temperature and ration size on the growth and energetics of salmonids in captivity. *Comparative Biochemistry and Physiology* **73B**, 81–91.

Elliott, J.M. and Davison, W. (1975). Energy equivalents of oxygen consumption in animal energetics. *Oecologia* **19**, 195–201.

El Sayed, M.M. (1984). Variations in the fatty acids of pelamys *Euthynnus alletteratus*. *Revue Internationale d'Océanographie Medicale* **75–76**, 77–97.

Elton, C. (1958). "Ecology of Invasions by Animals and Plants". Methuen, London, 181 pp.

Emeretli, I.V. (1981a). Activity of lactate dehydrogenase in tissues of Black Sea fishes (In Russian). *Ekologiya Morya* 1981 (7), 57–60.

Emeretli, I.V. (1981b). Activity of succinate dehydrogenase in tissue mitochondria of Black Sea fishes (In Russian). *Ekologiya Morya* 1981 (7), 60–63.

Emeretli, I.V. (1990). The activity of energy metabolism enzymes in Black Sea fishes (In Russian). *In* "Bioenergetics of Aquatic Organisms" (G.E. Shulman and G.A. Finenko, eds), pp. 178–189. Naukova Dumka, Kiev.

Emeretli, I.V. (1994). Dependence of the energy metabolism enzyme activity of Black Sea fishes on temperature at different periods of the annual cycle (In Russian). *Voprosy Ikhtiologii* **34**, 395–399.

Emeretli, I.V. (1996). Separate effect of hydrostatic pressure and hypoxia on the activity of tissue lactate dehydrogenase in fishes with different mobility (In Russian). *Gydrobiologii Zhurnal* **32**, 68–72.

Epple, A. (1982). Functional principles of vertebrate endocrine systems. *Verhandlungen der Deutschen Zoologische Gesellschaft* **75**, 117–126.

Epstein, V.M.(1992). "Science and Education as a Developing System" (In Russian). Kharkov, 28pp.

Everson, I. (1970). The population dynamics and energy budget of *Notothenia neglecta* of Signy Island, South Orkney Islands. *Bulletin of the British Antarctic Survey* **23**, 25–50.

Evgenyeva, T.P. and Kocherezhkina, E.V. (1994). Impact of different toxicants on the histogenesis of muscle tissue of Russian sturgeon fry (In Russian). *Doklady Akademii Nauk SSSR* **336**, 555–558.

Evgenyeva, T.P., Basurmanova, O.K. and Shekhter, A.B. (1989). Degenerative changes in white muscle of Russian sturgeon (In Russian). *Doklady Akademii Nauk SSSR* **307**, 462–466.

Faktorovich, K.A. (1958). On the disturbance of fat metabolism in the liver of rainbow trout cultivated using artificial foods (In Russian). *In* "Proceedings of the Conference on Fish Physiology" (G.S. Karzhinkin and G.A. Malyukina, eds), pp.237–243.

Faktorovich, K.A. (1967). On the character of fat metabolism in the liver of some fishes of the genus *Salmo* in relation to differences in their biology (In Russian). *In* "Metabolism and Biochemistry of Fish" (G.S. Karzinkin, ed.), pp.112–117. Nauka, Moscow.

Farbridge, K.J. and Leatherland, J.F. (1987). Lunar cycles of coho salmon, *Oncorhynchus kisutch*. II. Scale amino acid uptake, nucleic acids, metabolic reserves and plasma thyroid hormones. *Journal of Experimental Biology* **129**, 179–189.

Farkas, T. (1984). Adaptation of fatty acid composition to temperature – a study on carp (*Cyprinus carpio*) liver slices. *Comparative Biochemistry and Physiology* **79B**, 531–535.

Farkas, T. and Herodek, S. (1964). The effect of environmental temperature on the fatty acid composition of crustacean plankton. *Journal of Lipid Research* **5**, 369–375.

Farkas, T. and Roy, R. (1989). Temperature mediated restructuring of phosphatidyl ethanolamines in livers of freshwater fishes. *Comparative Biochemistry and Physiology* **93B**, 217–222.

Farkas, T., Csengeri, I., Majoros, F. and Olah, J. (1980). Metabolism of fatty acids of fish. III. Combined effect of environmental temperature and diet on formation and deposition of fatty acids in the carp. *Aquaculture* **20**, 29–40.

Farmer, G.J. and Beamish, F.W.H. (1969). Oxygen consumption of *Tilapia nilotica* in relation to swimming speed and salinity. *Journal of the Fisheries Research Board of Canada* **26**, 2807–2821.

Febry, R. and Lutz, P. (1987). Energy partitioning in fish: the activity related cost of osmoregulation in a euryhaline cichlid. *Journal of Experimental Biology* **128**, 63–85.

Fedorova, G.V. (1974). The patterns of absorption, excretion and biological effect of C^{14} in freshwater fish at different stages of ontogenesis (In Russian). *Izvestiya Gosudarstvennogo Nauchno-issledovatelskogo Instituta Ozernogo i Rechnogo Rybnogo* **91**, 1–188.

Fedoseeva, O.N. and Ovcharkina, M.G. (1993). The analysis of some methods of calculation of the diel ration for fish larvae (In Russian). *Gydrobiologicheskii Zhurnal* **29**(2), 25–30.

Ferguson, C.F. and Raymont, J.K.B. (1974). Biochemical studies on marine zooplankton. XII. Further investigations on *Euphausia superba*. *Journal of the Marine Biological Association UK* **54**, 719–725.

Ferron, A. and Legget, W.C. (1994). An appraisal of condition measures for marine fish larvae. *Advances in Marine Biology* **30**, 217–304.

Fessler, J.L. and Wagner, H.H. (1969). Some morphological and biochemical changes in steelhead trout during the parr-smolt transformation. *Journal of the Fisheries Research Board of Canada* **26**, 2823–2841.

Filatov, V.N. and Shvydky, G.V. (1988). Seasonal dynamics of the physiological condition and size structure of pacific saury during the fishing season near the southern Purle Islands (In Russian). *Biologiya Morya* 1988 (5), 61–64.

Fischer, R.U., Standora, E.A. and Spotila, J.R. (1987) Predator-induced changes in thermoregulation of bluegill, *Lepomis marcochirus*, from a thermally-altered reservoir. *Canadian Journal of Fisheries and Aquatic Science* **44**, 1629–1634.

Fischer, Z. (1970). The elements of energy balance in grass carp. *Polskie Archiwum Hydrobiologii* **17**, 421–434.

Fisher, C.R. (1996). Ecophysiology of primary production at deep-sea vents and seeps. *Biosystematics and Ecology Series* **11**, 313–336.

Fisher, S. and Weber, P.C. (1984). Prostaglandin 1–3 is formed in vivo in man's dietary eicosapentaenoic acid. *Nature, Lond.* **307**, 165–168.

Florkin, M. and Schoffeniels, E. (1965). Euryhalinity and the concept of physiological radiation. *In* "Studies in Comparative Biochemistry" (K.A. Munday, ed.), pp. 6–40. Pergamon Press, Oxford.

Flowerdew, M.W. and Grove, D.J. (1980). An energy budget for juvenile thick-lipped mullet. *Journal of Fish Biology* **17**, 395–410.

Fomovsky, M.A. (1981). Influence of the temperature factor of the aquatic environment on thermal metabolism in fish (In Russian). *In* "Proceedings of fifth All-Union Limnological Conference, Irkutsk" (G.I. Galazy, ed.), pp.84–85.

Fontaine, M. (1948). On the role played by internal factors in certain migrations of fish. A critical study of different methods of investigation. *Journal du Conseil Permanent International pour l'Exploration de la Mer* **15**, 284–294.

Fontaine, M. (1969). Control by endocrines of reproduction in teleost fish. *Verhandlungen der Internationalen Vereinigung für Theoretische und Angewandte Limnologie* **17**, 611–624.

Forrester, C.R. and Alderdice, D.F. (1966). Effects of salinity and temperature on embryonic development of the Pacific cod (*Gadus macrocephalus*). *Journal of the Fisheries Research Board of Canada* **23**, 319–340.

Foskett, J.K., Bern, H.A., Machen, T.E. and Conner, M. (1983). Chloride cells and the hormonal control of teleost fish osmoregulation. *Journal of Experimental Biology* **106**, 244–281.

Fouda, M.M. and Miller, P.J. (1979). Alkaline phosphatase activity in the skin of common goby in relation to cycles in scale and body protein. *Journal of Fish Biology* **15**, 263–273.

Fox, C.F. (1972). The structure of cell membranes. *Scientific American* **226**, 30–38.

Franklin, C.E., Davison, W. and Carey, P.W. (1991). The stress response of an antarctic teleost to an acute increase in temperature. *Journal of Thermal Biology* **16**, 173–177.

Franklin, C.E., Forster, M.E. and Davison, W. (1992). Plasma cortisol and osmoregulatory changes in sockeye salmon transferred to sea water. Comparison between successful and unsuccessful adaptation. *Journal of Fish Biology* **41**, 113–122.

Freed, G. (1965). Changes in activity of cytochrome oxidase during adaptation of goldfish to different temperatures. *Comparative Biochemistry and Physiology* **14**, 651–665.

Frolkis, V.V. (1975). "Ageing and Biological Potentialities of Organisms" (In Russian). Nauka, Moscow, 272pp.

Frolova, L.K. (1960). Some aspects of radioactive cobalt behaviour in the fish organism (In Russian). *Zhurnal Obshchey Biologii* 21, 301–305.

Fry, F.E.J. (1957). The aquatic respiration of fish. *In* "The physiology of fishes", Vol. 4 (M.E. Brown, ed.), pp. 1–63. Academic Press, New York.

Fry, F.E.J. (1958). Temperature compensation. *Annual Review of Physiology* 20, 207–224.

Fry, F.E.J. (1971). The effect of environmental factors on the physiology of fish. *In* "Fish Physiology" (W.S. Hoar and D.J. Randall, eds), Vol. 6, pp.1–99. Academic Press, London and New York.

Furspan, P., Prange, H.D. and Greenwald, L. (1984). Energetics and osmoregulation in the catfish. *Comparative Biochemistry and Physiology* 77A, 773–778.

Gabos, M., Pora, E.A. and Race, L. (1973). Effect of thyroxine (T_4), TSH and thiouracil TU (thiourea) treatment on the oxygen consumption of the carp. *Studii si Cercetâri de Biologie, Seria Zoologie* 25, 39–43.

Gal'chenko, V.F., Galkin, S.V., Lein, A.Y., Moskalev, L.I. and Ivanov, M.V. (1988). Role of bacterial symbionts in nutrition of invertebrates from areas of active underwater hydrothermal systems. *Oceanology* (English transl. of *Oceanologiya*) 28, 786–794.

Galkovskaya, G.A. and Sushchenya, L.M. (1978). "Growth of Aquatic Animals at Variable Temperature" (In Russian). Nauka i Tekhnika, Minsk, 141 pp.

Galvin, P., Sadusky, D., McGregor, D. and Cross, T. (1995). Population genetics of Atlantic cod using amplified single locus minisatellite VNTR analysis. *Journal of Fish Biology* 47(suppl. A), 200–208.

Gandzyura, V.P. (1985). Phosphorus content in fish inhabiting artificial reservoirs on the Dnieper River (In Russian). *Gydrobiologicheskii Zhurnal* 21, 84–87.

Gas-Baby, N., Laffont, J. and Labat, R. (1967). Physiological and histological proof of the regeneration of the cardiac branch of the vagus nerve in the carp. *Journal de Physiologie, Paris* 59, 39–42.

Gatz, A.J. (1973). Speed, stamina and muscles in fishes. *Journal of the Fisheries Research Board of Canada* 30, 325–328.

Gebruk, A.V., Galkin, S.V., Vereschaka, A.L., Moskalev, L.I. and Southward, A.J. (1997). Ecology and biogeography of the hydrothermal vent fauna of the Mid-Atlantic Ridge. *Advances in Marine Biology* 32, 93–144.

Gee, J.H. and Holst, H.M. (1992). Buoyancy regulation by the sticklebacks *Culaea inconstans* and *Pungitius pungitius* in response to different salinities and water densities. *Canadian Journal of Zoology* 70, 1590–1594.

Gemelli, L., Martino, G. and Tota, B. (1980). Oxidation of lactate in the compact and spongy myocardium of tuna fish. *Comparative Biochemistry and Physiology* 65B, 321–326.

Georgiev, G.S. (1971). Carotenoids and vitamin A content in *Salmo irideus* eggs and their significance in the initial periods of the embryogenesis. *Folia Balcanica* 2(9), 11pp.

Gerald, V.M. (1976). The effect of starvation on energy turnover and protein metabolism in *Ophiocephalus punctatus*. *Hydrobiologia* 49, 131–201.

Geraskin, P.P. and Lukyanenko, V.J. (1972). The species specificity of fractional composition of blood haemoglobin in the Acipenseridae (In Russian). *Zhurnal Obshchey Biologii* 33, 478–483.

Gerbilsky, N.L. (1941). The method of hypophysial injections and its significance for fish culture (In Russian). *In* "The Method of Hypophysial Injections and its Significance for Reproduction of Fish Stocks" (N.L. Gerbilsky, ed.), pp.5–36. Leningrad University Press, Leningrad.

Gerbilsky, N.L. (1956). Special features and the tasks of ecological histophysiology regarded as a trend of histological research (In Russian). *Arkhiv Anatomii, Gistologii i Embriologii* **2**, 114–122.

Gerbilsky, N.L. (1958). The problem of migration impulse in relation to the analysis of intraspecific biological groups (In Russian). "Proceedings of Conference on Fish Physiology", pp. 142–152. Akademii Nauk SSSR, Moscow.

Gerking, S.D. (1952).The protein metabolism of sunfishes of different ages. *Physiological Zoology* **25**, 358–372.

Gerking, S.D. (1966). Annual growth cycle, growth potential and growth compensation in the bluegill sunfish in northern Indiana lakes. *Journal of the Fisheries Research Board of Canada* **23**, 1923–1956.

Gerking, S.D. (1972). Revised food consumption estimate of bluegill sunfish population in Wyland Lake, Indiana, USA. *Journal of Fish Biology* **4**, 301–308.

Gershanovich, A.D., Markevich, N.B. and Dergaleva, Zh.T. (1984). On the employment of condition factor in ichthyological studies (In Russian). *Voprosy Ikhtiologii* **24**, 740–752.

Gershanovich, A.D., Pegasov, V.A. and Shatunovsky, M.J. (1987). "Ecology and Physiology of the Fry of the Acipenseridae" (In Russian). Agropromizdat, Moscow, 215 pp.

Gershanovich, A.D., Lapin, J.I. and Shatunovsky, M.I. (1991). Characteristics of lipid metabolism in fish (In Russian). *Uspekhi Sovremennoy Biologii* **111**, 207–209.

Ginatulina, L.K., Shedko, S.V., Miroshnichenko, J.L. and Ginatulin, A.A. (1988). Divergence of mitochondrial DNA sequences in pacific salmon (in Russian). *Zhurnal Evolutsionnoy Biokhimii i Physiologii* **24**, 477-483.

Ginetsinsky, A.G. (1963). "Physiological Mechanisms of Water and Salt Equilibrium" (In Russian). Akademia Nauk SSSR, Moscow, Leningrad, 427 pp.

Ginetsinsky, A.G. and Lebedinsky, A.V. (1956). "Textbook of Normal Physiology" (In Russian). Medgiz, Moscow, 536 pp.

Girard, J.-P. (1976). Salt excretion by the perfused head of trout adapted to sea water and its inhibition by adrenaline. *Journal of Comparative Physiology* **111**, 77–91.

Gleebe, B.D. and Leggett, W.C. (1981). Temporal intrapopulation differences in energy allocation and use by American shad during spawning migration. *Canadian Journal of Fisheries and Aquatic Science* **38**, 795–805.

Glushankova, M.A. (1967). Ambient temperature and thermoresistance of actomyosin, alkaline phosphatase and adenylate kinase of poikilothermal animals (In Russian). *In* "Variability of Cell Resistance in Animals at Auto- and Phylogenesis" (B.P. Ushakov, ed.), pp. 126–141. Nauka, Leningrad.

Goldstein, L. and Forster, R.P. (1970). Nitrogen metabolism in fishes. *In* "Comparative Biochemistry of Nitrogen Metabolism" (J. Campbell, ed.). Vol. 2, pp. 496–515. Academic Press, New York.

Golovachev, S.A. (1985). Setting of the norms for the contents of polyunsaturated fatty acids in artificial foods (In Russian). *In* "Biology of Water Bodies of Western Ural" (Kudersky, ed.), pp. 158–167. Perm University Press.

Golovko (Kulikova), N.I. (1964). Electrophoretic study of blood serum proteins in Black Sea large and small horse-mackerel (In Russian). *Trudy Azovsko-chernomorskogo Nauchno-issledovatelskogo Instituta Morskogo Rybnogo Khozyaistva I Okeanografii* **22**, 73–94.

Goncharova, A.V. (1978). On the method of determination of alkaline phosphatase activity of fish scale (In Russian). *Informatzione Byulletin Biologiya Vnutrennich Vod* 1978 (4), 71–74.

Gordon, M.S. (1968). Oxygen consumption of red and white muscles from tuna fishes. *Science, New York* **159**, 87–90.

Goromosova, S.A. and Shapiro, A.Z. (1984). "Basic Features of the Biochemistry of Energy Metabolism in Mussels" (In Russian). Legkaya i pishchevaye promyshlennost, Moscow, 118 pp.

Gosh, R.I. (1985). "Energy Metabolism in Reproductive Cells and Embryos of Fish (In Russian). Naukova Dumka, Kiev, 148 pp.

Gosh, R.I. (1989). Energy metabolism of fish spermatozoa (In Russian). *Gydrobiologicheskii Zhurnal* **25**, 61–71.

Gowri, C. and Joseph, K.T (1968). Characterisation of acid-soluble collagen from a 'live fish' or air-breathing fish of India group. *Leather Science* **15**, 300–305.

Graf, I.A. (1982). The ultrastructure of skeletal muscle in Black Sea fishes of different natural mobility (In Russian). In "Biology of Shelf Zones of the World" Vol. 2, pp.156–157. Ocean, Vladivostock.

Graham, J.B., Koehrn, F.J. and Dickson, K.A. (1983). Distribution of relative proportions of red muscle in scombrid fishes: consequences of body size and relationships to locomotion and endothermy. *Canadian Journal of Zoology* **61**, 2087–2096.

Graham, M. (1956). "Sea Fisheries: Their Investigation in the United Kingdom". Edward Arnold, London.

Grassle, J.F. (1986). The ecology of deep-sea hydrothermal vent communities. *Advances in Marine Biology* **23**, 301–362.

Graves, J.E. and Somero, G.N. (1982). Electrophoretic and functional enzymic evolution in four species of eastern Pacific barracudas from different thermal environments. *Evolution* **36**, 97–106.

Greene, C.W. (1919). Biochemical changes in the muscle tissue of king salmon during the fast of the spawning migration. *Journal of Biological Chemistry* **39**, 435–456.

Greer-Walker, M. (1970). Growth and development of skeletal muscle fibres of the cod. *Journal du Conseil International pour l'Exploration de la Mer* **33**, 228–244.

Greer-Walker, M. and Pull, G.A. (1973). A survey of red and white muscle in marine fish. *Journal of Fish Biology* **7**, 295–300.

Greze, V.N. (1979). Bioproductive system of the Black Sea and its functional characteristics (In Russian). *Gydrobiologicheskii Zhurnal* **15**, 3–9.

Greze,V.N., Fedorina, A.I. and Chmyr, V.D. (1973). "The Production of Basic Components of Zooplanktonic Nutriment for Plankton-eating Fishes in the Black Sea" (In Russian). Naukova Dumka, Kiev, 245 pp.

Grubinko, V.V. (1991). The role of glutamine in maintaining nitrogen homeostasis in fish (In Russian). *Gydrobiologicheskii Zhurnal* **27**, 46–56.

Grubinko, V.V. and Arsan, O.M. (1995). The role of glucose alanine cycle in the adaptation to ammonia in fish (In Russian). *Doklady Natsionalnoy Akademii Nauk Ukrainy* 1995 (1), 107–109.

Gubin, I.E., Shkorbatov, G.L. and Vasenko, L.G. (1972). Oxidative activity of carp, brain homogenates under different temperature influences (In Russian). *In* "Energy Aspects of the Growth and Metabolism of Aquatic Animals" (G.E. Shulman, ed.), pp.36–40. Karelian Branch of the Academy of Science of USSR, Petrogovodsk.

Guppy, M., Hulbert, W.C. and Hochachka, P.W. (1979). Metabolic sources of heat and power in tuna muscle. II. Enzyme and metabolite profiles. *Journal of Experimental Biology* **82**, 303–320.

Guryanova, S.D. (1980). Lipid composition of some tissues of turbot infested by helminths (In Russian). *In* "Biochemistry of Freshwater Fish of Karelia" (V.S. Sidorov, ed.), pp. 36–40. Karelian Branch of the Academy of Sciences of the USSR, Petrozavodsk.

Gusar, A.G. and Getmantsev, V.A. (1985). "Black Sea Sprat (Distribution, Behaviour, Biological Grounds for Light Fishing)" (In Russian). All-union Institute of Scientific and Technical Information, Moscow, 229 pp.

Gusar, A.G., Shchepkin, V.Ya., Shulman, G.E. and Getmantsev, V.A. (1987). Functional characteristics allowing attraction of Black Sea sprats by light (In Russian). *Doklady Akademii Nauk SSSR* **294**, 1267–1270.

Gustavson, K.H. (1953). Hydrothermal stability and intermolecular organisation of collagens from mammalian and teleost skins. *Svensk Kemisk Tidsskrift* **65**, 70–76.

Hagar, A.F. and Hazel, J.R. (1985). Changes in desaturase activity and the fatty acid composition of microsomal membranes from liver tissue of thermally-acclimating rainbow trout. *Journal of Comparative Physiology* **156B**, 35–42.

Haines, T.A. (1973). An evaluation of RNA–DNA ratio as a measure of long-term growth in fish populations. *Journal of the Fisheries Research Board of Canada* **30**, 195–199.

Haines, T.A. (1980). Seasonal patterns of muscle RNA–DNA ratio and growth in black crappie. *Environmental Biology of Fishes* **5**, 67–70.

Hakanson, J.L. (1989). Analysis of lipid component for determining the condition of anchovy larvae *Engraulis mordax*. *Marine Biology* **102**, 143–151.

Hamada, A. and Maeda, W. (1983). Oxygen uptake due to specific dynamic action of the carp. *Japanese Journal of Limnology* **44**, 225–239.

Hamoir, G., Focant, B. and Distèche, M. (1972). Proteinic criteria of differentiation of white, cardiac and various red muscles in carp. *Comparative Biochemistry and Physiology* **41 B**, 665–674.

Hansen, H.J.M. (1987). Comparative studies on lipid metabolism in various salt-transporting organs of the European eel (*Anguilla anguilla*). Mono-unsaturated phosphatidyl ethanolamine as a key substance. *Comparative Biochemistry and Physiology* **88B**, 323–332.

Hansen, H.J.M., Olsen, A.G. and Rosenkilde, P. (1995). Formation of phosphatidyl ethanolamine as a putative regulator of salt transport in the gills and oesophagus of the rainbow trout (*Oncorhynchus mykiss*). *Comparative Biochemistry and Physiology* **112B**, 161–167.

Hartley, S.E. (1995). Mitochondrial DNA analysis distinguishes between British population of the whitefish. *Journal of Fish Biology* **47**(suppl. A), 145–155.

Haschemeyer, A.E. (1969). Oxygen consumption of temperature-acclimated toad-fish (*Opsanus tau*). *Biological Bulletin, Marine Biological Laboratory, Woods Hole* **136**, 28–33.

Haschemeyer, A.E.V. (1980). Temperature effects on protein metabolism in cold-adapted fishes. *Antarctic Journal of the United States* **15**, 147–149.

Haschemeyer, A.E.V. and Smith, M.A.K. (1979). Protein synthesis in liver, muscle and gill of mullet (*Mugil cephalus*) in vivo. *Biological Bulletin, Marine Biological Laboratory, Woods Hole* **156**, 93–102.

Haschemeyer, A.E.V., Persell, R. and Smith, M.A.K. (1979). Effect of temperature on protein synthesis in fish of the Galapagos and Perlas Islands. *Comparative Biochemistry and Physiology* **64B**, 91–95.

Hatanaka, M., Kosaka, M. and Sato, Y (1956). Growth and food consumption in plaice. *Tohoku Journal of Agricultural Research* **7**, 151–162.

Hayashi, K. and Takagi, T. (1978). Seasonal variations in lipids and fatty acids of Japanese anchovy. *Bulletin of the Faculty of Fisheries, Hokkaido University* **28**, 38–47.

Hayes, F.R., Wilmot, I.R. and Livingstone, D.A. (1951). The oxygen consumption of the salmon egg in relation to development and activity. *Journal of Experimental Zoology* **116**, 377–395.

Hayes, L.W., Tinsley, I.I. and Lowry, R.R. (1973). Utilisation of fatty acids by the developing steelhead sac-fry *Salmo gairdneri*. *Comparative Biochemistry and Physiology* **45B**, 695–707.

Hazel, J.R. (1972). The effect of temperature acclimation upon succinic dehydrogenase activity from the muscle of common goldfish: lipid reactivation of the soluble enzyme. *Comparative Biochemistry and Physiology* **43B**, 863–882.

Hazel, J.R. (1979). Influence of thermal acclimation of membrane lipid composition of rainbow trout. *American Journal of Physiology* **236**, 91–101.

Hazel, J.R. and Livermore, R.C. (1990). Fatty-acyl coenzyme A pool in liver of rainbow trout, *Salmo gairdneri*. Effects of temperature acclimation. *Journal of Experimental Zoology* **256**, 31–37.

Hazel, J.R. and Prosser, C.L. (1970). Interpretation of inverse acclimation to temperature. *Zeitschrift für Vergleichende Physiologie* **67**, 217–228.

Hazel, J.R. and Prosser, C.L. (1974). Molecular mechanisms of temperature compensation in poikilotherms. *Physiological Reviews* **54**, 620–677.

Healy, M.C. (1972). Bioenergetics of a sand goby population. *Journal of the Fisheries Research Board of Canada* **29**, 187–194.

Hebb, C., Morris, D. and Smith, M.W. (1969). Choline acetyl transferase activity in the brain of goldfish acclimated to different temperatures. *Comparative Biochemistry and Physiology* **28**, 29–36.

Helly, J.J. (1976). The effect of temperature and thermal distribution on glycolysis in two rockfish species (*Sebastes*). *Marine Biology* **37**, 89–95.

Hemmingsen, E.A. and Douglas, E.L. (1970). Respiratory characteristics of the haemoglobin-free fish *Chaenocephalus aceratus*. *Comparative Biochemistry and Physiology* **33**, 733–744.

Hempel, G. and Blaxter, J.H.S. (1961). The experimental modification of meristic characters in herring (*Clupea harengus* L.). *Journal du Conseil Permanent International pour l'Exploration de la Mer* **26**, 336–346.

Henderson, I.W. and Chester Jones, I. (1974). Actions of hormones on osmoregulatory systems of fish. *Fortschritte der Zoologie*, 391–418.

Henderson, R.J. and Almater, S.M. (1989). Seasonal changes in the lipid composition of herring (*Clupea harengus*) in relation to gonad maturation. *Journal of the Marine Biological Association UK* **69**, 323–384.

Henderson, R.J. and Tocher, D.R. (1987). The lipid composition and biochemistry of freshwater fish. *Progress in Lipid Research* **26**, 281–347.

Henderson, R.J., Sargent, J.R. and Hopkins, C.C. (1984). Changes in the content and fatty acid composition of lipid in an isolated population of the capelin during sexual maturation and spawning. *Marine Biology* **78**, 255–263.

Henderson, R.J., Bell, M.V. and Sargent, J.R. (1985). The conversion of polyunsaturated fatty acids to prostaglandins by tissue homogenates of the turbot, *Scophthalmus maximus*. *Journal of Experimental Marine Biology and Ecology* **85**, 93–99.

Hennessy, J.P. and Siebenaller, J.F. (1987). Pressure-adaptive differences in proteolytic inactivation of M4-lactate dehydrogenase analogues from marine fishes. *Journal of Experimental Zoology* **241**, 9–15.

Higashi, H., Kaneko, T., Ishii, S., Mashida, I. and Shugichashi, T. (1964). The influence of food fats on fish. *Bulletin of the Japanese Society of Scientific Fisheries* **30**, 778–785.

Hilditch, T.P. and Williams, P.N. (1964). "The Chemical Constitution of Natural Fats". Chapman and Hall, London, 745 pp.

Hinterleitner, S., Platzer, U. and Wieser, W. (1987). Development of the activities of oxidative, glycolytic and muscle enzymes during early larval life in three families of freshwater fish. *Journal of Fish Biology* **30**, 315–326.

Hirao, S., Yamada, J. and Kikuchi, R. (1955). Relation between chemical constituents of rainbow trout eggs and the hatching rate. *Bulletin of the Japanese Society of Scientific Fisheries* **21**, 240–243.

Hirshfield, M.F. (1980). An experimental analysis of reproductive effort and cost in the Japanese medaka. *Ecology* **61**, 282–292.

Hoar, W.S. (1953). Factors which control and regulate the time of migration in fish. *Biological Reviews of the Cambridge Philosophical Society* **28**, 246–286.

Hoar, W.S. and Cottle, M.K. (1952). Some effects of temperature acclimatization on the chemical constitution of goldfish tissues. *Canadian Journal of Zoology* **30**, 49–54.

Hochachka, P.W. (1962). Glycogen stores in trout tissues before and after stream planting. *Journal of the Fisheries Research Board of Canada* **19**, 127–136.

Hochachka, P.W. (1969). Intermediate metabolism in fishes. *In* "Fish Physiology" (W.S. Hoar and D.J. Randall, eds), Vol. 1, pp. 351–383. Academic Press, London and New York.

Hochachka, P.W. (1975). Why study proteins of abyssal organisms? *Comparative Biochemistry and Physiology* **52B**, 1–2.

Hochachka, P.W., Fields, J. and Mustafa, T. (1973). Animal life without oxygen: basic biochemical mechanisms. *American Zoologist* **13**, 543–555.

Hochachka, P.W. and Hayes, F.R. (1962). The effect of temperature acclimation on pathways of glucose metabolism in the trout. *Canadian Journal of Zoology* **20**, 261–270.

Hochachka, P.W. and Mustafa, T. (1972). Invertebrate facultative anaerobiosis. *Science, New York* **178**, 1056–1060.

Hochachka, P.W. and Somero, G.N. (1973). "Strategies of Biochemical Adaptation". W.B. Saunders Company, Philadelphia, 398pp.

Hochachka, P.W. and Somero, G.N. (1977). Biochemical adaptation to the environment. *In* "Fish Physiology", (W.S. Hoar and D.J. Randall, eds), Vol. 6 pp. 99–156, Academic Press, London and New York.

Hochachka, P.W. and Somero, G.N. (1984). "Biochemical Adaptation". Princeton University Press, Princeton, New Jersey.

Hofer, R., Forstner, H. and Rettenwander, R. (1982). Duration of gut passage and its dependence on temperature and food consumption in roach. *Rutilus rutilus* L.: laboratory and field experiments. *Journal of Fish Biology* **20**, 289–299.

Holdway, D.A. and Beamish, F.W.H. (1985). The effect of growth rate, size and season on oocyte development and maturity of Atlantic cod (*Gadus morhua*). *Journal of Experimental Marine Biology and Ecology* **85**, 3–19.

Holeton, G.F. (1974). Metabolic cold adaptation of polar fish: fact or artefact? *Physiological Zoology* **47**, 137–152.

Holeton, G.F. (1980). Oxygen as an environmental factor of fish. In "Environmental Physiology of Fishes" (M.A. Ali, ed.), pp. 7–32, NATO Advanced Study Institute Series, Plenum Press, New York.

Holub, B.J., Piekarski, J. and Leatherland, J.F. (1977). Differential biosynthesis of molecular species of 1,2-diacyl-glycerols and phosphatidyl cholines in cold and warm acclimated goldfish. *Lipids* **12**, 316–318.

Hooper, S.N., Paradis, M. and Ackman, R.G. (1973). Distribution of trans-6-hexadecenoic acid, 7-methyl-7-hexadecenoic acid and common fatty acids in lipids of the ocean sunfish. *Lipids* **8**, 509–516.

Houde, E.D. and Schekter, R.C. (1983). Oxygen uptake and comparative energetics among eggs and larvae of three subtropical marine fishes. *Marine Biology* **72**, 283–293.

Hudson, R.C.L. (1973). On the function of the white muscle in teleosts at intermediate swimming speeds. *Journal of Experimental Biology* **58**, 509–522.

Hughes, G.M. and Johnston, I.A. (1978). Some responses of the electric ray (*Torpedo marmorata*) to low ambient oxygen tensions. *Journal of Experimental Biology* **73**, 107–117.

Hulbert, W.C., Guppy, M., Murphy, B. and Hochachka, P.W. (1979). Metabolic sources of heat and power in tuna muscle. 1. Muscle fine structure. *Journal of Experimental Biology* **82**, 289–301.

Hulet, W.H., Masel, S.J., Jodrey, L.H. and Wehr, R.G. (1967). The role of calcium in the survival of marine teleosts in dilute sea water. *Bulletin of Marine Science of the Gulf and Caribbean* **17**, 677–688.

Idler, D.R. and Bitners, I. (1958). Biochemical studies on sockeye salmon during spawning migration. *Canadian Journal of Biochemistry and Physiology* **36**, 793–798.

Idler, D.R. and Clemens, W.A. (1953). The energy expenditures of Fraser River sockeye salmon during the spawning migration to Chilko and Stuart Lakes. International Pacific Salmon Fishery Commission, Progress Report, 1–80.

Idler, D.R. and Freeman, H.C. (1965). A demonstration of an impaired hormone metabolism in moribund Atlantic cod (*Gadus morhua* L). *Canadian Journal of Biochemistry and Physiology* **43**, 620–623.

Idler, D.R. and Truscott, B. (1972). Corticosteroids in fish. *In* "Steroids in Non-mammalian Vertebrates" (D.R. Idler, ed.), pp. 127–253. Academic Press, London, New York.

Ikeda, T. (1974). Nutritional ecology of marine zooplankton. *Memoirs of the Faculty of Fisheries. Hokkaido University* **22**, 1–97.

Ilyin, V.S. and Titova, G.V. (1963). In vitro effect of insulin on the activity of liver glucose-6-phosphate dehydrogenase and glucose-6-phosphatase (In Russian). *Biochimika* **28**, 987–991.

Ip, Y.K., Lee, C.G.L., Low, W.P. and Lam, T.J. (1991). Osmoregulation in the mudskipper, *Boleophthalmus boddaerti*. 1. Responses of branchial cation activated and anion stimulated adenosine triphosphatases to changes in salinity. *Fish Physiology and Biochemistry* **9**, 63–68.

Ipatov, V.V. (1970). Dynamics of blood serum proteins in Baltic cod in relation to reproductive products, maturation and season (In Russian). *Voprosy Ikhtiologii* **10**, 892–896.

Ipatov, V.V. (1976). The dynamics of nucleic acids in Baltic cod during the reproductive period (In Russian). *In* "Proceedings of the third All-Union Conference on Fish Ecological Physiology" (V.D. Romanenko, ed.), p. 128. Naukova Dumka, Kiev.

Ipatov, V.V. and Kondratyeva, N.M. (1986). Bioenergy aspects of ecology of Baltic sprat. *Fischerei-Forschung, Rostock* **24**, 52–54.

Ipatov, V.V. and Lukyanenko, V.I. (1979). Fish serum proteins: heterogeneity, structure and functions (In Russian). *Uspekhi Sovremennoy Biologii* **88**,108–124.

Irving, D.O. and Watson, K. (1976). Mitochondrial enzymes of tropical fish: a comparison with fish from cold waters. *Comparative Biochemistry and Physiology* **54B**, 81–92.

Isuev, A.P. and Musaev, B.S. (1989). Comparative description of fatty acid composition of lipids at early stages of ontogenesis in carp, bighead, chum, Caspian salmon and Russian sturgeon (In Russian). *Voprosy Ikhtiologii* **29**, 342–345.

Itina, N.A. (1970). Structure and function of muscle fibrils in lower vertebrates (In Russian). *Uspekhi Sovremennoy Biologii* **70**, 286–304.

Ivanov, L. (1983). Population parameters and methods from literature on sprat-catching in the west part of the Black Sea (In Bulgarian). *Izvestiya Instituta po Ribno Resursi, Varna* **20**, 7–46.

Ivanov, S.L. (1929). Climatic variability of plant chemistry (In Russian). *In* "Technical Encyclopedia," Vol.3, pp.84–85, Moscow.

Ivanova, M.N. (1980). On the life duration of *Osmerus eperlanus* of White Lake (In Russian). *Voprosy Ikhtiologii* **20**, 481–489.

Ivlev, V.S. (1938). On the transformation of energy during the growth of invertebrates (In Russian). *Buylleten Moskovskogo Obshchestva Ispytatelei Prirody, Moskva, Otdel Biologiya* **47**, 267–277.

Ivlev, V.S. (1939). Energy balance in carp (In Russian). *Zoologicheskii Zhurnal* **18**, 315–326.

Ivlev, V.S. (1955). "Experimental Ecology of Fish Nutrition" (In Russian). Pishchepromizdat, Moscow, 252 pp. [English translation (1961). "Experimental Ecology of the Feeding of Fishes." Yale University Press, Newhaven, Connecticut.]

Ivlev, V.S. (1959). Experience of assessment of the evolutionary significance of energy metabolic levels (In Russian). *Zhurnal Obshchey Biologii* **20**, 94–103.

Ivlev, V.S. (1960). On the utilisation of food by plankton-eating fishes (In Russian). *Voprosy Ikhtiologii* **2**, 158–168.

Ivlev, V.S. (1962). Active energy metabolism in young Baltic salmon (*Salmo salar* L.). *Voprosy Ikhtiologii* **2**, 158–168.

Ivlev, V.S. (1963). The study of animal distribution in graded environmental conditions as a method of ecologo-physiological analysis (In Russian). *Trudy Sevastopolskoi Biologicheskoi Stanntsii* **16**, 277–282.

Ivlev, V.S. (1964). Metabolic rates and locomotion velocities in some Black Sea larvae (In Russian). *Voprosy Ikhtiologii* **4**, 118–124.

Ivlev, V.S. (1966). Elements of physiological hydrobiology (In Russian). *In* "Physiology of Marine Animals" (V.S. Ivlev, ed.), pp. 3–45. Nauka, Moscow.

Ivleva, E.V. (1989a). Seasonal dynamics of the functional activity of the thyroid gland in Black Sea fishes (In Russian). *Zhurnal Evolutsionnoy Biochimii i Physiologii* **25**, 467–473.

Ivleva, E.V. (1989b). Morphology of the thyroid gland in five species of Black Sea fish (In Russian). *Zhurnal Evolutsionnoy Biochimii i Physiologii* **25**, 644–647.

Ivleva, I.V. (1981). "Environmental Temperature and the Rate of Energy Metabolism in Aquatic Animals" (In Russian). Naukova Dumka, Kiev, 231 pp.

Jamieson, A. (1974). Genetic 'tags' for marine fish stocks. In "Sea Fisheries Research" (F.R. Harden Jones, ed.), pp.91–99. Elek Science, London.

Jamieson, A. and Birley, A.J. (1989). The distribution of transferrin alleles in haddock stocks. *Journal du Conseil, The ICES Journal of Marine Science* **45**, 248–262.

Jangaard, P.M., Ackman, R.G. and Sipos, J.C. (1967). Seasonal changes in fatty acid composition of cod liver, flesh, roe and milt lipids. *Journal of the Fisheries Research Board of Canada* **24**, 613–627.

Jankowski, H.D. and Korn, H. (1965). The influence of the adaptation temperature on the mitochondria content of fish muscle. *Naturwissenschaften* **52**, 642–643.

Jannasch, H.W. (1995). Microbial interactions with hydrothermal fluids. *In* "Seafloor Hydrothermal Systems" (S.E. Humphris, R.A. Zierenberg, L.S. Mullineaux and R.E. Thompson, eds), pp. 273–296. American Geophysical Union, Washington, DC.

Jannasch, H.W. (1997). Two new hyperthermophilic Archaea from mid-Atlantic hydrothermal vents. *RIDGE Events* **8**(2), 22–23

Jedryczkowski, W. and Fischer, Z. (1973). Preliminary report on the metabolism of the silver eel. *Polskie Archiwum Hydrobiologii* **20**, 507–516.

Jeffries, H.P. (1972). Fatty acid ecology of a tidal marsh. *Limnology and Oceanography* **17**, 433–440.

Job, V. and Gerald, V.M. (1976). Food uptake and energy turnover during forced activity in the fish *Channa punctatus*. *Monitore Zoologico Italiano* **10**, 239–252.

Jobling, M. (1980). Effects of starvation on proximate chemical composition and energy utilisation of plaice. *Journal of Fish Biology* **17**, 325–334.

Jobling, M. (1981). The influence of feeding on the metabolic rate of fishes: a short review. *Journal of Fish Biology* **18**, 385–400.

Jobling, M. (1983). A short review and critique of methodology used in fish growth and nutrition studies. *Journal of Fish Biology* **23**, 685–703.

Jobling, M. and Davies, S.P. (1980). Effects of feeding on metabolic rate and the specific dynamic action in plaice. *Journal of Fish Biology* **16**, 629–638.

Johnson, D.W. (1973). Endocrine control of hydromineral balance in teleosts. *American Zoologist* **13**, 799–818.

Johnston, C.E., Gray, R.W., McLennan, A. and Paterson, A. (1987). Effects of photoperiod, temperature and diet on the reconditioning response, blood chemistry and gonad maturation of Atlantic salmon kelts (*Salmo salar*) held in fresh water. *Canadian Journal of Fisheries and Aquatic Science* **44**, 702–711.

Johnston, I.A. (1975). Studies on the swimming musculature of the rainbow trout. II. Muscle metabolism during severe hypoxia. *Journal of Fish Biology* **7**, 459–467.

Johnston, I.A. (1977). A comparative study of glycolysis in red and white muscles of the trout and mirror carp. *Journal of Fish Biology* **11**, 575–588.

Johnston, I.A. (1981a). Quantitative analysis of muscle breakdown during starvation in the marine flatfish *Pleuronectes platessa*. *Cell and Tissue Research* **214**, 369–386.

Johnston, I.A. (1981b). Structure and function of fish muscles. *Symposia of the Zoological Society, London* **48**, 71–113.

Johnston, I.A. (1982). Physiology of muscle in hatchery-raised fish. *Comparative Biochemistry and Physiology* **73B**, 105–124.

Johnston, I.A. (1983). Comparative studies of contractile proteins from the skeletal and cardiac muscles of lower vertebrates. *Comparative Biochemistry and Physiology* **76A**, 439–445.

Johnston, I.A. (1985). Temperature, muscle energetics and locomotion in inshore Antarctic fish. *Oceanis* **11**, 125–142.

Johnston, I.A. (1993). Phenotypic plasticity of fish muscle to temperature change. *In* "Fish Ecophysiology" (C. Rankin and S.B. Jensen, eds), pp.322–340. Chapman and Hall, London.

Johnston, I.A. and Bernard, L.M. (1982). Ultrastructure and metabolism of skeletal muscle fibres in the tench; effects of long-term acclimation to hypoxia. *Cell and Tissue Research* **227**, 179-199.

Johnston, I.A. and Bernard, L.M. (1983). Utilisation of the ethanol pathway in carp following exposure to anoxia. *Journal of Experimental Biology* **104**, 73–78.

Johnston, I.A. and Camm, J.-P. (1987). Muscle structure and differentiation in pelagic and demersal stages of the Antarctic teleost *Notothenia neglecta*. *Marine Biology* **94**, 183–190.

Johnston, I.A. and Goldspink, G. (1973). Some effects of prolonged starvation on the metabolism of the red and white myotomal muscles of the plaice. *Marine Biology* **19**, 348–353.

Johnston, I.A. and Horne, Z. (1994). Immunocytochemical investigations of muscle differentiation in the Atlantic herring (*Clupea harengus*: Teleostei). *Journal of the Marine Biological Association UK* **74**, 79–91.

Johnston, I.A. and Maitland, B. (1980). Temperature acclimation in crucian carp, morphometric analyses of muscle fibre ultrastructure. *Journal of Fish Biology* **17**, 113–125.

Johnston, I.A. and Moon, T.W. (1981). Fine structure and metabolism of multiple innervated fast muscle fibres in teleost fish. *Cell and Tissue Research* **219**, 93–109.

Johnston, I.A. and Tota, B. (1974). Myofibrillar ATPase in the various red and white muscles of the tunny (*Thunnus thynnus* L.) and the tub gurnard (*Trigla lucerna* L.). *Comparative Biochemistry and Physiology* **49B**, 367–373.

Johnston, I.A. and Walesby, N.J. (1977). Molecular mechanisms of temperature adaptation in fish myofibrillar adenosine triphosphatases. *Journal of Comparative Physiology* **119**, 195–206.

Johnston, I.A., Frearson, N. and Goldspink, G. (1973). The effects of environmental temperature on the properties of myofibrillar adenosine triphosphatase from various species of fish. *Biochemical Journal* **133**, 735–738.

Johnston, I.A., Davison, W. and Goldspink, G. (1975). Adaptations in Mg^{2+}-activated myofibrillar ATPase activity induced by temperature acclimation. *FEBS Letters* **50**, 293–295.

Johnston, I.A., Davison, W. and Goldspink, G. (1977). Energy metabolism of carp swimming muscles. *Journal of Comparative Physiology*, **114**, 203–216.

Johnston, I.A., Eddy, F.B. and Malory, M.O. (1983). The effect of temperature on muscle pH, adenylate and phosphogen concentrations in *Oreochromis alcalicus grahami*, a fish adapted to an alkaline hot-spring. *Comparative Biochemistry and Physiology* **23**, 717–724.

Johnston, I.A., Guderley, H, Franklin, C.E., Crockford, T. and Kamunde, C. (1994). Are mitochondria subject to evolutionary temperature adaptation? *Journal of Experimental Biology* **195**, 293–306.

Johnston, P.V. and Roots, B.I. (1964). Brain lipids, fatty acids and temperature acclimation. *Comparative Biochemistry and Physiology* **11** B, 303–309.

Jones, N.R. (1959). The free amino acids of fish. II. Fresh skeletal muscle from lemon sole (*Pleuronectes microcephalus*). *Journal of the Science of Food and Agriculture* **10**, 282–286.

Jones, P.L. and Sidell, B.D. (1982). Metabolic responses of striped bass to temperature acclimation. II. Alterations in metabolic carbon sources and distributions of fibre types in locomotory muscle. *Journal of Experimental Zoology* **219**, 163–171.

Jürss, K. (1982). Äthanol – ein Endprodukt des anaeroben stoffwechsels von Fischen. *Biologische Rundschau* **20**, 178 (only).

Kazansky, B.N. (1963). Raising fish progeny in different seasons to provide repeated cycles of fish breeding culture (sturgeons) (In Russian). *In* "Sturgeon Culture in Water Bodies of the USSR", pp.56–65. Academi Nauk SSSR, Moscow.

Kalabukhov, N.I. (1950). "Ecologo-physiological Features of Animals and the Environment" (In Russian). Kharkov University Press, Kharkov, 270 pp.

Kalyuznaya, T.I. (1982). Seasonal variations in physiological condition of full-grown Korfokaragin herring (In Russian). *Biologiya Morya* 1982(3), 46–51.

Kamler, E., Zuromska, H. and Nissinen, T. (1982). Bioenergetical evaluation of environmental and physiological factors determining egg quality and growth in *Coregonus albula*. *Polskie Archiwum Hydrobiologii* **29**, 71–121.

Kanazawa, A., Teshima, S. and Ono, K. (1979). Relationship between essential fatty acid requirements of aquatic animals and capacity for bioconversion of linolenic acid to highly unsaturated fatty acids. *Comparative Biochemistry and Physiology* **63B**, 295–301.

Kaplansky, S.Ya. (1945). Proteins of blood plasma and their role in the metabolic processes of the organism (In Russian). *Uspekhi Sovremennoy Biologii* **19**, 324–338.

Karamushko, L.I. (1993). Influence of nutrition on metabolic rates and specific dynamic effect in cod, wolf-fish and plaice (In Russian). *Zoologicheskii Zhurnal* **72**, 106–115.

Karamushko, L.I. and Shatunovsky, M.I. (1993). Quantitative problems of temperature effect in relation to energy metabolism rate in *Gadus morhua*, *Anarhichas lupus* and *Pleuronectes platessa* (In Russian). *Voprosy Ikhtiologii* **33**, 111–120.

Karmanova, I.G., Titkov, E.S. and Popover, D.I. (1976). Species specific daily periodicity of locomotion and rest in Black Sea fishes (In Russian). *Zhurnal Evolutsionnoy Biokhimii i Physiologii* **12**, 486–488.

Karpevich, A.F. (1975). "Theory and Practice of the Acclimation of Aquatic Organisms" (In Russian). Pishchevaya Promyshlennost, Moscow, 432 pp.

Karpevich, A.F. (1976). On the change in the reaction of digestive juices in marine fishes during digestion (In Russian). *Physiologicheskii Zhurnal SSSR* **21**, 100–123.

Karpevich, A.F. (1985a). Potential properties of aquatic organisms as a reserve for maintaining mariculture efficiency (In Russian). *In* "Biological Principles of Aquaculture in the Seas of the European part of the USSR" (V.E. Sokolov and O.A. Scarlato, eds), pp.17–33. Nauka, Moscow.

Karpevich, A.F. (1985b). Changes in the structure and bio-productivity of aquatic ecosystems (In Russian). *Voprosy Ikhtiologii* **25**, 3–15.

Karpov, A.R. and Andreeva, A.P. (1992) The character of kinetics of lactate dehydrogenase reaction in some species of the Gadidae family (In Russian). *Voprosy Ikhtiologii* **32**, 144–149.

Karzinkin, G.S. (1935). Learning more about the fish production of aquatic bodies. II. Study of nutritional physiology of mirror carp fingerlings (In Russian). *Trudy Limnologicheskoi Stantsii Kosine* **19**, 21–52.

Karzinkin, G.S. (1952). "The Principles of Biological Productivity of Water Bodies" (In Russian). Pishchepromizdat, Moscow, 372 pp.

Karzinkin, G.S. (1962) "Application of Radioactive Isotopes to Fish Culture" (In Russian). Pishchepromizdat, Moscow, 71 pp.

Kasumyan, A.O. and Taurik, L.P. (1993). Behaviour reaction of sturgeon fry to amino acids (In Russian). *Voprosy Ikhtiologii* **33**, 691–700.

Kayama, M. (1986). Fish farming and aquaculture. Can we modify fish fat with more EPA? *Journal of the Faculty of Applied Science, Hiroshima University* **25**, 19–28.

Kayama, M. and Ikeda, Y. (1975). Study of lipids, especially wax esters of micronecton fish from Sagami and Suruga gulfs. *Journal of Japanese Oil Chemists' Society*, **24**, 435–440.

Kayama, M. and Nevenzel, J. (1974). Wax ester biosynthesis by midwater marine animals. *Marine Biology* **24**, 79–285.

Kayama, M., Tsuchia, J. and Nevenzel, J.C. (1963). Incorporation of linolenic-1-C14 acid into eicosa-pentaenoic and docosahexaenoic acids in fish. *Journal of the American Oil Chemists' Society* **40**, 499–502.

Kelly, P.B., Reiser, R. and Hood, D.W. (1958). The effect of diet on the fatty acid composition of several species of freshwater fish. *Journal of the American Oil Chemists' Society* **35**, 503–505.

Kelso, J.R.M. (1973). Seasonal energy changes in walleye and their diet in West Blue Lake, Manitoba. *Transactions of the American Fisheries Society* **102**, 363–368.

Kemp, P. and Smith, M.F. (1970). Effect of temperature acclimatization on the fatty acid composition of goldfish intestinal lipids. *Biochemical Journal* **117**, 9–15.

Kerr, S.R. (1982). Estimating the energy budgets of actively predatory fishes. *Canadian Journal of Fisheries and Aquatic Science* **39**, 371–379.

Khailov, K.M. (1971). "Ecological Metabolism in the Sea" (In Russian). Naukova Dumka, Kiev. 252 pp.

Khailov, K.M., Kovardakov, S.A. and Prazukin, A.B. (1986). Substantiation of chosen parameters for estimating the condition of aquatic ecosystems for multipurpose use (In Russian). *Vodnyey Resoursy* **1986**(6), 65–74.

Khakimullin, A.A. (1988). The gas exchange rate in cultured fry of Siberian sturgeon during muscle performance (In Russian). *Voprosy Ikhtiologii* **28**, 282–288.

Khaskin, V.V. (1975). "The Energetics of Heat Production and Adaptation to Cold" (In Russian). Nauka, Novosibirsk, 199 pp.

Khaskin, V.V. (1981). Energy metabolism (In Russian). *In* "Ecological Physiology of Animals" (A.D. Slonim, ed.), Vol. 2, pp.379–407. Nauka, Leningrad.

Khlebovich, V.V. (1986). "Acclimation of Animal Organisms" (In Russian). Nauka, Leningrad, 135 pp.

Khotkevich (Yuneva), T.V. (1974). Metabolism of fish with different functional activity (In Russian). *Doklady Akademii Nauk SSSR* **219**, 505–507.

Khotkevich (Yuneva), T.V. (1975). The character of accumulation and excretion of radioactive tracer by the proteins of scorpion fish and pickerel in the pre-wintering and prespawning seasons (In Russian). *Biologiya morya* 1975 (1), 68–72.

Khristophorov, O.L. Changes in the gonads and hypophysis of Arctic cod caused by aging (In Russian). *Proceedings of the All-Union Institute of Marine Fishery and Oceanography* **115**, 160–171.

Kiceniuk, J.W. and Jones, D.R. (1977). The oxygen transport system in trout during sustained exercise. *Journal of Experimental Biology* **69**, 247–260.

Kim, E.D. (1974). Annual and age-induced changes in the cholesterol and phospholipid content of full-grown reproductive products of carp (In Russian). *In* "Different Quality of Early Ontogenesis in Fish" (V.J. Vladimirov, ed.), pp.114–127. Naukova Dumka, Kiev.

Kim, Yen, Kadura, S.N., Byatcharina, L.I., Khrapounov, S.N. and Berdyshev, G.D. (1986). Basic nuclear proteins of testes and spermatozoids in cypriniforms and perciforms. (In Russian). *Gydrobiologicheskii Zhurnal* **6**, 70–75. [English translation (1987). *Hydrobiological Journal* **22**, 70–75.]

King, J.R., Backer, S. and Farner, D.S. (1963). A comparison of energy reserves during autumnal and vernal periods in the whitecrowned sparrow. *Ecology* **44**, 513–521.

Kinne, O. and Kinne, E.M. (1962). Rates of development in embryos of a cyprinodont fish exposed to different temperature–salinity–oxygen combinations. *Canadian Journal of Zoology* **40**, 231–253.

Kinsella, J.E. (1966). General metabolism of the hexapod embryo with particular reference to lipids. *Comparative Biochemistry and Physiology* **19**, 291–304.

Kinumaki, T., Sugii, K., Iida, H. and Takahashi, T. (1972). Addition of fat-soluble vitamins to the feeding-stuffs for parent rainbow trout with particular reference to the effect on the vitamin levels of eggs and fry. *Bulletin of the Tokai Regional Research Laboratory* **71**, 133–160.

Kirpichnikov, V.S. (1958). Resistance to cold and winter tolerance of carp, sazan and their hybrids (In Russian). In "Proceedings of Conference on Fish Physiology" (G.S. Karzinkin and G.A. Malyukina, eds), pp.261–270. Akademiya Nauk SSSR, Moscow.

Kirpichnikov, V.S. (1978). "The Genetic Foundations of Fish Selection" (In Russian). Nauka, Leningrad, 250 pp.

Kirpichnikov, V.S. (1987). "Genetics and Selection of Fish" (In Russian). Nauka, Leningrad, 520 pp.

Kitchell, J.F., Stewart, D.J. and Weininger, D. (1977). Applications of a bioenergetics model to yellow perch and walleye. *Journal of the Fisheries Research Board of Canada* **34**, 1922–1935.

Kititsina, L.A. and Kurovskaya, L.A. (1991). Physiologo-biochemical changes in the organism of carp in ectoparasitic invasions (In Russian). *Gydrobiologicheskii Zhurnal* **27**, 65–71.

Kizevetter, I.V. (1942). Techno-chemical characteristics of Far Eastern commercial fishes (In Russian). *Trudy TINRO* **21**, 3–225.

Kizevetter, I.V. (1948). On the changes in the body chemical composition of sockeye salmon (In Russian). *Trudy TINRO* **28**, 28–32.

Klaro, P. and Lapin, V.I. (1971). Changes of some biochemical parameters in the organs and tissues of *Lutianus synagris* of Barabañu Bay during the maturation of the reproductive products (In Russian). *Voprosy Ikhtiologii* **11**, 877–891.

Kleckner, N.W. and Sidell, B.D. (1985). Comparison of maximal activities of enzymes from tissues of thermally acclimated and naturally acclimatized chain pickerel. *Physiological Zoology* **58**, 18–20.

Klekowski, R.Z., Opalinski, K.W. and Rakusa-Suszczewski, S. (1973). Respiration of Antarctic amphipod *Paramoera walkeri* during the winter season. *Polish Archives of Hydrobiology* **20**, 301–308.

Kleymenov, I.Ya. (1971). "Nutritive Value of Fish" (In Russian). Pishchevaya promyshlennost, 151 pp.

Klovach, N.V. (1983). Effect of temperature and season on the level of metabolism in the Azov Sea atherine (In Russian). *Biologicheskii Nauki* 1983(7), 58–63.

Klyachko, O.S. and Ozernyuk, N.D. (1991). Temperature adaptations of metabolism: the influence of temperature on kinetic parameters of lactate dehydrogenase (K_m) in fish of different species during growth (In Russian). *Doklady Academii Nauk SSSR* **319**, 1252–1255.

Klyachko, O.S., Polosukhina, E.S. and Ozernyuk, N.D. (1992). Functional distinctions of muscle lactate dehydrogenase in fishes adapted to different environmental temperature (In Russian). *Doklady Akademii Nauk SSSR* **325**, 1246–1251.

Klyashtorin, L.B. (1982). "Aquatic Respiration and Oxygen Demand of Fish" (In Russian). Legkaya i Pishchevaya promyshlennost, 168 pp.

Klyashtorin, L.B. (1996). Climate and fishery outlook in the Pacific region (In Russian). *Rybnoye Khozyaistvo* 1996(4), 37–42.

Klyashtorin, L.B. and Smirnov, B.P. (1983). On the estimation of energy expenditure on respiration by fish (In Russian). *In* "Physiological Principles of Respiration in Marine and Diadromous Fishes" (J.A. Shekanova, ed.), pp.76–81. Legkaya i pishchevaya promyshlennost, Moscow.

Klyashtorin, L.B. and Smirnov, B.P. (1990). Estimation of the readiness for marine migration in cultured sockeye salmon fry (In Russian). *Rybnoye Khozyaistvo* 1990 (2), 42–45.

Knipprath, W.G. and Mead, J.F. (1966). Influence of temperature on the pattern of muscle and organ lipids of rainbow trout. *Fish Industry Review* 1966(3), 23–27.

Knipprath, W.G. and Mead, J.F. (1968). The effect of the environmental temperature on the fatty acid composition and on the in vivo incorporation of 1-^{14}C-acetate in goldfish. *Lipids* **3**, 121–128.

Koch, F. and Weiser, W. (1983). Partitioning of energy in fish: can reduction of swimming activity compensate for the cost of production? *Journal of Experimental Biology* **107**, 141–146.

Kokshaysky, N.V. (1974). "An Essay on the Biological Aero- and Hydrodynamics (Flight and Swimming of Animals)" (In Russian). Nauka, Moscow, 254 pp.

Kondrashova, M.N. and Chagovets, N.R. (1971). Succinic acid in the skeletal muscle at high-rate muscle performance and at rest (In Russian). *Doklady Akademii Nauk SSSR* **198**, 243–246.

Kondrashova, M.N. and Mayersky, E.I. (1978). Interrelation between hormonal and mitochondrial regulation (In Russian). *In* "Regulation of Energy Metabolism and the Physiological State of the Organism" (M.N. Kondrashova, ed.), pp. 217–229. Nauka, Moscow.

Kondratyeva, T.P. (1977). Changes in total protein content and the composition of protein fractions in the blood serum of some Black Sea fish during the spawning season (In Russian). *Gydrobiologicheskii Zhurnal* **13**, 75–79.

Kondratyeva, T.P. and Astakhova, L.P. (1994). Morphological and biochemical characteristics of white and red muscle in fish differing in physiological condition (In Russian). *In* "Proceedings of the 1st Congress of Hydrological Society of Ukraine" (V.D. Romanenko, ed.), p. 237. Naukova Dumka, Kiev.

Konovalov, Yu.D. (1984). "Proteins and their Reactive Groups in Early Ontogenesis of Fish" (In Russian), Naukova Dumka, Kiev, 194 pp.

Konovalov, Yu.D. (1989). Total protein, albumins and globulins in the eggs of silver carp and big-head of different ecological groups and characteristics of the fish culture and biology (In Russian). *Gydrobiologicheskii Zhurnal* **25**, 103.

Konovets, I.N., Grubinko, V.V., Arsan, O.M. and Kulik, V.A. (1993). Functioning of the adaptive systems for detoxication of ammonia in carp under the impact of temperature (In Russian). *Gydrobiologicheskii Zhurnal* **29**, 47–52.

Konovets, I.N., Kulik, V.A., Arsan, O.M., Grubinko, V.V. and Gavriley, D.V. (1994). The impact of lead on nitrogen metabolism in carp at varying temperature of the aquatic environment (In Russian). *Gydrobiologicheskii Zhurnal* **30**, 78–86.

Konstantinov, A.S. (1993). Effect of termperature variations on the growth, energetics and physiological state of young fish (In Russian). *Izvestiya Rossiskaya Akademii Nauk, Ser. Biologiya* (1), 55–63.

Konstantinov, A.S. and Sholokhov, A.M. (1990). Influence of temperature oscillation on the growth rate, energetics and physiological condition of Russian sturgeon fry (In Russian). *Ekologya* 1990 (4), 69–75.

Konstantinov, A.S. and Yakovchuk, A.M. (1993). Species specific metabolites as a factor of limitation of fish density (In Russian). *Voprosy Ikhtiologii* **33**, 829–833.

Konstantinov, A.S., Zdanovich, V.V. and Tikhomirov, D.G. (1989). Influence of temperature oscillation on metabolic rates and energetics of fish fry (In Russian). *Voprosy Ikhtiologii* **29**, 1019–1027.

Korzhuev, P.A. (1936). Effect of high temperature on trypsin of warm- and cold-blooded animals (In Russian). *Physiologicheskii Zhurnal SSSR* **21**, 433–437.

Korzhuev, P.A. (1964). "Haemoglobin" (In Russian). Nauka, Moscow, 287 pp.

Koshelev, B.V. (1984). "Ecology of Fish Reproduction" (In Russian). Nauka, Moscow, 309 pp.

Kostetsky, E.Ya. (1985). Phospholipid content of molluscs and crustaceans (In Russian). *Biologiya Morya* 1985 (12), 52–61.

Kostetsky, E.Ya., Sanina, N.M. and Naumenko, N.V. (1992). Influence of fatty acid composition on the structure of the thermogram of phosphatidyl choline phase conversion of the sea cucumber, *Cucumaria frandatrix* (In Russian). *Zhurnal Evolutsionnoy Biokhimii i Physiologii* **28**, 426–433.

Kostylev, E.F. (1973). Changes in the biochemical composition of plankton with depth (In Russian). *Biologiya Morya* 1973 (3), 35–47.

Kotov, A.M. (1976). Seasonal dynamics of haemotological characteristics of some Black Sea fishes and changes generated by experimental poisoning with oil products (In Russian). *Gydrobiologicheskii Zhurnal* **12**, 63–68.

Kotov, A.M. (1979). Morpho-physiologo-biochemical characteristics of blood in some Black Sea fishes (In Russian). *Trudy VNIRO* **129**, 86–93.

Kotsar, N.I. (1976). Adaptive reactions of carp to varying free carbon dioxide concentrations in the water (In Russian). *Dopovi Akademii Nauk Ukrainskoi RSR, Kiev Series B* **9**, 827–830.

Kovalevskaya, L.A. (1956). The energetics of fish during motion (In Russian). *Trudy Morskogo Gidrofizicheskogo Instituta AN SSSR* **7**, 161–165.

Kozlov, A.N. (1972). Some features of fat metabolism of marbled *Notothenia* in the prespawning period (In Russian). *Trudy VNIRO* **85**, 117–128.

Kozlov, A.N. (1975). Some features of energy balance in three species of *Notothenia* (In Russian). *Trudy VNIRO* **96**, 92–100.

Kramer, D.L. and McClure, M. (1981). The transit cost of aerial respiration in the catfish *Corydoras aeneus*. *Physiological Zoölogy* **54**, 189–194.

Krayushkina, L.S. (1983). Development of the osmoregulatory function at early ontogenesis of salmonidae (In Russian). *In* "Biological Principles of Salmon Culture in Water Bodies of the USSR" (O.A. Scarlato, ed.), pp. 56–72. Nauka, Moscow.

Kreps, E.M. (1977). Biochemical adaptations of marine animals (In Russian). *Biologiya Morya* 1977 (5), 6–15.

Kreps, E.M. (1981). "Lipids of Cell Membranes" (In Russian). Nauka, Leningrad, 339 pp.

Kreps, E.M., Avrova, N.F., Zabelinsky, S.A., Kruglova, E.E., Levitina, M.V., Obukhova, E.L., Pomozanskaya, L.F., Pravdina, N.J., Chebotareva, M.A. and Chirkovskaya, E.V. (1977). On the biochemical evolution of brain in vertebrates (In Russian). *Zhurnal Evolutsionnoy Biokhimii i Physiologii* **13**, 556–569.

Kreps, E.M., Tyurin, V.A., Gorbunov, N.V., Maksimovich, A.A., Polyakov, V.N., Plyusnin, V.V. and Kagan, V.E. (1986). Activation of peroxidase oxidation of lipids during migrational stress in gorbuscha: possible mechanism of adaptation (In Russian). *Doklady Akademii Nauk SSSR* **286**, 1009–1012.

Kreps, E.M., Tyurin, V.A., Chelomin, V.P., Gorbunov, N.V., Nalivaeva, N.N., Tyurina, Yu.Yu, Avrova, N.F. and Kagan, V.E. (1987). The study of mechanisms of inactivation of peroxidative oxidation of lipids in synaptosomes of marine teleosts (In Russian). *Zhurnal Evolutsionnoy Biokhimii i Physiologii* **23**, 461–467.

Kriksunov, E.A. and Shatunovsky, M.I. (1979). Some aspects of structural variability in the smelt population (In Russian). *Voprosy Ikhtiologii* **19**(5), 55–62.

Krivobok, M.N. (1953). Food utilisation by the fry of carp at the Azovo-Dolgy fish farm (In Russian). *Trudy VNIRO* **24**, 102–116.

Krivobok, M.N. (1964). On the role of liver in maturation of the ovaries of Baltic herring (In Russian). *Voprosy Ikhtiologii* **4**, 483–494.

Krivobok, M.N. and Storozhuk, A.Ya. (1970). Effect of size and age of female Volga sturgeons on the weight and chemical composition of mature eggs (In Russian). *Voprosy Ikhtiologii* **10**, 1012–1017.

Krivobok, M.N. and Tarkovskaya, O.I. (1960). Determination of the term of spawning migrations of Baltic herring based on the study of its fat metabolism (In Russian). *Trudy VNIRO* **42**, 171–189.

Krogh, A. (1931). Dissolved substances as food for aquatic organisms. *Biological Reviews of the Cambridge Philosophical Society* **6**, 412–442.

Krogh, A. (1939). Osmotic regulation in aquatic animals. London, 242 pp.

Krogius, F.V. (1978). On the significance of genetic and ecological factors in population dynamics of sockeye salmon (*Oncorhynchus nerka*) from Lake Dalneye (In Russian). *Voprosy Ikhtiologii* **18**, 211–221.

Krüger, F. (1962). Über die mathematische Darstellung des tierischen Wachstums. *Naturwissenschaften* **49**, 454.

Krueger, H.M., Saddler, J.B., Chapman, G.A., Tinsley, I.J. and Lowry, R.R. (1968). Bioenergetics, exercise and fatty acids of fish. *American Zoologist* **8**, 119–129.

Krumschnabel, G. and Lackner, R. (1993). Stress responses in rainbow trout, *Oncorhynchus mykiss* alevins. *Comparative Biochemistry and Physiology* **104A**, 777–783.

Kryvi, H. (1977). Ultrastructure of the different fibre types in axial muscles of the sharks *Etmopterus spinax* and *Galeus melastomus*. *Cell and Tissue Research* **184**, 287–300.

Kryvi, H., Flood, P.R. and Gulyaev, D. (1980). The ultrastructure and vascular supply of the different fibre types in the axial muscle of the sturgeon. *Cell and Tissue Research* **212**, 117–126.

Kryzhanovsky, S.G. (1949). Ecologo-morphological regularities of cyprinid, cobitid and siluroid development (In Russian). *Trudy Instituta Morfologii Zhivotnyk Akademiya Nauk SSSR* **1**, 5–332.

Kudryavtseva, G.V. (1990). Ecologio-physiological features and the role of the pentose-phosphate shunt for carbohydrate metabolism in adaptations of lower aquatic vertebrates (Cyclostomes and Pisces) (In Russian). *Uspekhi Sovremennoy Biologii* **109**, 171–182.

Kudryavtseva, G.V. (1991). Prostaglandins: ecological potential of the effect (In Russian). *Uspehki Sovremennoy Biologii* **111**, 698–706.

Kukharev, N.N., Rebik, S.T. and Trushin, Yu.K. (1988). On the problem of trophic activity of schooling pelagic and demersal fish inhabiting hypoxic waters of the western Arabian Sea (In Russian). *In* "Nutrition of Marine Fish and Utilisation of the Nutritive Base as an Element of Fisheries Prognostication", pp.124–125. Murmansk.

Kulikova, N.I. (1967). On the blood serum lipoproteins of fish (In Russian). *In* "Metabolism and Biochemistry of Fish", pp. 292–296. Nauka, Moscow.

Kulinsky, V.I. and Olkhovsky, I.A. (1992). Two alternative strategies – resistance and tolerance – developing under unfavourable conditions. Role of hormones and receptors (In Russian). *Uspekhi Sovremennoy Biologii* **112**, 697–714.

Kurant, V.Z. and Arsan, O.M. (1991). Impact of zinc on protein and nucleic acid content of carp tissues (In Russian). *Gydrobiologicheskii Zhurnal* **27**, 45–48.

Kurant, V.Z., Yakovenko, B.V. and Yavonenko, A.F. (1983). Age changes in the content of nucleic acids and proteins in carp tissues. *Gydrobiologicheskii Zhurnal* **19**, 75–78.

Kutty, M.N. (1968). Respiratory quotients in goldfish and rainbow trout. *Journal of the Fisheries Research Board of Canada* **25**, 1689–1728.

Kutty, M.N. (1972). Respiratory quotient and ammonia excretion in *Tilapia mossambica*. *Marine Biology* **16**, 126–133.

Kutty, M.N. and Mohamed, M.P. (1975). Metabolic adaptations of mullet with special reference to energy utilisation. *Aquaculture* **5**, 253–270.

Kuzmina, V.V. (1985). Temperature adaptations of enzymes accounting for the membrane digestion in freshwater bony fishes (In Russian). *Zhurnal Obshchey Biologii* **46**, 824–837.

Kuzmina, V.V. (1992). Activity of digestive enzymes in the intestinal mucous membrane of Black Sea bony fishes varying in ecology (In Russian). *Voprosy Ikhtiologii* **32**(2), 141–148.

Kychanov, V.M. (1981). On the serum proteins, lipo- and glycoproteins of white fish in the process of gonad maturation (In Russian). *Voprosy Ikhtiologii* **21**, 489–497.

Kychanov, V.M. (1984). Colloid resistance of fish serum proteins (In Russian). *Voprosy Ikhthyiologii* **24**, 302–306.

Kychanov, V.M. and Volodina, N.A. (1985). Ecologo-physiological aspects of cultural reproduction of white fish (In Russian). *Gydrobiologicheskii Zhurnal* **20**, 45–49.

Lapin, V.I. (1973). Comparative study of fatness and qualitative composition of lipids in White and Black Sea flounders (In Russian). *Biologicheskii Nauki* 1973 (3), 41–48.

Lapin, V.I. and Basaanzhov, G. (1989). Seasonal rhythmicity of physiologio-biochemical processes in Mongolian grayling (In Russian). *Voprosy Ikhtiologii* **29**, 831–841.

Lapin, V.I. and Shatunovsky, M.I. (1981). The composition, physiological and ecological significance of fish lipids (In Russian). *Uspekhi Sovremennoy Biologii* **92**, 380–393.

Lapina, N.N. (1991). Effect of the warm effluent from an atomic power station on the physiological characteristics of female bleak during the spawning season (In Russian). *Ekologiya* 1991 (4), 43-50.

Lapina, N.N. and Lapin, V.I. (1982). The dynamics of physiology-biochemical characteristics of generative synthesis in some Cyprinidae females (In Russian). *Voprosy Ikhtiologii* **22**, 285- 293.

Lapkin, V.V. (1979). Annual rhythmicity of the vital activity of fishes of temperate latitudes: a thermodynamic approach (In Russian). *Voprosy Ikhtiologii* **19**, 782–792.

Lasker, R. (1962). Efficiency and rate of yolk utilisation by developing embryos and larvae of the Pacific sardine. *Journal of the Fisheries Research Board of Canada* **19**, 867–875.

Lasker, R. (1970). Utilisation of zooplankton energy by a Pacific sardine population in the California current. *In* "Marine Food Chains" (J.H. Steele, ed.), pp. 265–284. Edinburgh.

Lasker, R. and Theilacker, G.H. (1962). The fatty acid composition of the lipids of some Pacific sardine tissues in relation to ovarian maturation and diet. *Lipid Research* **3**, 60–64.

Laurence, G.G. (1975). Laboratory growth and metabolism of the winter flounder from hatching through metamorphosis at three temperatures. *Marine Biology* **32**, 223–229.

Lavéty, J., Afolabi, O.A. and Love, R.M. (1988). The connective tissues of fish. IX. Gaping in farmed species. *International Journal of Food Science and Technology* **23**, 23–30.

Leatherland, J.F. (1994). Reflections on the thyroidology of fishes: from molecules to humankind. *Guelph Ichthyology Reviews* **2**, 1–67 .

Leatherland, J.F., McKeown, B.A. and John, T.M. (1974). Circadian rhythm of plasma prolactin, growth hormone, glucose and free fatty acid in juvenile kokanee salmon, *Oncorhynchus nerka*. *Comparative Biochemistry and Physiology* **47A**, 821–828.

Lebedev, N.V. (1940). The possibility of forecasting the term of migration of Azov Sea anchovy (In Russian). *Zoologicheskii Zhurnal* **19**, 646–670.

Le Cren, E.D. (1958). Observations on the growth of perch (*Perca fluviatilis* L.) over twenty two years with special reference to the effects of temperature and changes in population density. *Journal of Animal Ecology* **27**, 287–334.

Lee, C.G. and Ip, Y.K. (1987). Environmental effect on plasma thyroxine (T4), 3,5,3'-triiodo-ʟ-thyronine (T3), prolactin and cyclic adenosine 3'-5'-monophosphate (cAMP) content in the mudskippers *Periophthalmus chrysospilos* and *Boleophthalmus boddaerti*. *Comparative Biochemistry and Physiology* **87A**, 1009–1014.

Lee, D.J., Roehm, I.N., Yu, T.C. and Sinnhuber, R.O. (1967). Effect of 3-fatty acids on the growth rate of rainbow trout, *Salmo gairdneri*. *Journal of Nutrition* **92**, 93–98.

Lee, R.F., Nevenzel, J. and Paffenhafer, G.A. (1972). The presence of wax esters in marine planktonic copepods. *Naturwissenschaften* **59**, 406–411.

Lee, R.F., Phleger, C.F. and Horn, M.H. (1975). Composition of oil in fish bones: possible function in neutral buoyancy. *Comparative Biochemistry and Physiology* **50B**, 13–16.

Lee, R.W. and Meier, A.H. (1967). Diurnal variations of the fattening response to prolactin in the golden-top minnow, *Fundulus chrysotus*. *Journal of Experimental Zoology* **166**, 307–315.

Lehninger, A.L. (1972). "Biochemistry. The Molecular Basis of Cell Structure and Function". Worth Publishers, New York.

Leibson, L.G. (1972). Metabolism and its endocrine regulation in fish of different locomotion activity (In Russian). *Zhurnal Evolutsionnoy Biochimii i Physiologii* **8**, 280–288.

Leslie, G.M. and Buckley, G.T. (1976). Phospholipid composition of goldfish (*Carassius auratus*) liver and brain and temperature-dependence of phosphatidyl choline synthesis. *Comparative Biochemistry and Physiology* **53B**, 335–337.

Levanidov, I.P. (1950). Chemical composition of the flesh of herring in coastal waters of Western Sakhalin (In Russian). *Rybnoye Khozyaistvo* 1950 (2), 37–41.

Levanidov, I.P. (1968). Classification of fish based on the fat and protein content of the body (In Russian). *Rybnoye Khozyaistvo* 1968 (9), 50–51.

Lewander, K., Dave, G., Johansson, M.-L., Larsson, Å. and Lidman, U. (1974). Metabolic and haematological studies on the yellow and silver phases of the European eel, *Anguilla anguilla*. – I. Carbohydrate, lipid, protein and inorganic ion metabolism. *Comparative Biochemistry and Physiology* **47B**, 571–581.

Lewis, R.W. (1962). Temperature and pressure effects on the fatty acids of some marine ectotherms. *Comparative Biochemistry and Physiology* **6**, 75–89.

Limansky, V.V. (1970). Characteristics of the blood groups of European anchovy inhabiting the Black and Azov Seas (In Russian). *Trudy VNIRO* **69**, 213–219.

Lindemann, R.L. (1942). The trophic–dynamic aspect of ecology. *Ecology* **23**, 399–418.

Liu, E.H., Smith, M.H., Godt, M.J.W., Chesser, R.K., Latheo, A.K. and Henzler, D.J. (1985). Enzyme levels in natural mosquito-fish populations. *Physiological Zoology* **58**, 242–252.

Liu, R.K. and Walford, R.L. (1966). Increased growth and life-span with lowered ambient temperature in the annual fish. *Cynolebias adloffi*. *Nature, Lond.* **212**, 1277–1278.

Lizenko, E.I., Nefedova, Z.A. and Sidorov, V.S. (1980). Ecological pattern of lipid composition of eggs of some fish species (In Russian). *In* "Biochemistry of the Fishes of Karelia" (V.S. Sidorov and E.I. Lizenko, eds), pp. 6–15. Petrozavodsk.

Lizenko, E.I., Nefedova, Z.A., Titova, V.F. and Sterligova, O.P. (1983). Lipid composition and biological role of different groups of lipids in eggs and milt of freshwater fish (In Russian). *In* "Comparative Biochemistry of Aquatic Animals" (V.S. Sidorov and R.U. Vysotskaya, eds), pp. 28–42. Petrozavodsk.

Lizenko, E.I., Zagorskikh, O.M. and Solovyev, L.G. (1991). Comparative characteristics of lipid composition of organs of the River Enisey and Caspian sturgeons (In Russian). *In* "Biochemical Features of Fish Diseases", pp. 36–41. Karelian Scientific Centre Publishing, Petrozavodsk.

Love, R.M. (1957). The biochemical composition of fish. *In* "The Physiology of Fishes", (M.E. Brown, ed.), Vol. 1, pp. 401–418. Academic Press, New York.

Love, R.M. (1958). Studies on the North Sea cod. III. – Effects of starvation. *Journal of the Science of Food and Agriculture* **9**, 617–620.

Love, R.M. (1960). Water content of cod (*Gadus callarias* L.) muscle. *Nature, Lond.* **185**, 692.

Love, R.M. (1962). The measurement of 'condition' in North Sea cod. *Journal du Conseil International pour l'Exploration de la Mer* **27**, 34–42.

Love, R.M. (ed.) (1970). "The Chemical Biology of Fishes" Vol. 1, 547 pp. Academic Press, London, New York.

Love, R.M. (1974). Colour stability in cod (*Gadus morhua* L.) from different grounds. *Journal du Conseil International pour l'Exploration de la Mer* **35**, 207–209.

Love, R.M. (1975). Variability in Atlantic cod (*Gadus morhua*) from the northeast Atlantic: a review of seasonal and environmental influences on various attributes of the flesh. *Journal of the Fisheries Research Board of Canada* **32**, 2333–2342.

Love, R.M. (1979). The post-mortem pH of cod and haddock muscle and its seasonal variation. *Journal of the Science of Food and Agriculture* **30**, 433–438.

Love, R.M. (ed.) (1980). "The Chemical Biology of Fishes", Vol. 2. Academic Press, London, New York, 943 pp.

Love, R.M. (1988). "The Food Fishes, Their Intrinsic Variation and Practical Implications". Farrand Press, London, 276 pp.

Love, R.M. (1997). Biochemical dynamics and the quality of fresh and frozen fish. *In* "Fish Processing Technology", 2nd edn (G.M. Hall, ed.), pp.1–31. Blackie Academic and Professional, London.

Love, R.M. and Black, D. (1990). Dynamics of stored energy in North Sea cod (*Gadus morhua* L.) and cultured rainbow trout (*Salmo gairdneri* Richardson). *In* "Animal Nutrition and Transport Processes. I. Nutrition in Wild and Domestic Animals" (J. Mellinger, ed.), Comparative Physiology, Vol. 5, pp.193–202. Karger, Basel.

Love, R.M. and Lavéty, J. (1977). Wateriness of white muscle: a comparison between cod (*Gadus morhua*) and jelly cat (*Lycichthys denticulatus*). *Marine Biology* **43**, 117–121.

Love, R.M. and Muslemuddin, M. (1972). Protein denaturation in frozen fish. XII. The pH effect and cell fragility determinations. *Journal of the Science of Food and Agriculture* **23**, 1229–1238.

Love, R.M., Lavéty, J. and Garcia, N.G. (1972). The connective tissues of fish. VI. Mechanical studies on isolated myocommata. *Journal of Food Technology* **7**, 291–301.

Love, R.M., Robertson, I., Lavéty, J. and Smith, G.L. (1974). Some biochemical characteristics of cod (*Gadus morhua* L.) from the Faroe Bank compared with those from other fishing grounds. *Comparative Biochemistry and Physiology* **47B**, 149–161.

Love, R.M., Hardy, R. and Nishimoto, J. (1975). Lipids in the flesh of cod (*Gadus morhua* L.) from Faroe Bank and Aberdeen Bank in early summer and autumn. *Memoirs of the Faculty of Fisheries, Kagoshima University* **24**, 123–126.

Love, R.M., Munro, L.J. and Robertson, I. (1977). Adaptation of the dark muscle of cod to swimming activity. *Journal of Fish Biology* **11**, 431–436.

Lovern, J.A. (1934). Fat metabolism in fishes. V. The fat of the salmon in its young freshwater stages. *Biochemical Journal* **28**, 1961–1963.

Lovern, J.A. (1937). Variation in the chemical composition of herring. *Journal of the Marine Biological Association UK* **22**, 281–293.

Lovern, J.A. (1942). The composition of the depot fats of aquatic animals. *Special Reports of the Food Investigation Board, Lond.* **51**, 1–72.

Lovern, J.A. (1964). The lipids of marine organisms. *Oceanography and Marine Biology, Annual Review* **2**, 169–191.

Lowe-McConnell, R.H. (1979). Ecological aspects of seasonality in fishes of tropical waters. *Symposia of the Zoological Society of London* **44**, 327–359.

Lozina-Lozinsky, L.K. and Zaar, E.I. (1987). The features of the reaction of stenothermal organisms to variable temperatures (In Russian). *Zhurnal Obshchey Biologii* **48**, 538–548.

Lühmann, M. (1953). Über die Fettspeicherung bei Ostsee-heringen und ihre Beriehungzum Fortplanzungs-Zyklus. *Kieler Meeresforschung* **9**, 213–227.

Lühmann, M. (1955). Über die Fettspeicherung bei der Kleinen maräne. *Archiv für Fischereiwissenschaft* **6**, 119–131.

Lukyanenko, V.I. (1967). "Fish Toxicology" (In Russian). Pishchevaya Promyshlennost, Moscow, 216 pp.

Lukyanenko, V.I. (1971). "Immunobiology of Fish" (In Russian). Pishchevaya Promyshlennost, Moscow, 366 pp.

Lukyanenko, V.I. (1976) "Fish Immunobiology" (In Russian). Pishchevaya promyshlennost, Moscow, 366 pp.

Lukyanenko, V.I. (1987). "Ecological Aspects of Ichthyotoxicology" (In Russian). Agropromizdat, Moscow, 240 pp.

Lukyanenko, V.I. (1989). "Immunobiology of Fish" (In Russian). Agropromizdat, Moscow, 269 pp.

Lukyanenko, V.I. (1990). On the general conception of the preservation of water bodies from pollution (In Russian). *Vestnik Akademii Nauk SSSR* 1990 (4), 75–81.

Lukyanenko, V.I. (1992). How can we save Caspian sturgeon? *Vestnik Rossiyskoy Akademii Nauk* 1992 (2), 55–74.

Lukyanenko, V.I., Geraskin, P.P. and Bal, N.V. (1978). Heterogeneity and polymorphism of haemoglobin in two species of the genus *Huso* (In Russian). *Doklady Akademii Nauk SSSR* **237**, 994–998.

Lukyanenko, V.I., Vasilyev, A.S. and Lukyanenko, V.V. (1991). "Heterogeneity and Polymorphism of Fish Haemoglobin" (In Russian). Nauka, St Petersburg, 392 pp.

Lunde, G. (1973). Analysis of trace elements, phosphorus and sulphur in the lipid and the non-lipid phase of halibut and tunny. *Journal of the Science of Food and Agriculture* **24**, 1029–1038.

Lushchak, V.I. (1994). The role of the redistribution of phosphorylating and glycolytic enzymes in the adaptation of Aquatic organisms to environmental conditions (In Russian). *Gydrobiologicheskii Zhurnal* **30**(6), 50–58.

Luts, G.I. (1986). "The Ecology and Fishery of Azov Kilka" (In Russian). Rostov Publishing House, Rostov-on-Don, 88 pp.

Luts, G.I. and Rogov, S.F. (1978). Dynamics of fat content and its formation in kilka and anchovy stocks in the Azov Sea depending on winter temperatures (In Russian). *Gydrobiologicheskii Zhurnal* **14**, 31–35.

Lynen, F. (1972). *In* "Current Trends in the Biochemistry of Lipids" (J. Ganguly and R.M.S. Smellie, eds), pp. 5–25, Academic Press, London.

Lyzlova, E.M. and Serebrennikova, T.P. (1983). The study of the enzymes of carbohydrate and amino acid metabolism in fish and lamprey (In Russian). *Zhurnal Evolutsionnoy Biokhimii i Physiologii* **19**, 222–225.

McBride, J.R., Fagerlund, U.H.M., Smith, M. and Tomlinson, N. (1965). Post-spawning death of Pacific salmon (*Oncorhynchus nerka*) maturing and spawning in captivity. *Journal of the Fisheries Research Board of Canada* **22**, 775–782.

MacDonald, A.G. (1975). "Physiological Aspects of Deep Sea Biology", Cambridge University Press, Cambridge, 450 pp.

MacFadyen, E. (1963). "Animal Ecology: Aims and Methods", 2nd edn. Pitman, London, 344 pp.

McFarland, W.N. and Munz, F.W. (1958). A re-examination of the osmotic properties of the Pacific hagfish, *Polistotrema stouti*. *Biological Bulletin, Marine Biological Laboratory, Woods Hole* **114**, 348–356.

McKay, M.C., Lee, R.F. and Smith, M.A.K. (1985). The characterisation of the plasma lipoproteins of the channel catfish, *Ictalurus punctatus*. *Physiological Zoölogy* **58**, 693–704.

MacKeown, B.A., Leatherland, J.F. and John, T.M. (1975). The effect of growth hormone and prolactin on the mobilisation of free fatty acids and glucose in the kokanee salmon, *Oncorhynchus nerka*. *Comparative Biochemistry and Physiology* **50B**, 425–430.

MacKinnon, J.C. (1973). Analysis of energy flow and production in an unexploited marine flatfish population. *Journal of the Fisheries Research Board of Canada* **30**, 1717–1728.

McNabb, R.A. and Pickford, G.E. (1970). Thyroid function in male killifish, *Fundulus heteroclitus*, adapted to high and low temperatures and to fresh water and sea water. *Comparative Biochemistry and Physiology* **33**, 783–792.

McVean, J.C. and Montgomery, J.C. (1987). Temperature compensation in myotomal muscle: Antarctic versus temperate fish. *Environmental Biology of Fish* **19**, 27–33.

Maetz, J. (1971). Fish gills: mechanisms of salt transfer in fresh water and sea water. *Philosophical Transactions of the Royal Society of London B* **262**, 209–249.

Maetz, J. (1973). Transport mechanisms in seawater adapted fish gills. *Alfred Benson Symposia* **5**, 427–444.

Maetz, J. (1974). Aspects of adaptation to hypo-osmotic and hyper-osmotic environments. *In* "Biochemical and Biophysical Perspectives in Marine Biology" (D.C. Malins and J.R. Sargent, eds), pp. 1–167. Academic Press, London, New York.

Maksimovich, A.A. (1988). The characteristics of carbohydrate metabolism in Pacific salmon under total starvation (In Russian). *Izvestiya Akademii Nauk SSSR, Seriya Biologiya* 1988 (4), 500–508.

Maksimovich, A.A. (1989). The structural and functional changes of neurosecretory nuclei in pink salmon hypothalamus during spawning migration (In Russian). *Izvestiya Akademii Nauk SSSR, Seriya Biologiya* 1989(5), 674–681.

Malikova, E.M. (1962). Biochemical changes in juvenile salmon during smoltification (In Russian). *In* "Biology of Inland Water Bodies of Baltic Countries" (E.M. Malikova, ed.), pp.100–104. Academy of Sciences of USSR, Moscow, Leningrad.

Malikova, E.M. (1967). Some aspects of reproduction related to improving fish stocks in Latvia (In Russian). *Voprosy Ikhtiologii* **7**, 961–966.

Malinovskaya, M.V. (1988). The pathways of carbohydrate metabolism in fish and their temperature adaptation (In Russian). *Gydrobiologicheskii Zhurnal* **24**, 29–39.

Malins, D.C. and Barone, A. (1970). Glyceryl ether metabolism: regulation of buoyancy in dogfish *Squalus acanthias*. *Science, New York* **67**, 79–80.

Malins, D.C. and Wekell, J.C. (1969). The lipid biochemistry of marine organisms. *Progress in the Chemistry of Fats and Other Lipids* **10**, 339–363.

Malyarevskaya, A.Ya. (1959). "Nitrogen Metabolism in Carp" (In Russian). Academy of Sciences of Ukrainian SSR, Kiev, 88 pp.

Malyarevskaya, A.Ya. (1979). "The Metabolism of Fish under Anthropogenic Eutrophication of Water Bodies" (In Russian). Naukova Dumka, Kiev, 254 pp.

Malyarevskaya, A.Ya. and Karasina, F.M. (1991). Dynamics of heavy metals and total thiamine accumulation in fish (In Russian). *Gydrobiologicheskii Zhurnal* **27**, 69–74.

Malyarevskaya, A.Ya., Bilyk, T.I., Sherstyuk, V.V. and Tugay, V.A. (1985). Seasonal dynamics of macro-erg compounds in fish muscle (In Russian). *Gydrobiologicheskii Zhurnal* **21**, 55–62.

Malyukina, G.A. (1966). Some aspects of the physiology of schooling behaviour of fish (In Russian). *Trudy VNIRO* **60**, 201–212.

Mancera, J.M., Perez-Figares, J.M. and Fernandez-Llebrez, P. (1993). Osmoregulatory responses to abrupt salinity changes in the euryhaline sea bream (*Sparus aurata* L.). *Comparative Biochemistry and Physiology* **106A**, 245–250.

Mancera, J.M., Perez-Figares, J.M. and Fernandez-Llebrez, P. (1994). Effect of cortisol on brackish water adaptation in the euryhaline gilthead seabream (*Sparus aurata* L.). *Comparative Biochemistry and Physiology* **107A**, 397–402.

Mankowski, W., Elwertowski, J. and Maciejczyk, J. (1961). Produkcja planktonu w Poludniowym Baltyku a biologia saprota. *Gospodarka Rybna* **13**, 8–9.

Mann, K.H. (1965). Energy transformations by a population of fish in the River Thames. *Journal of Animal Ecology* **34**, 253–275.

Mann, K.H. (1969). The dynamics of aquatic ecosystems. *Advances in Ecological Research* **6**, pp.1–83, Academic Press, London, New York.

Manteifel, B.P. (1980). "Ecology of Animal Behaviour" (In Russian). Nauka, Moscow, 220 pp.

Manteifel, B.P. (1984). The main tasks and means of regulation of fish behaviour to provide well-balanced fishery and fish culture (In Russian). *In* "Modern Problems of Ichthyology", pp.256–264, Nauka, Moscow.

Manteifel, B.P. (1987). "The Ecological and Evolutionary Aspects of Animal Behaviour" (In Russian). Nauka, Moscow. 272 pp.

Marshall, N.B. (1972). Swimbladder organisation and depth ranges of deepsea teleosts. *Symposia of the Society for Experimental Biology* **26**, 261–272.

Marti, Yu.Yu. (1932). On the feasibility of forecasting the character of fish migration (In Russian). *Rybnoye Khozyaistvo* 1932 (7), 43–44.

Marti, Yu.Yu. (1956). "The Main Stages of the Life-history of Atlanto-Scandinavian Herrings" (In Russian). Polar Institute of Marine Fishery and Oceanography, Moscow, 70 pp.

Marti, Yu.Yu. (1980). "Migration of Marine Fishes" (In Russian). Pishchevaya Promyshlennost, Moscow, 248 pp.

Mathur, G.B. (1967). Anaerobic respiration in the cyprinoid fish *Rasbora daniconius*. *Nature, Lond.* **214**, 318–319

Matishev, G.G. and Pavlova, L.G. (1994). Degradation of marine ecosystems of northern Europe provoked by the exploitation of bioresources and methods for their restoration (In Russian). *Izvestiya Rossiskaya Akademii Nauk, Seriya Biologya* 1994 (1), 119–126.

Matyukhin, V.A. (1973). "Bioenergetics and Physiology of Fish Swimming" (In Russian). Nauka, Novosibirsk, 154 pp.

Matyukhin, V.A., Alikin, Yu.S., Belchenko, L.A., Stolbov, A.Ya., Turetsky, V.I. and Khaskin, V.V. (1984). The energetics of swimming (In Russian). *In* "Study of the Energetics of Fish Swimming at Low Velocities" (E.J. Galazy and M.A. Medvedev, eds), pp. 6–38, Nauka, Novosibirsk.

Masmanidi, N.D. (1974). On the symptoms of oil poisoning of hydrobionts (In Russian). *Rybnoye Khozyaistvo* 1974 (9), 28–29.

Matthews, R.A. and Thorpe, J.E. (eds) (1995). "Molecular Biology in Fish, Fisheries and Aquaculture", *Journal of Fish Biology* **47**(suppl. A), Academic Press, London.

Mayzaud, P. and Dallot, S. (1973). Respiration et excrétion azoteé du zooplankton. 1. Evolution des niveaux metaboliques de quelques espèces de Méditerranée occidentale. *Marine Biology*, **19**, 307–314.

Mead, J.F. and Kayama, M. (1967). Lipid metabolism in fish. *In* "Fish Oils" (M.E. Stansby, ed.), pp. 289–299. Avi Publishing Company, Westport, Connecticut.

Mead, J.F., Kayama, M. and Reiser, R. (1960). Biogenesis of polyunsaturated fatty acids in fish. *Journal of the American Oil Chemists' Society* **37**, 438–450.

Medford, B.A. and Mackay, W.C. (1978). Protein and lipid content of gonads, liver and muscle of northern pike in relation to gonad growth. *Journal of the Fisheries Research Board of Canada* **35**, 213–219.

Mednikov, B.M. (1973). Thermolability of the development of poikilothermal organisms and its molecular mechanisms (In Russian). *Uspechi Sovremennoy Biologii* **76**, 279–395.

Mednikov, B.M., Reshetnikov, Yu S. and Shubina, E.A. (1977). The study of the relationships between white fish by the method of DNA molecule hybridisation (In Russian). *Zoologicheskii Zhurnal* **56**, 333–341.

Meister, R., Zwingelstein, G. and Jouanneteau, J. (1973). Salinity and fatty acid composition of phosphoglycerides in the tissues of the eel (*Anguilla anguilla*). *Annales de l'Institut Michel Pacha Laboratoire Maritime de Physiologie* **6**, 58–71.

Melnichuk, G.L. (1970). Nutritive requirements and energy balance in carp fry of Kremenchug reservoir (In Russian). *Gydrobiologicheskii Zhurnal* **6**, 50–56.

Meyen, V.A. (1939). On the annual cycle of changes in the ovaries of bony fishes (In Russian). *Izvestiya Akademii Nauk SSSR* 1939 (3), 229–418.

Meyen, V.A., Karzinkin, G.S., Ivlev, V.S., Lipin, A.N. and Sheina, M.P. (1937). Utilisation of the natural nutritive base of the pond by two-year carp (In Russian). *Zoologicheskii Zhurnal* **16**, 209–223.

Meyer-Rochow, V.B. and Pyle, C.A. (1980). Fatty acid analysis of lens and retina of two Antarctic fish and of the head and body of the antarctic amphipod *Orchomene plebs*. *Comparative Biochemistry and Physiology* **65B**, 395–398.

Meyerson, F.Z. (1981). "Adaptation, Stress and their Prevention" (In Russian). Nauka, Moscow, 200 pp.

Mezhgherin, V.A. (1970). Energy structure of zoological systems (In Russian). *Ecologiya* 1970 (6), 52–61.

Mezhgherin, V.A. (1990). On the possibility of the newest formulation of evolutionary doctrine (In Russian). *Zhurnal Obshchey Biologii* **51**, 185–195.

Mikryakov, V.R. (1978). Urgent problems of fish immunology (In Russian). *Trudy Instituta Biologii Yuzhnich Morei AN SSSR* **32**, 116–133.

Mikryakov, V.R. and Silkina, N.I. (1982). Bactericide features and lipid composition of the blood serum of the rockling (In Russian). *Voprosy Ikhtiologii* **22**, 698–702.

Miller, N.G., Hill, M.W. and Smith, M.W. (1976). Positional and species analysis of membrane phospholipids extracted from goldfish adapted to different environmental temperatures. *Biochimica et Biophysica Acta* **55**, 644–654.

Milman, L.S. and Yurovitsky, Yu.G. (1973). "Mechanisms of Enzymatic Regulation of Carbohydrate Metabolism at Early Embryogenesis" (In Russian). Nauka, Moscow, 235 pp.

Mils, E.L. and Forney, J.I. (1981). Energetics, food consumption and growth of young yellow perch in Oneida Lake, New York. *Transactions of the American Fisheries Society* **110**, 479–488.

Minyuk, G.S. (1991). Relationship between the fat content and age of Black Sea sprat (In Russian). *Ekologiya Morya* **37**, 76–79.

Minyuk, G.S., Shulman, G.E., Shchepkin, V.Ya. and Yuneva, T.V. (1997). "Black Sea sprat: the Relationship between Lipid Dynamics, Biology and Fishery" (In Russian). EKOSI-Hydrophysica, Sevastopol, 139 pp.

Miroshnichenko, M.L., Bonch-Osmolovskaya, E.A. and Alekseev, V.A. (1990). Extremely thermophilic bacteria from Kraternaya Bight. *Biologiya Morya* **15**, 206–210.

Miura, T., Suzuki, N., Nagoshi, M. and Yamamura, K. (1976). The rate of production and food consumption of the Biwa masu, *Oncorhynchus rhodurus*, population in Lake Biwa. *Researches on Population Ecology* **17**, 135–154.

Mochek, A.D. (1987). "Ethological Structure of the Coastal Communities of Marine Fishes" (In Russian). Nauka, Moscow, 270 pp.

Moffat, A.M. (1996). Ecophysiology of mysids (Crustacea: peracarida) in the river Tamar estuary, Ph.D. thesis, University of Plymouth.

Mohamed, M.P. (1981). Metabolism of *Tilapia mossambica* with emphasis on hypoxia. *Indian Journal of Experimental Biology* **19**, 1098–1100.

Mohamed, M.P. and Kutty, M.N. (1983a). Respiratory quotient and ammonia quotient in goldfish, *Carassius auratus*, with special reference to ambient oxygen. *Proceedings of the Indian National Science Academy* **49B**, 303–310.

Mohamed, M.P. and Kutty, M.N. (1983b). Influence of hypoxia on metabolism and activity in *Puntius sarana*. *Proceedings of the Indian Academy of Science, Animal Science* **92**, 215–220.

Mohamed, M.P. and Kutty, M.N. (1986). Influence of ambient oxygen and random swimming activity on metabolic rates and quotients in the freshwater mullet, *Rhinomugil corsula*. *Proceedings of the Indian Academy of Science, Animal Science* **95**, 67–76.

Moiseeva, E.B. (1969). Morphofunctional characteristics of the round goby hypophysis in relation to the reproductive cycle (In Russian). *Arkhiv Anatomii, Histologii i Embryologii* **56**, 89–96.

Mommsen, T.P., French, C.J. and Hochachka, P.W. (1980). Sites and patterns of protein and amino acid utilisation during the spawning migration of salmon. *Canadian Journal of Zoology* **58**, 1785–1799.

Monin, Yu.G., Goncharevskaya, O.A. and Stolbov, A.Ya. (1989). The relationship between water–salt homeostasis and tissue respiration under varying muscle loading in horse-mackerel (In Russian). *Voprosy Ikhtiologii* **29**, 842–847.

Moore, J.W. and Potter, I.C. (1976). Aspects of feeding and lipid deposition and utilisation in the lamprey. *Journal of Animal Ecology* **45**, 699–712.

Morawa, F.W. (1955). Wachstum, Wachstums bedingungen und Aufurechsplütse des Sprattes in der Ostsee. *Zeitschrift für Fischerei und deren Hilfswissenschaften* **4**, 101–136.

Morozova, A.L. (1973). The carbohydrate-phosphorus metabolism in muscles of fish of different ecology (In Russian). *Proceedings of All-Union Hydrobiological Society* **18**, 128–136.

Morozova, A.L., Astakhova, L.P. and Silkina, E.N. (1978a). Carbohydrate metabolism in fish during swimming (In Russian). *In* "Elements of Physiology and Biochemistry in Total and Active Metabolism of Fish" (G.E. Shulman, ed.), pp. 122–144. Naukova Dumka, Kiev.

Morozova, A.L., Astakhova, L.P. and Silkina, E.N. (1978b). Aspects of the post-swimming restitution of fish (In Russian). *In* "Elements of Physiology and Biochemistry of Total and Active Metabolism in Fish", (G.E. Shulman, ed.), pp.175–184. Naukova Dumka, Kiev.

Morris, R.W. (1967). High respiratory quotients of two species of bony fish. *Physiological Zoology* **40**, 409–423.

Morris, R.W. and Schneider, M.J. (1969). Brain fatty acids of an antarctic fish, *Trematomus bernacchii*. *Comparative Biochemistry and Physiology* **28**, 1461–1465.

Mosse, P.R.L. and Hudson, R.C.L. (1977). The functional roles of different muscle fibre types identified in the myotomes of marine teleosts: a behavioural, anatomical and histochemical study. *Journal of Fish Biology* **11**, 417–430.

Motais, R. (1959). On the seasonal growth of an abyssal teleost, measured by the phosphatase activity in the scales. *Compte Rendu Hebdomadaire de l'Academie des Sciences, Paris* **248**, 311–312.

Motais, R., Romeu, F.G. and Maetz, J. (1965). Mechanism of euryhalinity. Comparative study of the flounder (euryhaline) and ballan wrasse during transfer into fresh water. *Compte Rendu des Séances de la Societé de Biologie* **261**, 801–804.

Muir, B.S. and Niimi, A.J. (1972). Oxygen consumption of the euryhaline fish aholehole with reference to salinity, swimming, and food consumption. *Journal of the Fisheries Research Board of Canada* **29**, 67–77.

Mukhina, R.I. (1958). Utilisation of artificial and natural food by the fingerlings of carp (In Russian). *Trudy Vsesoyuznogo nauchno-issledovatelskogo Instituta Prudovogo Rybnogo* **9**, 98–113.

Murat, J.C. (1976). Studies on the mobilisation of tissue carbohydrates in the carp. Thesis, Docteur d'Etat, University of Toulouse.

Muravskaya, Z.A. (1978). The nitrogen metabolism in fish during swimming (In Russian). *In* "Elements of Physiology and Biochemistry in Total and Active Metabolism of Fish" (G.E. Shulman, ed.), pp. 87–89. Naukova Dumka, Kiev.

Muravskaya, Z.A. and Belokopytin, Yu.S. (1975). Influence of the swimming velocity on nitrogen excretion and oxygen consumption in pickerel (In Russian). *Biologiya Morya* **1975** (5), 39–44.

Musatov, A.P. (1993). Van't Hoff temperature coefficient for energy metabolism in lower vertebrates (In Russian). *Gydrobiologicheskii Zhurnal* **29**, 77–80.

Nag, A.C. and Nursall, J.R. (1972). Histogenesis of white and red muscle fibres of the trunk muscle of a fish, *Salmo gairdneri*. *Cytobios* **6**, 227–246.

Nagorny, A.V. (1940). "The Problem of Ageing and Longevity" (In Russian). Kharkov University Press, Kharkov, 446 pp.

Nagorny, A.V. (1947). Regularities of the individual evolution of the animal organism. Correlation between the assimilative and dissimilative processes at different stages of individual evolution (In Russian). *Trudy Kkhar'kovskogo Universitets* **25**, 39–61.

Nakagawa, H. and Tsuchiya, Y. (1971). Study of rainbow trout eggs. III. Determination of lipid composition of fatty drops and lipoproteins. *Journal of the Faculty of Fisheries and Animal Husbandry, Hiroshima University* **10**, 11–19.

Nakano, H.Y., Ando, Y. and Shirata, S. (1985). Changes of acid phosphatase, total protein, DNA and RNA during the early development of chum salmon (*Oncorhynchus keta*). *Bulletin of the Hokkaido Regional Fisheries Research Laboratory* **50**, 71–77.

Narasimhan, P.V. and Sundararaj, B.I. (1971). Circadian variations in carbohydrate parameters in two teleosts, *Notopterus notopterus* and *Colisa fasciata*. *Comparative Biochemistry and Physiology* **39** B, 89–99.

Natochin, Yu.V. (1974). "Physiology of Kidney: Formulae, Calculations" (In Russian). Nauka, Leningrad, 162 pp.

Natochin, Yu.V. (1976). "The Ion Regulatory Function of Kidney" (In Russian). Nauka, Leningrad, 265 pp.

Natochin, Yu.V. (1988). Some principles of the evolution of functions at cell, organ and organismic levels with reference to the kidney and water-salt homeostasis. *Zhurnal Obshchey Biologii* **49**, 291–305.

Natochin, Yu.V. and Menshutkin, V.V. (1993). Problems of the evolution of functions in physiology, ecology and techniques (In Russian). *Zhurnal Evolutionnoy Biokhimii i Physiologii* **29**, 434–446.

Natochin, Yu.V., Krayushkina, L.S. and Maslova, M.N. (1975a). Enzymatic activity in gills and kidneys and the endocrine factors of ion metabolism regulation in smolt and spawning sockeye salmon (In Russian). *Voprosy Ikhtiologii* **15**, 131–141.

Natochin, Yu.V., Lukyanenko, V.I., Lavrova, E.A. and Metallov, G.F. (1975b). The isosmotic type of regulation in the sturgeon, Acipenser güldenstadti, during the marine period of life (In Russian). *Zhurnal Evolutsionnoy Biochimii i Physiologii*, **9**, 583–587.

Nebeker, A.V. and Brett, J.R. (1976). Effects of air-supersaturated water on survival of Pacific salmon and steelhead smolts. *Transactions of the American Fisheries Society* **105**, 338–342.

Nebeker, A.V., Bouck, G.R. and Stevens, D.G. (1976). Carbon dioxide and oxygen-nitrogen ratios as factors affecting salmon survival in air-supersaturated water. *Transactions of the American Fisheries Society* **105**, 425–429.

Nechaev, I.V. (1989). Noradrenaline and serotonin in the brain of fishes of different social structure (In Russian). *Doklady Akademii Nauk SSSR* **305**, 755–758.

Nechaev, I.V., Pavlov, D.S., Labas, Yu.A. and Legky, B.P. (1991). Dynamics of catecholamines in early ontogeny and the development of behavioural reactions in juvenile *Aequides pulcher* fry. *Voprosy Ikhtiologii* **31**, 822–838.

Needham, J. (1963). "Chemical Embryology". Academic Press, London, New York, 2021 pp.

Nefedova, Z.A. and Ripatti, P.O. (1990). Lipids of the egg membrane at some stages of salmon embriogenesis (In Russian). *In* "Biochemistry of Ecto- and Endothermic Organisms under Normal and Pathological Conditions", pp.50–56, Petrozavodsk.

Neifakh, A.A. and Timofeeva, M.Ya. (1977). "Molecular Biology of Developmental Processes"(In Russian). Nauka, Moscow, 312 pp.

Neighbors, M.A. (1988). Triacylglycerols and wax esters in the lipids of deep midwater teleost fishes of the Southern California Bight. *Marine Biology* **98**, 15–22.

Nemova, N.N. (1991). Properties and the physiological role of intracellular proteinases in fish tissues (In Russian). *Uspekhi Sovremennoy Biologii* **111**, 948–954.

Nemova, N.N. and Sidorov, V.S. (1990). Dynamics of cathepsin activity in gonads of female fish during maturation (In Russian). *Voprosy Ikhtiologii* **30**, 516–519.

Nemova, N.N., Sidorov, V.S., Grigoryeva, L.I., Valueva, T.A., Mosalov, V.V., Koyvyaryainen, E.I. and Lukyanenko, V.V. (1992). Intracellular proteinases in Russian sturgeon organs during muscular disintegration (In Russian). *Voprosy Ikhtiologii* **32**, 197–200.

Nevenzel, I. and Menon, N. (1980). Lipids of midwater marine fish: family Gonostomatidae. *Comparative Biochemistry and Physiology* **65B**, 351–355.

Nevenzel, J.C. (1970). Occurrence, function and biosynthesis of wax esters in marine organisms. *Lipids* **5**, 308–319.

Nevenzel, J.C., Rodegker, W. and Mead, J.F. (1965). The lipids of *Ruvettus pretiosus* muscle and liver. *Biochemistry* **4**, 1589–1594.

Nevenzel, J.C., Rodegker, W. and Mead, J.F. (1966). Lipids of the living coelacanth. *Science, New York* **152**, 1753–1755.

Nevenzel, J.C., Rodegker, W., Robinson, J.S. and Kayama, M. (1969). The lipids of some lantern fishes (family Myctophidae). *Comparative Biochemistry and Physiology* **31**, 25–36.

Newell, R.C. (1970). "Biology of Intertidal Animals". Logos Press, London, 555 pp.

Newsholme, E.A. and Stort, C. (1973). "Regulation in Metabolism". John Wiley and Sons, London, 349 pp.

Nichols, J.R. and Fleming, W.R. (1990). The effects of salinity, hypophysectomy and hormone administration on gill RNA metabolism of the euryhaline teleost *Fundulus kansae*. *Comparative Biochemistry and Physiology* **95A**, 121–126.

Nicol, J.A.C., Arnott, H.J., Mizuno, C.R., Ellison, E.C. and Chipault, J.R. (1972). Occurrence of glyceryl tridocosahexaenoate. *Lipids* **7**, 171–177.

Niimi, A.J. and Beamish, F.W.H. (1974). Bioenergetics and growth of largemouth bass in relation to body weight and temperature. *Canadian Journal of Zoology* **52**, 447–456.

Nikitin, V.N. (1966). On some basic factors of ontogenesis (In Russian). *In* "Major Problems of Age, Physiology and Biochemistry" (V.N. Nikitin, ed.), pp. 3–31. Meditsina, Moscow.

Nikitin, V.N. (1982). Theory of ageing (In Russian). *In* "Biology of Ageing" (V.V. Frolki's, ed.), pp. 153–174. Nauka, Leningrad.

Nikitin, V.N. and Babenko, N.A. (1987). Lipids and lipid metabolism during ontogenesis (In Russian). *Uspechi Sovremennoy Biologii* **104**, 331–345.

Nikolsky, G.V. (1963). "The Ecology of Fishes" (L. Birkett, transl.), Academic Press, London, New York, 352 pp.

Nikolsky, G.V. (1965). "Theory of Fish Stock Dynamics" (In Russian). Nauka, Moscow, 382 pp.

Nikolsky, G.V. (1974). "Fish Ecology" (In Russian). Vysshaya Shkola, Moscow, 357 pp.

Nikolsky, V.N., Shulman, G.E. and Gordina, A.D. (1988). Estimation of the possible share of fry in the production of Black Sea anchovy population (In Russian). "Proceedings of III All-Union Conference of Marine Biology" (A.L. Morozova, ed.), pp. 56–57. Naukova Dumka, Kiev.

Nordlie, F.G. and Leffler, C.W. (1975). Ionic regulation and the energetics of osmoregulation in *Mugil cephalus* L. *Comparative Biochemistry and Physiology* **51A**, 125–131.

Nordlie, F.G., Walsh, S.J., Haney, D.C. and Nordlie, T.F. (1991). The influence of ambient salinity on routine metabolism in the teleost *Cyprinodon variegatus* Lacépède. *Journal of Fish Biology* **38**, 115–122.

Novikov, G.G. (1993). Character of growth energetics in bony fishes of different ecological groups at different temperatures (In Russian). *Izvestiya Akademii Nauk SSSR, Ser. Biologiya* 1993 (1), 21–28.

Novikova, N.S. (1949). On the possibility of determination of the daily ration of fish under natural conditions (In Russian). *Vestnik Moskovskova Gos-Universiteta* 1949(9), 115–134.

Novozhilova, A.N. (1960). Zooplankton of the Azov Sea in 1957 (In Russian). *Trudy Azovski Instituta Rybnogo Khozyaistva* **1**, 143–166.

O'Boyle, R.N. and Beamish, F.W.H. (1977). Growth and intermediary metabolism of larval and metamorhosing stages of the landlocked sea lamprey, *Petromyzon marinus*. *Environmental Biology of Fishes* **2**, 1103–120.

Ochiai, T. and Fuji, A. (1980). Energy transformation by blenny population of Usu Bay, southern Hokkaido. *Bulletin of the Faculty of Fishes, Hokkaido University* **31**, 314–326.

Odum, E.P. (1959). "Fundamentals of Ecology", 2nd edn. Saunders, Philadelphia, London, 546 pp.

Ogawa, M. (1975). The effects of prolactin, cortisol and calcium-free environment of water influx in isolated gills of Japanese eel, *Anguilla japonica*. *Comparative Biochemistry and Physiology* **52A**, 539–543.

Ogorodnikova, L.G. and Lebedinskaya, I.N. (1984). Glycogen content, phosphorylase and glucose-6-phosphatase activity in fast and slow muscles of carp (In Russian). *Zhurnal Evolutsionnoy Biokhimii i Physiologii* **20**, 12–15.

Olson, R.J. and Boggs, C.H. (1986). Apex predation by yellowfin tuna (*Thunnus albacares*): independent estimates from gastric evacuation and stomach contents, bioenergetics and cesium concentrations. *Canadian Journal of Fisheries and Aquatic Science* **43**, 1760–1755.

Orbeli, L.A. (1958). Principal tasks and methods of evolutionary physiology (In Russian). *In* "Evolution of Nervous System Functions" (L.A. Orbeli, ed.), pp. 7–17. Medgiz, Leningrad.

Orel, G., Vio, E., Princi, M., Piero, D.D. and Aleffi, F. (1986). Stati di anossia dei fondali, popolamenti bentonici e pesca. *Nova Thalassia* **8**(Suppl. 3), 267–280.

Orel, G., Vio, E. and Aleffi, F. (1989). Biocenosi bentoniche e loro modificazioni in seguito a stress anossici. *In* "Atti Convenzione Nazionale Ancona 4 Aprile" (P.V. Curzi and F. Tombolini, eds), pp. 59–63.

Orlova, E.L., Berestovsky, E.G., Antonov, S.G. and Yaragina, N.A. (1990). Some features of Barents Sea cod feeding in the 1980s (In Russian). *Voprosy Ikhtiologii* **30**, 634–643.

Orton, J.H. (1929). Reproduction and death in invertebrates and fishes. *Nature, Lond.* **123**, 14–15.

Ostroumova, I.N. (1979). Physiologo-biochemical estimation of fish condition in fish culture (In Russian). *In* "Modern Problems of Ecological Physiology of Fish" (N.S. Stroganov, ed.), pp. 50–67. Pishchevaya Promyshlennost, Moscow.

Ostroumova, I.N. (1983). Protein requirement of fish and fish fry as related to developmental stages of the digestive function (In Russian). *Izvestiya Gosudarstvennogo Nauchno-issledovatelskogo Instituta Ozernogo i Rechnogo Rybnogo* **194**, 3–19.

Ota, T. and Yamada, M. (1975). Fatty acids of four fresh-water fish lipids. *Bulletin of the Faculty of Fisheries, Hokkaido University* **26**, 277–288.

Oven, L.S. (1976). "The Oogenesis and Character of Spawning of Marine Fishes" (In Russian). Naukova Dumka, Kiev, 132 pp.

Owen, J.M., Adron, J.W., Middleton, C. and Cowey, C.B. (1975). Elongation and desaturation of dietary fatty acids in turbot *Scophthalmus maximus* L. and rainbow trout, *Salmo gairdnerii* Rich. *Lipids* **10**, 528–531.

Owen, T.G. and Hochachka, P.W. (1974). Purification and properties of dolphin muscle aspartate and alanine transaminases and their possible roles in the energy metabolism of diving mammals. *Biochemical Journal* **143**, 541–553.

Ozernyuk, N.D. (1985). "Energy Metabolism of Fish During Early Ontogenesis" (In Russian). Nauka, Moscow, 176 pp.

Ozernyuk, N.D. (1988). Principle of energy minimum in ontogenesis and stability of development processes (In Russian). *Zhurnal Obshchey Biologii* **49**, 552–562.

Ozernyuk, N.D. (1993). Principles of minimalizing metabolism and optimal conditions of species development (In Russian). *Izvestiya Akademii Nauk SSSR, Seriya Biologiya*, 1993 (1), 8–15.

Ozernyuk, N.D., Bulgakova, Yu.V., Demin, V.I., Androsova, I.M. and Stelmashchuk, E.V. (1993). Mechanisms of evolutionary and ontogenetic temperature adaptations of metabolism in poikilotherms (In Russian). *Izvestiya Akademii Nauk SSSR Seriya Biologiya*, 1993, 703–705.

Pagliarani, A., Pirini, M., Trigari, G. and Ventrella, V. (1986). Effect of diets containing different oils on brain fatty acid composition in sea bass (*Dicentrarchus labrax*). *Comparative Biochemistry and Physiology* **83B**, 277–282.

Paloheimo, J.E. and Dickie, L.M. (1966). Food and growth of fishes. III. Relations between food, body size and growth efficiency. *Journal of the Fisheries Research Board of Canada* **23**, 1209–1948.

Paloheimo, J.E. and Plowright, R.C. (1979). Bioenergetics, population, growth and fisheries management. *In* "Environmental Bio-monitoring, Assessment, Prediction and Management", pp.241–268. International Corporation Publishing House, Fairland.

Pandey, N.B. and Mushi, J.S. (1976). Role of thyroid glands in regulation of metabolic rate in an air-breathing siluroid fish, *Heteropneustes*. *Journal of Endocrinology* **69**, 421–425.

Parina, E.V. (1967). "Age and Metabolism of Proteins" (In Russian). Kharkov State University, Kharkov, 204 pp.

Parina, E.V. and Kaliman, P.A. (1978). "Mechanisms of Enzyme Regulation in Ontogenesis" (In Russian). Kharkov University Press, Kharkov, 202 pp.

Parker, S.J. and Specker, J.L. (1990). Salinity and temperature effects on whole-animal thyroid hormone levels in larval and juvenile striped bass, *Morone saxatilis*. *Fish Physiology and Biochemistry* **8**, 507–514.

Parnova, P.G. (1986). Lipids of insecta. Long-chain polyunsaturated fatty acids and their functional role (In Russian). *Zhurnal Evolutsionnoy Biokhimii i Physiologii* **22**, 74–83.

Parnova, P.G. and Svetashev, V.I. (1985). Polyunsaturated fatty acids in the lipid composition of water insects (In Russian). *Zhurnal Evolutsionnoy Biokhimii i Physiologii* **21**, 242–247.

Parrish, B.B. and Saville, A. (1965). The biology of the northeast Atlantic herring populations. *Oceanography and Marine Biology Annual Review* **3**, 323–373.

Parry, G.D. (1983). The influence of the cost of growth on ectotherm metabolism. *Journal of Theoretical Biology* **101**, 453–477.

Paton, D.N. (1898). Report on investigations on the life history of the salmon in fresh water. *Journal of Physiology* **22**, 17–25.

Patton, J.S. (1975). The effect of pressure and temperature on phospholipid and triglyceride fatty acids of fish white muscle; a comparison of deepwater and surface marine species. *Comparative Biochemistry and Physiology* **52B**, 105–110.

Patton, S. and Thomas, A.J. (1971). Composition of lipid foams from swim bladders of two deep ocean fish species. *Journal of Lipid Research* **12**, 331–335.

Patton, S. and Trams, E.G. (1973). Salmon heart triglycerides during spawning migration. *Comparative Biochemistry and Physiology* **46B**, 851–855.

Pavlov, D.S. (1970). "Optomotor Reaction and Fish Behaviour in a Water Stream" (In Russian). Nauka, Moscow, 147 pp.

Pavlov, D.S. (1979). "Biological Aspects of Fish Behaviour Control in a Water Stream" (In Russian). Nauka, Moscow, 319 pp.

Pavlov, I.P. (1932). "Twenty five Years' Experience of Objective Investigation of High Nervous Activity (Behaviour) of Animals" (In Russian). Akademiya Nauk SSSR, Moscow, 240 pp.

Pearse, C. and Achtenberg, H. (1917). Habits of yellow perch in Wisconsin lakes. *Bulletin of the US Bureau of Fisheries* **36**, 297–366.

Pegel, V.A. and Remouev, V.A. (1967). On the role of the environment in the formation and manifestation of thermoregulatory reactions in fish (In Russian). *In* "Metabolism and Biochemistry of Fish" (G.S. Karzinkin, ed.), pp. 198–205. Nauka, Moscow.

Pekkarinen, M. (1980). Seasonal variations in lipid content and fatty acids in the liver, muscle and gonads of eel pout, *Zoarces viviparus* (teleostei) in brackish water. *Annales Zoologici Societatis Zoologicae Botanicae Fennicae Vanamo* **17**, 249–254.

Penczak, T. (1985). A method of estimating total food consumed by fish populations. *Hydrobiologia* **123**, 241–244.

Penczak, T., Zalewski, M. and Malinski, M. (1976). Production of pike, roach and chub in a selected fragment of the Pilic River (Barbel region). *Polskie Archiwum Hydrobiologii*, **23**, 139–153

Penny, K.K. and Goldspink, G. (1981). Temperature adaptation by the myotomal muscle of fish. *Journal of Thermal Biology* **6**, 297–306.

Pentegov, B.P., Mentov, Yu.N. and Kurnaev, E.F. (1928). Physico-chemical characteristics of the spawning and migration starvation of keta salmon (In Russian). *Izvestiya Tikhookeanskogo Nauchno-promyslovoj Stantsii* **2**(1), 1–68.

Pérez-Pinzón, M.A. and Lutz, P.L. (1991). Activity-related cost of osmoregulation in the juvenile snook (*Centropomus undecimalis*). *Bulletin of Marine Science* **48**, 58–66.

Pertseva, M.N. (1981). On the significance of membrane lipids for functioning of the hormone-sensitive adenylate cyclase system (In Russian). *Zhurnal Evolutsionnoy Biokhimii i Physiologii* **27**, 281–286.

Petipa, T.S. (1981). "Copepod Trophodynamics in Marine Planktonic Communities" (In Russian). Naukova Dumka, Kiev, 241 pp.

Phillips, A.M. (1969). Nutrition, digestion and energy utilisation. In "Fish Physiology" (W.S. Hoar and D.J. Randall, eds), Vol. 1, pp.391–423. Academic Press, London and New York.

Phillips, A.M., Podoliak, H.A., Dumas, R.F. and Thoesen, R.W. (1958). The nutrition of trout. *Fisheries Research Bulletin, State of New York Conservation Department, Albany* **22**, 1–87.

Phillips, A.M., Podoliak, H.A., Poston, H.A., Livingston, D.L., Booke, H.E. and Hammer, G.L. (1963). Cortland Hatchery Report 31 for the year 1962. *Fisheries Research Bulletin, State of New York Conservation Department, Albany* **26**, 1–93.

Phleger, C.F. (1975). Bone lipids of Kona Coast reef fish: skull buoyancy in the hawkfish, *Cirrhites pinnulatus*. *Comparative Biochemistry and Physiology* **52B**, 101–104.

Phleger, C.F. and Benson, A.A. (1971). Cholesterol and hyperbaric oxygen in swimbladders of deep sea fishes. *Nature, Lond.* **230**, 122.

Phleger, C.F. and Holtz, R.B. (1973). The membranous lining of the swimbladder in deep sea fishes. 1. Morphology and chemical composition. *Comparative Biochemistry and Physiology* **45B**, 867–873.

Phleger, C.F., Patton, J., Grimes, P. and Lee, R.F. (1976). Fish-bone oil: percent total body lipid and carbon-14 uptake following feeding of 1–14C-palmitic acid. *Marine Biology* **35**, 85–89.

Pickford, G.E., Grant, F.B. and Umminger, B.L. (1969). Studies on the blood serum of the euryhaline cyprinodont fish, *Fundulus heteroclitus*, adapted to fresh or to salt water. *Transactions of the Connecticut Academy of Arts and Sciences* **43**, 25–70.

Pierce, R.J. and Wissing, T.E. (1974). Energy cost of food utilisation in the bluegill. *Transactions of the American Fisheries Society* **103**, 38–45.

Pinder, L.J. and Eales, J.G. (1969). Seasonal buoyancy changes in Atlantic salmon parr and smolt. *Journal of the Fisheries Research Board of Canada* **26**, 2093–2100.

Pionetti, J.-M., Carriere, S. and Quessada, J. (1986). Evolution des reserves energetiques lors de l'entrée dans la vie trophique de poissons marins. *Oceanis* **12**, 251–260.

Pirsky, L.I., Morozova, T.I. and Berdyshev, G.D. (1969). Age related changes in the amount and composition of nucleic acids in liver of yellowfin sole (In Russian). *Gydrobiologicheskii Zhurnal* **5**, 118–120.

Plack, P.A., Woodhead, A.D. and Woodhead, P.M.J. (1961). Vitamin A compounds in the ovaries of the cod from the Arctic. *Journal of the Marine Biological Association UK* **41**, 617–630.

Plisetskaya, E.M. (1975). "Hormonal Regulation of the Carbohydrate Metabolism in Lower Vertebrates" (In Russian). Nauka, Leningrad, 209 pp.

Plisetskaya, E.M., Soltitskaya, L.P. and Leibson, L.G. (1977). Insulin in blood of catadromous lampreys and fish during spawning migration (In Russian). In "Evolution of Endocrinology of the Pancreas", pp.127–133. Nauka, Leningrad.

Poddubny, A.L. (1971). "Ecological Topography of Fish Population in Reservoirs" (In Russian). Nauka, Leningrad, 309 pp.

Podlesnykh, A.V. and Ardashev, A.A. (1990). Influence of the stocking density and cortisol introduction on metabolic transformation of chum (In Russian). *Ekologiya* 1990 (6), 73–75.

Podsevalov, B.N. and Perova, L.I. (1973). Seasonal changes in the fat content of mackerel and scad flesh (In Russian). *Trudy VNIRO* **54**, 6–9.

Polenov, A.L. (1968). "Hypothalamic Neurosecretion" (In Russian). Nauka, Moscow, 170 pp.

Polenov, A.L. (1983). Evolution of the hypothalamo-hypophysial neuroendocrine complex (In Russian). *In* "Evolutionary Physiology" (E.M. Kreps, ed.), Part 2, pp. 53–109. Nauka, Leningrad.

Polimanti, O. (1913). Über den Fettgehalt und die biologische bedeutung desselben für die Fisch und ihren Aufenhaltsort. *Biochemische Zeitschrift* **56B**, 214.

Polyakov, G.D. (1958). Underfeeding as a cause of mortality in the wintering finger-lings of carp (In Russian). "Proceedings of Conference of Ichthyological Committee of Academy of Sciences of USSR, Moscow", Vol. 8, pp. 255–260. Academy of Sciences, Moscow.

Polyakov, G.D. (1975). "Ecological Regulation of Variability of Fish Population" (In Russian). Nauka, Moscow. 159 pp.

Pomazanskaya, L.F., Chirkovskaya, E.V., Pravdina, N.I., Kruglova, E.E., Chebotareva, M.A. and Kreps, E.M. (1979). Phospholipids in the brain of fish and representatives of other classes of vertebrates (Comparative biochemical research) (In Russian). *In* "Physiology and Biochemistry of Marine and Freshwater Animals", pp.22–28, Nauka, Leningrad.

Poston, H.A., McCartney, T.H. and Pyle, E.A. (1969). The effect of physical condi-tioning upon the growth, stamina and carbohydrate metabolism of brook trout. *Fishery Research Bulletin, New York* **31**, 25–31.

Potts, W.T.W., Foster, M.A., Rudy, P.P. and Howells, G.P. (1967). Sodium and water balance in the cichlid teleost, *Tilapia mossambica*. *Journal of Experimental Biology* **47**, 461–470.

Potts, W.T.W., Foster, M.A. and Stather, J.W. (1970). Salt and water balance in salmon smolts. *Journal of Experimental Biology* **52**, 553–564.

Pozo, R., Pérez-Villarreal, B. and Saitua, E. (1992). Total lipids and omega-3 fatty acids from seven species of pelagic fish. *In* "Pelagic Fish: the Resource and its Exploitation" (J.R. Burt, R. Hardy and K.J. Whittle, eds), pp.142–148. Fishing News (Books), Cambridge.

Praag, D., Farber, S.J., Minkin, E. and Naftali, P. (1987). Production of eicosanoids by killifish gills and opercular epithelia and their effect in active transport of ions. *General and Comparative Endocrinology* **67**, 50–57.

Precht, H. (1964). Über die Bedeutung des Blutes für die Temperaturadaptazion von Fischen. *Zoologische Jahrbücher für Physiologie* **71**, 313–327.

Privolnev, T.I. (1948). Respiration of fish as a determining factor in fish distribution in the water body (In Russian). *Izvestia Gosudarstvennogo Nauchno-issledovatel-skogo Instituta Ozernogo I Rechnogo Rybnogo* **25**, 125–148.

Privolnev, T.I. and Brizinova, P.N. (1964). Melting-temperature of fish oils (In Russian). *Izvestia Gosudarstvennogo Nauchno-issledovatelskogo Instituta Ozernogo I Rechnogo Rybnogo* **58**, 45–57.

Prosser, C.L. (ed.) (1967). "Molecular Mechanisms of Temperature Adaptation". Publication no. 84, American Association for the Advancement of Science, Washington, DC.

Prosser, C.L. (1969). Principles and general concepts of adaptation. *Environmental Research* **2**, 404–416.

Prosser, C.L. (1979). "Comparative Animal Physiology," 3rd edn. W.B. Saunders Company, Philadelphia, 1011 pp.

Prosser, C.L. and Brown, F.A. (1962). "Comparative Animal Physiology", 2nd edn. W.B. Saunders Company, Philadelphia.

Protasov, V.R. (1978). "Fish Behaviour" (In Russian). Pishchevaya promyshlennost, Moscow, 296 pp.

Pryanishnikova, E.N., Zhulanova, Z.I. and Romantsev, E.F. (1975). Prostaglandins – a new class of biologically active compounds (In Russian). *Biologiya Nauki* 1975 (6), 27–41.

Puchkov, N.V. (1954). "Fish Physiology" (In Russian). Pishchepromizdat, Moscow, 372 pp.

Pütter, A. (1909). "Die Ernährung der Wassertiere und der Stoffhoushalt der Gewässer". Fischer, Jena. 168 pp.

Putman, R.W. and Freel, R.W. (1978). Haematological parameters of five species of marine fishes. *Comparative Biochemistry and Physiology* 61A, 585–588.

Quinn, T.P. (1988). Estimated swimming speeds of migrating adult sockeye salmon. *Canadian Journal of Zoology* 66, 2160–2163.

Rabinovich, A.L. and Ripatti, P.O. (1990). On the conformation properties and functions of docosohexaenoic acid (In Russian). *Doklady Akademii Nauk SSSR* 314, 752–756.

Radakov, D.V. (1972). "Schooling of Fish as an Ecological Phenomenon" (In Russian). Nauka, Moscow, 174 pp.

Radakov, D.V. and Solovyev, B.S. (1959). The first experience of using a submarine for the observation of herring behaviour (In Russian). *Rybnoye Khozyaistvo* 1959 (7), 16–21.

Rakitskaya, L.V. (1982). Seasonal dynamics of haematological characteristics in Black Sea fishes of different ecology (In Russian). *Ekologiya Morya* 1982 (10), 90–93.

Ramaswama, R. and Sushella, L. (1974). A mechanism of thermogenesis by modification of succinate dehydrogenase. *In* "Biomembranes. Architecture, Biogenesis, Bioenergetics and Differentiation", pp.261–277. New York.

Rambhasker, B. and Rao, K.S. (1987). Comparative haematology of ten species of marine fish from the Visakhapatman Coast. *Journal of Fish Biology* 30, 53–66.

Rao, G.M.M. (1969). Effect of activity, salinity and temperature on plasma concentrations of rainbow trout. *Canadian Journal of Zoology* 47, 131–134.

Rashcheperin, V.K. (1967). Ecology of the reproduction of round goby in the Azov Sea (In Russian). Synopsis of Candidate Dissertation, Kaliningrad, 19 pp.

Rass, T.S. (1948). On the life periods and regularities of fish development and growth. *Izvestiya Akademii Nauk SSSR Seriya Biologiya* 1948 (3), 295–305.

Ratnayake, W.N. and Ackman, R.G. (1979). Fatty alcohols in capelin, herring and mackerel oils and muscle lipids: 1. Fatty alcohol details linking dietary copepod fat with certain fish depot fats. *Lipids* 14, 795–803.

Ray, A.K. and Medda, A.K. (1975). Effect of variation of temperature on the protein and RNA contents of liver and muscle of Lata fish (*Ophicephalus punctatus*). *Science and Culture* 41, 532–534.

Rayner, M.D. and Keenan, M.J. (1967). Role of red and white muscle in the swimming of skipjack tuna. *Nature, Lond.* 214, 392–393.

Reay, G.A. (1957). Factors affecting initial and keeping quality. *Food Investigation Board, Department of Scientific and Industrial Research, Annual Report* 1957, p.3.

Reeve, E.B. and MacKinley, J.E. (1970). Measurement of albumin synthetic rate with bicarbonate-C^{14}. *American Journal of Physiology* 218, 498–509.

Reinhardt, S.B. and Van Vleet, E.S. (1986). Lipid composition of twenty-two species of Antarctic midwater zooplankton and fish. *Marine Biology* 91, 149–159.

Reshetnikov, Yu.S. (1980). "The Ecology and Taxonomy of White Fish" (In Russian). Nauka, Moscow, 301 pp.

Reshetnikov, Yu.S. and Klaro, R.M. (1976). The rhythmicity of biological processes in *Lutjanus synagris*, a tropical fish (In Russian). *Voprosy Ikhtiologii* **16**, 784–796.

Reshetnikov, Yu.S., Koshelev, B.V. and Mochek, A.D. (1984). Soviet–Cuban ichthyological researches in the coastal waters of Cuba. *Voprosy Ikhtiologii* **24**, 77–84.

Revina, N.I. (1964). Some elements of the nitrogen metabolism in horse-mackerel fingerlings (In Russian). *Trudy Azovo-Chernomorkogo Nauchno-issledovatelnogo Instituta Morskogo Ribnogo Khozyastva I Okeanografii* **22**, 133–136.

Reynolds, W.W. and Casterlin, M.E. (1980). The role of temperature in the environmental physiology of fishes. *In* "Environmental Physiology of Fishes" (M.A. Ali, ed.), pp. 497–518. NATO Advanced Study Institute Series, No. 35. Plenum Press, New York.

Reznichenko, P.N. (1980). "The Transformation and Change of Functional Mechanisms in Oncogenesis of Lower Vertebrates" (In Russian). Nauka, Moscow, 216 pp.

Ricker, W.E. (1979). Growth rates and models. *In* "Fish Physiology. Bioenergetics and Growth" (W.S. Hoar and D.J. Randall, eds), Vol. 8, pp.678–744. Academic Press, New York.

Ripatti, P.O., Mikhailova, N.V. and Antonova, V.V. (1993). Estimation of pools of fatty acids connected covalently with coenzyme A and proteins (In Russian). *In* "Biochemical Methods in Ecological and Toxicological Research", pp.156–168. Petrozavodsk.

Roberts, J.L. (1973). Effects of thermal stress on gill ventilation and heart rate in fishes. *In* "Responses of Fish to Environmental Changes" (W. Chavin, ed.), pp.64–86. Charles C. Thomas, Springfield, Illinois.

Robertson, J.D. (1954). The chemical composition of the blood of some aquatic chordates, including members of the Tunicata, Cyclostomata and Osteichthyes. *Journal of Experimental Biology* **31**, 424–442.

Robertson, O.H., Hane, S., Wexler, B.C. and Rinfret, A.P. (1963). The effect of hydrocortisone on immature rainbow trout (*Salmo gairdnerii*). *General and Comparative Endocrinology* **3**, 422–436.

Roche, J., Collet, J. and Mourgue, M. (1940). Phosphatase activity and the growth of dermal bones (scales) in fish (Selachii and Teleostei). *Enzymologia* **8**, 257–260.

Romanenko, V.D. (1978). "Liver and the Regulation of Intermediate Metabolism. Mammals and Fish" (In Russian). Naukova Dumka, Kiev, 183 pp.

Romanenko, V.D., Evtushenko, N.Yu. and Kotsar, N.I. (1980). "Carbon Dioxide Metabolism in Fish" (In Russian). Naukova Dumka, Kiev, 179 pp.

Romanenko, V.D., Arsan, O.M. and Solomatina, V.D. (1982). "Calcium and Phosphorus in the Vital Activity of Hydrobionts" (In Russian). Naukova Dumka, Kiev, 153 pp.

Romanenko, V.D., Arsan, O.M. and Solomatina, V.D. (1991). "The Mechanisms of Temperature Acclimation of Fish" (In Russian). Naukova Dumka, Kiev, 192 pp.

Rome, L.C., Loughna, P.T. and Goldspink, G. (1985). Temperature acclimation: improved sustained swimming performance in carp at low temperatures. *Science, New York* **228**, 194–196.

Rome, L.C., Alexander, R.M., Funke, R., Lutz, G. and Freedman, M. (1986). Muscle function in fish during swimming. *Biological Bulletin, Marine Biological Laboratory, Woods Hole* **171**, 502.

Roots, B.J. (1968). Phospholipids of goldfish brain: the influence of environmental temperature. *Comparative Biochemistry and Physiology* **25**, 457–466.

Ross, D.A. (1977). Lipid metabolism of the cod, *Gadus morhua* L. Ph.D. thesis, University of Aberdeen, Scotland.

Ross, D.A. and Love, R.M. (1979). Decrease in the cold-store flavour developed by frozen fillets of starved cod (*Gadus morhua* L.). *Journal of Food Technology* **14**, 115–122.

Rowe, D.K. and Thorpe, J.E. (1990). Suppression of maturation in male Atlantic salmon (*Salmo salar* L.) parr by reduction in feeding and growth during spring months. *Aquaculture* **86**, 291–313.

Rowe, D.K., Thorpe, J.E. and Shanks, A.M. (1991). Role of fat stores in the maturation of male Atlantic salmon (*Salmo salar* L.) parr. *Canadian Journal of Fisheries and Aquatic Science* **48**, 405–413.

Rudneva-Titova, I.I. (1994). Correlation between antioxidative enzyme activity and lipid peroxidate oxidation processes in the embryogenesis of Black Sea round goby (In Russian). *Ontogenesis* **25**, 13–20.

Rudneva-Titova, I.I. (1995). Lipid content and peroxidation of lipid in the blood serum of Black Sea cartilaginous and bony fishes (In Russian). *Zhurnal Evolutsionnoy Biokhimii i Physiologii* **31**, 14–20.

Rudneva-Titova, I.I. and Zherko, N.V. (1994). Influence of polychlorvinyl biphenyls on the activity of antioxidative enzymes and lipid peroxidation in muscles and liver of two Black Sea fish species (In Russian). *Biochimiya* **59**, 34–44.

Ruhland, M.L. (1969). Relationship between the activity of the thyroid gland and oxygen consumption in the cichlides (teleostei). *Experientia* **25**, 944–945.

Ryabushko, V.I. and Propp, L.N. (1985). Respiration and nitrogen and phosphorus excretion by echinoderms from the South China Sea (In Russian). *Biologiya Morya* (6), 42–46.

Rychagova, T.L. (1989). Dynamics of morphophysiological and biochemical characteristics of Caspian kilka (*Clupeonella engrauliformes*) over the annual cycle (In Russian). *Voprosy Ikhtiologii* **29**, 62–67.

Rychagova, T.L., Sedov, S.I. and Valentinova, S.Ya. (1987). Prediction of times of spring migration of caspian kilka (In Russian). *Rybnoye Khozyaistvo* **1987** (4), 54–55.

Ryzkov, L.P. (1976). "Morphophysiological Regularities and Transformation of Matter and Energy at Early Ontogenesis of Freshwater Fishes of the Salmon Family" (In Russian), Petrozavodsk, Karelia, 288 pp.

Rzhavskaya, F.M. (1976). "Fats of Fish and Marine Animals" (In Russian), Pishchevaya Promyshlennost, Moscow, 470 pp.

Saddler, J.B., Lowry, R.R., Krueger, H.M. and Tinsley, L. (1966). Distribution and identification of fatty acids from the coho salmon. *Journal of the American Oil Chemists' Society* **43**, 321–324.

Saez, L., Goicoschea, O., Amthauer, R. and Krauskopi, M. (1982). Behaviour of RNA and protein synthesis during the acclimation of the carp. Studies with isolated hepatocytes. *Comparative Biochemistry and Physiology* **72B**, 31–38.

Safyanova, T.E. and Demidov, V.F. (1955). The Black Sea anchovy response to artificial light during the breeding and feeding seasons (In Russian). *Trudy Azovsko-Chernomorskogo Nauchno-issledovatelnogo Instituta Morskogo Ribnogo Khozyastva i Okeanografii* **16**, 71–88.

Salmenkova, E.A. (1973). Genetics of fish isoenzymes (In Russian). *Uspekhi Sovremennoy Biologii* **75**, 217–235.

Samyshev, E.Z. (1992). Dynamics of phyto- and zooplankton (In Russian). *In* "Hydrometeorology and Hydrochemistry of the Seas of the USSR" (A.I. Simonov, A.I. Ryabinin and D.E. Gershanovich, eds), Vol. 4, no. 2, pp.173–178, Gydrometeoizdat, St Petersburg.

Sand, O., Petersen, J.M. and Korsgaard-Emmersen, B. (1980). Changes in some carbo-hydrate metabolising enzymes and glycogen in liver, glucose and lipid in serum during vitellogenesis and after induction by oestradiol-13β in the flounder. *Comparative Biochemistry and Physiology* **65B**, 327–332.

Sandnes, K, Ulgenes, Y., Braekkan, O.R. and Utne, F. (1984). The effect of ascorbic acid supplement in broodstock feed on reproduction of rainbow trout (*Salmo gairdneri*). *Aquaculture* **43**, 167–177.

Sargent, J.R. (1976). The structure, metabolism and function of lipids in marine organisms. *In* "Biochemical and Biophysical Perspectives in Marine Biology" (D.C. Malins and J.R. Sargent, eds). Vol. 3, pp.149–212. Academic Press, London.

Sargent, J.R. (1978). Marine wax esters. *Science Progress* **65**, 437–458.

Sargent, J.R. and Henderson, R.J. (1980). Lipid metabolism in marine animals. *Transaction of the Biochemical Society* **8**, 296–297.

Sargent, J.R., Gatten, R.R. and McIntosh, R. (1971). Metabolic relationships between fatty alcohol and fatty acids in the liver of *Squalus acanthias*. *Marine Biology* **10**, 346–355.

Sato, Y. and Tsuchiya, Y. (1970). Studies on the lipids of *Ruvettus pretiosus*. II. The composition of the unsaponifiable material and the purgative action of the oils on the mouse. *Tohoku Journal of Agricultural Research* **21**, 176–182.

Sautin, Yu.Yu. (1985). Somatotropin and prolactin forming activity of the hypophysis and some aspects of protein and lipid metabolism in carp in thermal and pond culture (In Russian). *Gydrobiologicheskii Zhurnal* 1985 (3), 92–98.

Sautin, Yu.Yu. (1989). The problem of regulation of the adaptive changes of lipoge-nesis, lipolysis and lipid transport in fish (In Russian). *Uspekhi Sovremennoy Biologii* **107**, 131–149.

Sautin, Yu.Yu. and Romanenko, V.D. (1982). Effect of photoperiod and temperature on somatotropic and lactotropic activity of the hypophysis of carp (In Russian). *Zhurnal Evolutsionnoy Biokhimii i Physiologii* **28**, 471–477.

Savina, M.V. (1992). "Mechanisms of Tissue Respiration Adaptations in the Evolution of Vertebrates" (In Russian). Nauka, St Petersburg, 200 pp.

Savina, M.V. and Plisetskaya, E.M. (1976). Glycogen synthesis in isolated tissues of the lamprey, *Lampetra fluviatilis* (In Russian). *Zhurnal Evolutsionnoy Biokhimii i Physiologii* **12**, 282–284.

Savina, M.V., Ivanova, T.I. and Egoyauz, M.A. (1993). Mitochondria of some poikilo-thermal vertebrates: oxidative phosphorylation and adenine nucleotides (In Russian). *Zhurnal Evolutsionnoy Biochimii i Physiologii* **29**, 113–119.

Savitz, J. (1969). Effects of temperature and body weight on endogenous nitrogen excretion in the bluegill sunfish. *Journal of the Fisheries Research Board of Canada* **26**, 1813–1821.

Schmidt, J. (1930). Racial investigations. X. The Atlantic cod (*Gadus callarias* L.) and local races of the same. *Comptes Rendus des Travaux Laboratoire Carlsberg* **18**, 72 pp.

Schmidt-Nielsen, K. (1975). "Animal Physiology. Adaptation and the Environment", Part 1. Cambridge University Press, London, New York. 699 pp.

Schmidt-Nielsen, K. (1979). "Animal Physiology. Adaptation and the Environment", Part 2. Cambridge University Press, London, New York.

Scholander, P.F., Flagg, W., Hoch, R.J. and Irving, L. (1953). Studies on the physi-ology of frozen plants and animals in the Arctic. *Journal of Cellular and Comparative Physiology*, Suppl. 1, 1–56.

Schünke, M. and Wodtke, (1983). Cold-induced increase of delta 9- and delta 6-desat-urase activities in endoplasmic membranes of carp liver. *Biochimica et Biophysica Acta* **734**, 70–75.

Schwartz, S.S., Dobrinsky, L.N. and Toparkova, L.Ya. (1965). The dynamic description of morpho-physiological features of animals (In Russian). *Byulleten Moskogo Obshchestva Ispytalteli Prirody, Moskva* **70**(5), 5–15.

Schwassmann, H.O. (1979). Biological rhythms, their adaptive significance. *In* "Environmental Physiology of Fishes" (M.A. Ali, ed.). pp.613–630. Plenum Press, New York, London.

Secombes, C., Lewis, A., Laird, L., Needham, E. and Priede, I. (1985). Role of autoantibodies in the autoimmune response to testis in rainbow trout (*Salmo gairdneri*). *Immunology* **56**, 409–415.

Selivonchick, D.P. and Roots, B.J. (1976). Variation in myelin lipid composition induced by change in environmental temperature of goldfish. *Journal of Chemical Biology* **1**, 131–135.

Selivonchick, D.P., Johnston, P.V. and Roots, B.J. (1977). Acyl and alkenyl group composition of brain subcellular fractions of goldfish acclimated to different environmental temperatures. *Neurochemical Research* **2**, 379–393.

Selkov, E.E. (1978). Temporal organisation of energy metabolism and the cell clock (In Russian). *In* "Regulation of Energy Metabolism and Physiological Condition of the Organism", pp.15–32, Nauka, Moscow.

Selye, H. (1973). The evolution of the stress concept. *American Scientist* **61**, 692–699.

Semenchenko, N.N. (1988). Mechanisms of self-regulation of sockeye salmon population abundance (In Russian). *Voprosy Ikhtiologii* **28**, 44–52.

Semenchenko, S.M. (1992). Dynamics of the energy metabolism during larval development of Baikal omul (In Russian). *Voprosy Ikhtiologii* **32**, 117–123.

Semenkova, T.B. (1984). Effect of hormonal factors on lipid level in liver of Siberian sturgeon in the Lena river (In Russian). *Voprosy Ikhtiologii* **24**, 158–164.

Senkevich, N.K. (1967). Relationship between alkaline phosphatase activity in the scales of some Azov and Black Sea fishes and the rate and term of their linear growth (In Russian). *In* "Metabolism and Biochemistry of Fish" (G.S. Karzinkin, ed.), pp.265–269, Nauka, Moscow.

Serebrennikova, T.P. (1981). The study of glycogen phosphorylase features in muscle tissues of horse-mackerel and plaice (In Russian). *Zhurnal Evolutsionnoy Biokhimii i Physiologii* **17**, 537–541.

Serebrennikova, T.P., Silkina, E.N., Shmelev, V.K., Morozova, A.L. and Nesterov, V.P. (1991). The regulation of glycogenolysis in skeletal muscle of annular bream during intensive performance as dependent on temperature (In Russian). *Zhurnal Evolutsionnoy Biokhimii i Physiologii* **27**, 427–431.

Sergeev, Yu.S. (1979). The method of quantitative estimation of metabolic feeding and production rate of fish (In Russian). *In* "Current Problems of Ecological Physiology of Fish" (N.S. Stroganov, ed.), pp. 185–192. Nauka, Moscow.

Sergeeva, N.T. (1985). Effect of imbalance of fatty acid and amino acid composition of food on lipid metabolism in rainbow trout (In Russian). *In* "Plastic Metabolism in Fish", pp.15–21, Kaliningrad.

Sergeeva, N.T., Zhdanov, Yu.I., Lempert, O.T. and Pisareva, N.A. (1987). Effect of essential fatty acids and vitamins in food on substance metabolism in rainbow trout (In Russian). *Rybnoye Khozyaistvo* 1987 (8), 41–44.

Severtsev, A.N. (1934). "Main Trends of Evolution. Morphobiological Theory of Evolution" (In Russian). Biomedgiz, Moscow, Leningrad, 149 pp.

Shackley, S.E. and King, P.E. (1978). Protein yolk synthesis in *Blennius pholis*. *Journal of Fish Biology* **13**, 179–193.

Shaklee, J.B., Christiansen, J.A., Sidell, B.D., Prosser, C.L. and Whitt, G.S. (1977). Molecular aspects of temperature acclimation in fish: contributions of changes in enzyme activities and isozymic patterns to metabolic reorganisation in the green sunfish. *Journal of Experimental Zoology* **201**, 1–20.

Shamardina, I.P. (1954). Changes in the respiratory rate of growing fish (In Russian). *Doklady Akademii Nauk SSSR* **98**, 689–692.

Sharp, G.D. (1973). An electrophoretic study of haemoglobins of some scombroid fishes and related forms. *Comparative Biochemistry and Physiology* **44B**, 381–388.

Shatunovsky, M.I. (1963). Dynamics of fatness and water content in muscles and gonads of Baltic flounder as related to gonad maturation (In Russian). *Voprosy Ikhtiologii* **3**, 652–667.

Shatunovsky, M.I. (1967). Alterations of biochemical composition of White Sea flounder liver and blood during the maturation of reproductive materials in summer and autumn (In Russian). *Vestnik Moskovkogo Gosudarstvennogo Universiteta, Seryiya Biologii, Pochvovedenie* 1967 (2), 22–30.

Shatunovsky, M.I. (1970). Characteristics of the qualitative composition of fats in eggs, fry and spawning females of the spring and autumn Baltic herring in the Gulf of Riga (the Baltic Sea) (In Russian). *Voprosy Ikhtiologii* **10**, 1026–1034.

Shatunovsky, M.I. (1978). Annual balances of matter and energy in some age groups of cod, haddock, Baltic herring and flounder (In Russian). *Trudy VNIRO* **120**, 13–19.

Shatunovsky, M.I. (1980). "Ecological Regularities in Marine Fishes" (In Russian). Nauka, Moscow, 284 pp.

Shatunovsky, M.I. and Rychagova, M.I. (1990). On some size and age-related changes in the metabolism of anchovy kilka, *Clupeonella engrauliformis* (In Russian). *Voprosy Ikhtiologii* **30**, 154–158.

Shaverdov, R.S. (1964). On the relationships between large and small horse-mackerel of the Black Sea (In Russian). *Voprosy Ikhtiologii* **4**, 82–91.

Shchepkin, V.Ya. (1972). A comparison between the liver and muscle lipids of horse-mackerel and scorpionfish (In Russian). *Nauchnye Doklady Vysshey Shkoly. Biologicheskiye Nauki* 1972 (2), 36–39.

Shchepkin, V.Ya. (1978). The study of succinate dehydrogenase activity of white skeletal, red muscles and liver in fish with different ecological and physiological peculiarities (In Russian). *Biologiya Morya, Kiev* **46**, 104–107.

Shchepkin, V.Ya. (1979). Seasonal dynamics of the lipid composition of liver and muscle in horse-mackerel and scorpionfish (In Russian). *Gydrobiologicheskii Zhurnal*, **15**, 77–84.

Shchepkin, V.Ya. and Minyuk, G.S. (1987). Dynamics of the lipid content of sprat muscle over the annual cycle (In Russian). *Ekologiya Morya* **27**, 61–64.

Shchepkin, V.Ya. and Minyuk, G.S. (1990). The levels of accumulated energy and the response to light in Black Sea sprat (In Russian). *In* "Bioenergetics of Aquatic Organisms" (G.E. Shulman and G.A. Finenko, eds), pp. 207–221, Naukova Dumka, Kiev.

Shchepkin, V.Ya., Belokopytin, Yu.S., Minyuk, G.S. and Shulman, G.E. (1994). Utilisation of proteins and lipids, and the oxygen consumption by pickerel in fast swimming (In Russian). *Gydrobiologicheskii Zhurnal* **30**, 58–61.

Shchepkina, A.M. (1980a). The lipid composition of tissues of Black Sea anchovy over the annual cycle and of the nematode infestation of larvae (In Russian). *Ekologiya Morya* **3**, 33–39.

Shchepkina, A.M. (1980b). The influence of helminths on the tissue lipid content of Black Sea anchovy and bullhead during the annual cycle. *Technical Reports National Marine Fisheries Service, NOAA* **25**, 49–51.

Shchepkina, A.M. (1990). The influence of helminths on the level of energy stored in the body of Black Sea mussels (In Russian). *In* "Bioenergetics of Water Organisms" (G.E. Shulman and G.A. Finenko, eds), pp. 72–78. Naukova Dumka, Kiev.

Shcherban, S.A. and Abolmasova, G.I. (1991). The growth of mussels at Laspi Bay in the Black Sea (In Russian). *Biologiya Morya* 1991 (2), 82–89.

Shcherbina, M.A. (1973). "The Digestion and Efficiency of Artificial Nutriment Utilisation in Carp" (In Russian). Pischevaya Promyshlennost, Moscow, 132 pp.

Shcherbina, M.A. (1989). On the endogenous feeding of carp fingerlings at different periods of wintering (In Russian). *Voprosy Ikhtiologii* **29**, 142–154.

Shcherbina, M.A., Burlachenko, I.V. and Sergeeva, N.T. (1988). On the chemical composition of eggs and the amino acid requirements of two species of Black Sea mullet (In Russian). *Voprosy Ikhtiologii* **28**, 132–137.

Shekhanova, I.A. (1983). "Radioecology of Fish" (In Russian). Legkaya i Pishchevaya Promyshlennost, Moscow, 208 pp.

Shekk, P.V. (1983). On the energy metabolism and food rations of grey mullet during wintering (In Russian). *In* "The Physiological Foundations of Marine and Diadromous Fish Reproduction" (J.A. Shelkanova, ed.), pp. 81–85. Legkaya i Pishchevaya Promyshlennost, Moscow.

Shekk, P.V., Kulikova, N.I. and Rudenko, V.I. (1990). Age-induced shifts in the response to cold in Black Sea golden mullet (In Russian). *Voprosy Ikhtiologii* **30**, 94–106.

Shelbourne, J.E. (1974). Population effects on the survival, growth and pigment of tank-reared plaice larvae. *In* "Sea Fisheries Research" (F.R. Harden Jones, ed.), pp.357–377. Elek Science, London.

Shelbourne, J.E. (1975). Marine fish cultivation: pioneering studies on the culture of the larvae of the plaice (*Pleuronectes platessa* L.) and the sole (*Solea solea* L.). *Fishery Investigations Lond. Ser. 2* **27**(9), 29 pp.

Shelukhin, G.K., Metallov, G.F. and Geraskin, P.P. (1989). Influence of temperature on sturgeon fry cultivated in Caspian water of varying salinity (In Russian). *In* "Ecological Biochemistry and Physiology of Fish", Part 2, pp. 231–233.

Shevchenko, V.V. (1977). Study of the efficient fishing of the North Sea haddock population using a production model (In Russian). *Trudy VNIRO* **121**, 93–99.

Shier, W.T., Lin, Y. and De Vries, A.L. (1972). Structure and mode of action of glycoproteins from an Antarctic fish. *Biochimica et Biophysica Acta* **263**, 406–413.

Shilov, I.A. (1977). "The Ecologo-Physiological Basis of Population Relationships in Animals" (In Russian). Nauka, Moscow, 263 pp.

Shilov, I.A. (1985). "Physiological Ecology of Animals" (In Russian). Vysshaya Shkola, Moscow, 328 pp.

Shimma, Y., Suzuki, R., Yamaguchi, M. and Akiyama, T. (1977). On the lipids of adult carp raised on fishmeal and SCP feed, and hatchability of their eggs. *Bulletin of the Freshwater Fisheries Laboratory, Tokyo* **27**, 35–48.

Shindo, K., Tsuchiya, T. and Matsumoto, J.J. (1986). Histological study on white and dark muscle of various fishes. *Bulletin of the Japanese Society of Scientific Fisheries* **52**, 1377–1399.

Shkorbatov, G.L. (1961). Intraspecies physiological variability in aquatic poikilotherms (In Russian). *Zoologicheskii Zhurnal* **40**, 1437–1452.

Shkorbatov, G.L., Gubin, I.E. and Dmitrenko, V.F. (1972). The oxidative and ATP-activity of carp muscle mitochondria as related to temperature adaptation (In Russian). *In* "Energy Aspects of Growth and Metabolism of Aquatic Animals" (G.E. Shulman, ed.), pp. 254–255.

Shlyakhov, V.A., Chashchin, A.K. and Kozkosh, N.I. (1990). Intensity of fishing and dynamics of the stock of Black Sea anchovy (In Russian). *In* "Biological Resources of the Black Sea" (K.S. Tracheva and Yu.K. Benko, eds), pp. 93–102. All-Union Institute of Marine Fishery and Oceanography, Moscow.

Shmalgausen, I.I. (1969). "Problems of Darwinism" (In Russian). Nauka, Leningrad, 494 pp.

Shmerling, M.D., Buzueva, I.I. and Filyushina, E.E. (1984). The electronic microscopic study of grayling skeletal muscles at rest and with muscular loads (In Russian). In "Studies of fish locomotion energetics", pp. 61–68. Nauka, Novosibirsk, Siberian Branch.

Shpet, G.I. (1971). "Increase of the Growth Rate and Productivity in Evaluating Animals" (In Russian). Urozhay, Kiev, 112 pp.

Shubnikov, D.A. (1959). On the applicability of data about fish fatness and blood components in scouting for Atlanto-Scandinavian herring in summer time (In Russian). *Rybnoye Khozyaistvo* 1959 (3), 12–14.

Shugaev, B.S. (1989). A mathematical model of selection, entailing energy metabolism of organisms as a tool for studying the evolutionary process (In Russian). *Zhurnal Obshchey Biologii* **50**, 199–206.

Shulman, G.E. (1957). Characteristics of the chemical composition of Azov anchovy during spring and wintering migrations (In Russian). *Rybnoye Khozyaistvo* 1957 (8), 68–70.

Shulman, G.E. (1960a). Dynamics of fat content of the fish body (In Russian). *Uspechi Sovremennoy Biologii* **49**, 225–239.

Shulman, G.E. (1960b). Dynamics of chemical composition of Azov Sea anchovy in relation to its biology (In Russian). *Trudy Azovsko-chernomorskogo Nauchno-issledovatelnogo Instituta morskogo Rybno Khozyastva i Oceanografii* **18**, 130–144.

Shulman, G.E. (1961). Correlation between fat and water contents of the fish and the technique of fatness calculation during an expedition (In Russian). *Trudy Azovsko-chernomorskogo Nauchno-issledovatelnogo Instituta morskogo Rybno Khozyastva i Oceanografii* **19**, 36–44.

Shulman, G.E. (1962). Elements of nitrogen balance and food rations of Azov anchovy (In Russian). *Doklady Akademii Nauk SSSR* **147**, 724–726.

Shulman, G.E. (1964a). The character of fat content dynamics in large Black Sea horse-mackerel in relation to its biology (In Russian). *Trudy Azovsko-chernomorskogo Nauchno-issledovatelnogo Instituta morskogo Rybno Khozyastva i Oceanografii* **22**, 101–106.

Shulman, G.E. (1964b). Seasonal changes of fat content in the body of 'small' Black Sea horse mackerel (In Russian). *Voprosy Ikhtiologii* **4**, 764–768.

Shulman, G.E. (1967). Dynamics of the fat stored in the liver of Azov round goby (In Russian). *Doklady Akademii Nauk SSSR* **175**, 710–719.

Shulman, G.E. (1972a). "Physiologo-Biochemical patterns in the Annual Cycle of Fish" (In Russian). Pishchevaya Promyshlennost, Moscow, 368 pp.

Shulman, G.E. (1972b). On the level of fat content in fish from different areas of the central Mediterranean (In Russian). *Voprosy Ikhtiologii* **12**, 133–140.

Shulman, G.E. (1974). "Life Cycles of Fish. Physiology and Biochemistry". Kulstad Press, John Wiley and Sons, New York, 253 pp.

Shulman, G.E. (1978a). The principles of physiologo-biochemical investigations of annual cycles of fish (In Russian). *Biologiya Morya, Kiev* **46**, 90–104.

Shulman, G.E. (1978b). Relation of protein growth and fat accumulation with weight and caloric content increase in fish of the genus *Engraulis* (In Russian). *Biologiya Morya, Vladivostok* **5**, 80–82.

Shulman, G.E. (1996). Physiologo-biochemical investigations of hydrobionts (In Russian). *Ekologiya Morya* **45**, 39–48.

Shulman, G.E. and Dobrovalov, I.S. (1979). The state of affairs in ecologo-physiological studies of Black Sea fishes (In Russian). *In* "The Foundations of

Biological Productivity of the Black Sea" (V.N. Greze, ed.), pp. 321–340, Naukova Dumka, Kiev.

Shulman, G.E. and Khotkevich, T.V. (Yuneva) (1977). Utilisation of fat in the teleost *Mullus barbatus ponticus* during swimming (In Russian). *Zhurnal Evolutsionnoy Biokhimii i Physiologii* **13**, 86–88.

Shulman, G.E. and Kokoz, L.M. (1968). Peculiarities of protein growth and fat accumulation in Black Sea fishes (In Russian). *Biologiya Morya, Kiev* **15**, 159–206.

Shulman, G.E. and Kulikova, N.I. (1966). On the specificity of protein composition of blood serum in fish (In Russian). *Uspekhi Sovremennoy Biologii* **62**, 42–60.

Shulman, G.E. and Nigmatullin, C.M. (1981). Changes in liver indices of the squid *Stenoteuthis oualaniensis* from the tropical Indian Ocean under experimental conditions (In Russian). *Ekologiya Morya* **5**, 95–103.

Shulman, G.E. and Shatunovsky, M.I. (1975). The basic principles of physiologo-biochemical studies of species (In Russian). *In* "Study of Species Productivity within the Distribution Range" (S.R. Volski's, ed.), pp.23–25. Mintas, Vilnius.

Shulman, G.E. and Shchepkina, A.M. (1983). Comparative evaluation of fish condition by morphological and physiologo-biochemical parameters (In Russian). *Rybnoye Khozyaistvo* 1983 (4), 25–27.

Shulman, G.E. and Urdenko, S.Yu. (1989). "Productivity of Fishes of the Black Sea" (In Russian). Naukova Dumka, Kiev, 188 pp.

Shulman, G.E. and Yakovleva, K.K. (1983). Hexaenoic acid and natural mobility of fish (In Russian). *Zhurnal Obshchey Biologii* **44**, 529–540.

Shulman, G.E. and Yuneva, T.V. (1990a). The role of docosohexaenoic acid in adaptations of fish (In Russian). *Gydrobiologicheskii Zhurnal* **26**, 43–51.

Shulman, G.E. and Yuneva, T.V. (1990b). Docosohexaenoic acid and unsaturation of lipids in fish (In Russian). *Gydrobiologicheskii Zhurnal* **26**, 50–55.

Shulman, G.E., Vengrzhin, E.P and Dubinina, V.N. (1957). Characteristics of the gas metabolism of Azov gobies in relation to the environment (In Russian). *Voprosy Ikhtiologii* **8**, 77–80.

Shulman, G.E., Revina, N.I. and Safyanova, T.E. (1970). Relationship between physiological condition and oogenesis feature of pelagic fish (In Russian). *Trudy VNIRO* **59**, 96–108.

Shulman, G.E., Shchepkin, V.Ya., Yakovleva, K.K. and Khotkevich (Yuneva), T.V. (1978). Lipids and their utilisation by fish while swimming (In Russian). *In* "Elements of Physiology and Biochemistry in Total and Active Metabolism of Fish" (G.E. Shulman, ed.), pp.100–121. Naukova Dumka, Kiev.

Shulman, G.E., Oven, L.S. and Urdenko, S.Yu. (1983). Evaluation of the generative production of Black Sea fishes (In Russian). *Doklady Akademii Nauk SSSR* **272**, 254–256.

Shulman, G.E., Abolmasova, G.I. and Muravskaya, Z.A. (1984). Physiologo-biochemical principles of investigation of ecological features of squid in epipelagic layers of the world ocean (in Russian). *Zhurnal Obshchey Biologii* **45**, 631–644.

Shulman, G.E., Shchepkin, V.Ya., Yakovleva, K.K., Minyuk, G.S., Getmantsev, V.A. and Levin, S.Yu. (1985). Formation of fishable stocks and the long-term variations of fatness in Black Sea sprat (In Russian). *Rybnoye Khozyaistvo* 1985 (5), 26–28.

Shulman, G.E., Belokopytin, Yu.S., Stolbov, A.Ya. and Yuneva, T.V. (1987). Ecologo-physiological studies of Black Sea sprat (In Russian). *Okeanologiya* **27**, 155–157.

Shulman, G.E., Shchepkin, V.Ya. and Minyuk, G.S. (1989). Determination of fat and lipid contents in Black Sea sprats (In Russian). *Rybnoye Khozyaistvo* 1989 (12), 86–87.

Shulman, G.E., Ostalovsky, E.M., Shershov, S.V. and Kryachko, V.I. (1990). Phospholipid composition of Black Sea fishes (In Russian). *In* "Bioenergetics of Aquatic Organisms" (G.E. Shulman and G.A. Finenko, eds). pp.189–196. Naukova Dumka, Kiev.

Shulman, G.E., Shchepkina, A.M. and Chesolin, M.B. (1992). Physiologo-biochemical analysis of the food supply of wing-armed squid at the dynamic zones of the Eastern Atlantic (In Russian). *Doklady Akademii Nauk SSSR* **322**, 813–816.

Shulman, G.E., Abolmasova, G.I. and Stolbov, A.Ya. (1993a). Protein utilisation in the energy metabolism of hydrobionts (In Russian). *Uspekhi Sovremennoy Biologii* **113**, 576–586.

Shulman, G.E., Stolbov, A.Ya. and Abolmasova, G.I. (1993b). A possible cause of the muscle destruction in Acipenseridae (In Russian). *Rybnoye Khozyaistvo* 1993 (4), p. 26.

Shulman, G.E., Chashchin, A.K., Minyuk, G.S., Shchepkin, V. Ya., Nikolsky, V.N., Dobrovolov, I.S., Dobrovolova, S.G. and Zhigulenko, A.S. (1994). Long-term monitoring of Black Sea sprat condition (In Russian). *Doklady Akademii Nauk* **335**(1), 124–126.

Shustov, Yu.A., Shurov, I.L. and Veselov, A.E. (1989). Influence of temperature on physical characteristics of lake salmon fry (In Russian). *Voprosy Ikhtiologii* **29**, 676–677.

Shvydky, G.V. (1986). The patterns of distribution of the shoals of ivasi of different fatness during feeding season (In Russian). *Rybnoye Khozyaistvo* 1986 (12), 22–24.

Shvydky, G.V. and Vdovin, A.N. (1991). Distribution of Okhotsk Sea walleye pollock, *Theragra chalcogramma*, of different fatness in summer time (In Russian). *Rybnoye Khozyaistvo* **1991** (9), 33–34.

Sidell, B.D. (1977). Turnover of cytochrome C in skeletal muscle of green sunfish during thermal acclimation. *Journal of Experimental Zoology* **199**, 233–270.

Sidell, B.D. (1980). Responses of goldfish (*Carassius auratus*) muscle to acclimation temperature: alterations in biochemistry and proportions of different fibre types. *Physiological Zoology* **53**, 98–107.

Sidell, B.D., Wilson, F.R., Hazel, J. and Prosser, C.L. (1973). Time course of thermal acclimation in goldfish. *Journal of Comparative Physiology* **84**, 119–127.

Sidorov, V.S. (1983). "Ecological Biochemistry of Fish Lipids". Nauka, Leningrad, 240 pp.

Sidorov, V.S. and Guryanova, S.D. (1981). The content of free amino acids in the liver of burbot and stickleback infested with pleurocercoid cestodes (In Russian). *Parasitologiya* **15**, 126–130.

Sidorov, V.S., Lizenko, E.I., Ripatti, P.O. and Bolgova, O.M. (1977). Lipids of fish. Total lipid content in organs of salmonidae and some other fishes (In Russian). *In* "Comparative Biochemistry of Fish and their Helminths", pp. 5–56. Petrozavodsk.

Sidorov, V.S., Bolgova, O.M., Yarzhombek, A.A. and Lizenko, E.I. (1985). Fatty acid composition of phosphatidyl choline in 'weak' and 'strong' year-old carp at the end of wintering (In Russian). *In* "Biochemistry of Fry of Freshwater fishes", pp. 5–14, Petrozavodsk.

Sidorov, V.S., Yurovitsky, Yu.G., Kirilyuk, S.D. and Taksheev, S.A. (1990). Principles and methods of ecologo-biochemical monitoring of water bodies (In Russian). *In* "Biochemistry of Ecto- and Endothermic Organisms in Normal and Pathological Situations", pp. 5–27, Karelian Scientific Centre, Petrozavodsk.

Sidorov, V.S., Kirilyuk, S.D. and Vysotskaya, R.Ch. (1991). Some results and problems of the study of biochemical changes in Volga sturgeon tissues during muscle destruction (In Russian). *In* "Biochemical Features of Fish Diseases" (V.S. Sidorov and R.U. Vysotskaya, eds), pp. 6–24, Petrozavodsk.

Sidorov, V.S.,Vysotskaya, R.U. and Nemova, N.N. (1993). Evolutionary aspects of ecologo-biochemical monitoring (In Russian). *In* "Biochemical Methods in Ecology and Toxicological Researches" (V. Sidorov, ed.), pp. 5–35, Karelian Scientific Centre Publishing, Petrozavodsk.

Siebenaller, J.F. (1984). Analysis of the biochemical consequences of ontogenetic vertical migration in a deep-living teleost fish. *Physiological Zoology* **57**, 598–608.

Silkina, E.N. (1990). The glycogen and lactate content in skeletal muscle of fish during short-term intensive swimming in summer and autumn (In Russian). *Zhurnal Evolutsionnoy Biokhimii i Physiologii* **26**, 68–72.

Sindermann, C.J. and Mairs, D.F. (1961). Blood properties of prespawning and postspawning anadromous alewives. *Fishery Bulletin US* **61**, 145–151.

Singer, S.J. and Nicholson, G.L. (1972). The fluid mosaic model of the structure of cell membranes. *Science, New York* **175**, 720–731.

Sinnhuber, R.O. (1969). The role of fats. *In* "Fish in Research" (O. Neuhaus and J.E. Halver, eds), pp. 245–259, Academic Press, New York.

Sinnhuber, R.O., Castell, J.D. and Lee, D.G. (1972). Essential fatty acid requirement of the rainbow trout, *Salmo gairdneri*. *Federation Proceedings* **31**, 1436–1441.

Skazkina, E.P. (1972). On the active metabolism in Azov Sea gobies (In Russian). *Trudy VNIRO* **85**, 138–144.

Skazkina, E.P. (1975). Energy metabolism of quinaldine-narcotised anchovy and kept singly or in groups (In Russian). *Doklady Akademii Nauk SSSR* **225**, 238–240.

Skazkina, E.P. and Danilevsky, N.N. (1976). On the utilisation of the nutritive base of the Black Sea by anchovy (In Russian). *Trudy VNIRO* **116**, 36–41.

Skazkina, E.P. and Kostyuchenko, V.A. (1968). Food rations of the Azov Sea round goby (In Russian). *Voprosy Ikhtiologii* **8**, 303–311.

Sklyarov, V.Ya., Gamygin, E.A. and Rhyzhkov, L.P. (1984). "Feeding of Fish" (In Russian). Legkaya i Pishchevaya Promyshlennost, Moscow, 120 pp.

Slatina, L.N. (1986). Diel metabolic rhythms in the annual cycle of Black Sea mussel (In Russian). *In* "Problems of Modern Biology", pp.187–194. Moscow University.

Slobodkin, L.B. (1962). Energy in animal ecology. *Advances in Ecological Research* **1**, 69–101.

Slobodkin, L.B. (1964).The strategy of evolution. *American Scientist* **52**, 342–357.

Slonim, A.D. (1971). "Ecological Physiology of Animals" (In Russian). Vysshaya Shkola, Moscow, 448 pp.

Slonim, A.D. (1979). Conception of physiological adaptations (In Russian). *In* "Ecological Physiology of Animals" (A.D. Slonim, ed.), Part 1, pp. 79–182, Nauka, Leningrad.

Smelova, I.V. (1970). Utilisation of foodstuff sulphur by fish (In Russian). *Trudy VNIRO* **49**, 170–173.

Smetanin, M.M. (1978). Growth of fish as a characteristic of population condition (In Russian). *Trudy Institut Biologii Vnuttennikh Vod AN SSSR* **32**(35), 43–54.

Smirnov, B.P. and Klyashtorin, L.B. (1989). The osmoregulatory capacities of chum fry during long-term freshwater breeding (In Russian). *Voprosy Ikhtiologii* **29**, 617–623.

Smirnov, B.P., Baryleina, I.A. and Klyashtorin, L.B. (1986). Dependence of standard metabolism on temperature in fry of the *Oncorhynchus* genus (In Russian). *Voprosy Ikhtiologii* **26**, 1003–1009.

Smirnov, V.S. (1967). On the problem of elementary relationships between weight and linear variability of animals (In Russian). *Zhurnal Obshchey Biologii* **28**, 644–647.

Smirnova, G.P. (1967). Dependence of the thermoresistance of *Xiphophorus helleri* on the amount of consumed food (In Russian). "Proceedings of All-Union Conference on Ecological Physiology of Fish" (G.S. Karzinkin, ed.), pp.114–115, Moscow University Press, Moscow.

Smith, K.L. (1982). Metabolism of two dominant epibenthic echinoderms measured at bathyal depth in the Santa Catalina basin. *Marine Biology* **72**, 249–256.

Smith, M.A.K. (1981). Estimation of growth potential by measurement of tissue protein synthetic rates in feeding and fasting rainbow trout. *Journal of Fish Biology* **19**, 213–220.

Smith, M.A.K., Matthews, R.W., Hudson, A.P. and Haschemeyer, A.E.V. (1980). Protein metabolism of tropical reef and pelagic fish. *Comparative Biochemistry and Physiology* **65B**, 415–418.

Smith, M.W. and Ellory, J.C. (1971). Temperature-induced change in Na transport and ATPase in goldfish. *Comparative Biochemistry and Physiology* **39A**, 209–218.

Smith, M.W., Colombo, V.E. and Munn, E.A. (1968). Effect of temperature on ionic activation of ATPase. *Biochemical Journal* **107**, 691–698.

Smith, P.E. and Eppley, R.W. (1982). Primary production and the anchovy population in the southern California Bight: comparison of time series. *Limnology and Oceanography* **27**, 1–17.

Smith, P.J., Birley, A.J., Jamieson, A. and Bishop, C.A. (1989). Mitochondrial DNA in the Atlantic cod, *Gadus morhua*: lack of genetic divergence between eastern and western populations. *Journal of Fish Biology* **34**, 369–373.

Smith, P.J., Jamieson, A. and Birley, A.J. (1990). Electrophoretic studies and the stock concept in marine teleosts. *Journal du Conseil permanent international pour l'Exploration de la Mer* **47**, 231–245.

Smith, R.L. (1973). Energy transformation by the sargassum fish, *Histio histio*. *Journal of Experimental Marine Biology and Ecology* **12**, 219–227.

Soengas, J.L., Otero, J. Fuentes, J., Andrés, M.D. and Aldegunde, M. (1991). Preliminary studies on carbohydrate metabolism changes in domesticated rainbow trout (*Oncorhynchus mykiss*) transferred to diluted sea water (12 p.p.t). *Comparative Biochemistry and Physiology* **98B**, 53–57.

Soengas, J.L., Barciela, P., Fuentes, J., Otero, J., Andrés, M.D. and Aldegunde, M. (1993). The effect of seawater transfer on liver carbohydrate metabolism of domesticated rainbow trout (*Oncorhynchus mykiss*). *Comparative Biochemistry and Physiology* **105B**, 337–343.

Soldatov, A.A. (1993). A comparative study of the oxygen binding function of blood of Black Sea gobies (In Russian). *Zhurnal Evolutsionnoy Biokhimii i Physiologii* **29**, 327–330.

Soldatov, A.A. (1996). The effect of hypoxia on red blood cells of flounder: a morphologic and autoradiographic study. *Journal of Fish Biology* **48**, 321–328.

Soldatov, A.A. and Maslova, M.N. (1989). Methaemoglobin concentration in fish blood over the annual cycle (In Russian). *Zhurnal Evolutionnoy Biokhimii i Physiologii* **25**, 454–459.

Solomatina, V.D., Pozharova, T.S. and Yudina, O.V. (1989). Tissue metabolism of phosphorus compounds in carp undergoing oxygen deficiency (In Russian). *Gydrobiologicheskii Zhurnal* **25**, 95.

Solomon, D.J. and Brafield, A.E. (1972). The energetics of feeding, metabolism and growth of perch. *Journal of Animal Ecology* **41**, 699–718.

Somero, G.N. and Siebenaller, J.F. (1979). Inefficient lactate dehydrogenases of deep-sea fishes. *Nature, Lond.* **282**, 100–102.

Somero, G.N., Giese, A.C. and Wohlschlag, D.E. (1968). Cold adaptation of the Antarctic fish *Trematomus bernacchii*. *Comparative Biochemistry and Physiology* **26**, 223–233.

Sorokin, K.F. (1982). "The Black Sea. The Nature and Resources" (In Russian). Nauka, Moscow, 216 pp.

Sorvachev, K.F. (1982). "The Foundations of Biochemistry of Fish Nutrition" (In Russian). Legkaya i Pishchevaya Promyshlennlost, Moscow, 247 pp.

Stanley, J.G. and Colby, P.J. (1971). Effects of temperature on electrolyte balance and osmoregulation in the alewife (*Alosa pseudoharengus*) in fresh and sea water. *Transactions of the American Fisheries Society* **100**, 624–638

Stanley, J.G. and Fleming, W.R. (1967). The effect of hypophysectomy on the electrolyte content of *Fundulus kansae* held in fresh water and in sea water. *Comparative Biochemistry and Physiology* **20**, 489–497.

Stansby, M.E. (1967). Fatty acid patterns in marine, freshwater and anadromous fish. *Journal of the American Oil Chemists' Society* **44**, p.64.

Steele, J.H. (1965). Some problems in the study of marine resources. *International Commission of Northwest Atlantic Fisheries, Special Publication* **6**, 463–476.

Stegeman, J.J. (1979). Temperature influence on basal activity and induction of mixed function oxygenase activity in *Fundulus heteroclitus*. *Journal of the Fisheries Research Board of Canada* **36**, 1400–1405.

Stevens, E.D. and Dizon, A.E. (1982). Energetics of locomotion in warm-bodied fish. *Annual Review of Physiology* **44**, 121–131.

Stevens, E.D., Lam, H.M. and Kendall, J. (1974). Vascular anatomy of the counter-current heat exchanger of skipjack tuna. *Journal of Experimental Biology* **61**, 145–153.

Stolbov, A.Ya. (1990). Tissue respiration and respiratory coefficients in Black Sea fishes at different periods of the annual cycle (In Russian). *In* "Bioenergetics of Aquatic Organisms" (G.E. Shulman, ed.), pp.160–166. Naukova Dumka, Kiev.

Stolbov, A.Ya. (1992). Energy metabolism of Black Sea sprat (In Russian). *Ekologiya Morya* **41**, 63–65.

Stolbov, A.Ya., Stavitskaya, E.N. and Shulman, G.E. (1995). Oxygen consumption and nitrogen excretion in Black Sea fishes of different ecological specialisation during oxygen deficiency (In Russian). *Gydrobiologicheskii Zhurnal* **31**, 73–78.

Stolbov, A.Ya., Stavitskaya, E.N. and Shulman, G.E. (1997). Dynamics of oxygen consumption and nitrogen excretion in Black Sea scorpion fish in short- and long-time hypoxia (In Russian). *Doklady Akademii Nauk* **356**, 569–571.

Storozhuk, A.Ya. (1971). Physiologo-biochemical methods in the studies of Atlanto-Scandinavian herring (In Russian). *Trudy VNIRO* **87**, 150–162.

Storozhuk, A.Ya. (1975). Seasonal dynamics of the physiological-biochemical condition of North Sea pollack (In Russian). *Trudy VNIRO,* 96, 114–120.

Stroganov, N.S. (1956). "Physiological Adjustment of Fish and the Environmental Temperature" (In Russian). Academy of Sciences, SSSR, Moscow, 154 pp.

Stroganov, N.S. (1960). Modern problems of aquatic toxicology (In Russian). *News of Moscow State University, Series Biology* (2), 3–17.

Stroganov, N.S. (1962). "Ecological Physiology of Fish" (In Russian). Moscow University Press, Moscow, 444 pp.

Stroganov, N.S. (1968). "Acclimatisation and Breeding of Sturgeons in Ponds (Ecologo-physiological and Biochemical Investigations)". Moscow University Press, Moscow, 377 pp.

Stroganov, N.S. (1979). Theoretical aspects of ecological physiology of fish in relation to growing toxicity of the water environment (In Russian). *In* "Modern Problems of Ecological Physiology of Fish" (N.S. Stroganov, ed.), pp.19–34. Moscow.

Studenikina, E.I., Volovik, S.P., Mirzoian, I.A. and Luts, G.I. (1991). The ctenophore *Mnemiopsis leidyi* in the Sea of Azov (In Russian). *Okeanologia* **31**, 981–985. [English translation (1992) *Oceanology* **31**, 722–725.]

Sukumaran, N. and Kutty, M.N. (1977). Oxygen consumption and ammonia excretion in the catfish with special reference to swimming speed and ambient oxygen. *Proceedings of the Indian Academy of Science* **86B**, 195–206.

Sullivan, K.M. and Smith, K.L. (1982). Energetics of sablefish under laboratory conditions. *Canadian Journal of Fisheries and Aquatic Science* **39**, 1012–1020.

Suppes, C., Tiemeier, O.W. and Deyoe, C.W. (1967). Seasonal variations of fat, protein and moisture in channel catfish. *Transactions of the Kansas Academy of Science* **70**, 349–358.

Sushkina, A.P. (1962). The rate of fat expenditure in *Calanus finmarchicus* and *C. glacialis* at different temperatures and their life histories (In Russian). *Zoologicheskii Zhurnal* **41**, 1004–1012.

Suyama, M. and Ogino, C. (1958). Changes in chemical composition during development of rainbow trout eggs. *Bulletin of the Japanese Society of Scientific Fisheries* **23**, 785–788.

Svetlichny, L.S., Yuneva, T.V., Shulman, G.E. and Houseman, J.A. (1994). Utilisation of protein in the energy metabolism of the cladoceran *Moine micrura* at varying dissolved oxygen contents (In Russian). *Doklady Akademii Nauk SSSR* **337**, 428–430.

Svetovidov, A.N. (1964). "The Fishes of the Black Sea" (In Russian). Nauka, Moscow, Leningrad, 552 pp.

Takahashi, R., Ichioka, K., Matano, M. and Zama, R. (1985). Seasonal variation of sardine muscle lipids and other components. *Bulletin of the Faculty of Fisheries, Hokkaido University* **36**, 248–257.

Takahashi, T. and Yokoyama, W. (1954). Physico-chemical studies on the skin and leather of marine animals – XII. The content of hydroxyproline in the collagen of different fish skins. *Bulletin of the Japanese Society of Scientific Fisheries* **20**, 525–529.

Takama, K., Love, R.M. and Smith, G.L. (1985). Selectivity in mobilisation of stored fatty acids by maturing cod, *Gadus morhua* L. *Comparative Biochemistry and Physiology* **80B**, 713–718.

Takeuchi, T. and Watanabe, T. (1976). Nutritive value of ω-3 highly unsaturated fatty acids in pollock liver oil for rainbow trout. *Bulletin of the Japanese Society of Scientific Fisheries* **42**, 907–919.

Tamozhnyaya, V.A. and Goromosova, S.A. (1985). Biochemical parameters of the metabolism of mussels under the influence of toxins (In Russian). *Ekologiya Morya* **21**, 64–68.

Tandler, A. and Beamish, F.W.H. (1979). Mechanical and biochemical components of apparent specific dynamic action in largemouth bass. *Journal of Fish Biology* **14**, 343–350.

Taranenko, N.F. (1964). Fat content of Azov anchovy as an index of reproductive capacity of the fish stock and the migration term (In Russian). *Trudy Azovsko-chernomorskogo Nauchno-issledovatelnogo Instituta morskogo Rybno Khozyaistva i Oceanografii* **22**, 137–147.

Telitchenko, M.M. (1974). Quality and safety of the environment (chemistry, toxicology and technology). Global aspects in the environmental context (In Russian). *Gydrobiologicheskii Zhurnal* **10**, 131–134.

Terroine, E.F. and Wurmser, R.L. (1922). Energy of growth. I. Development of *Aspergillus niger. Bulletin de la Société de Chimie Biologique* 4, 518–567.

Teskeredzic, E., Teskeredzic, Z., Tomec, M. and Modrusan, Z. (1989). A comparison of the growth performance of rainbow trout (*Salmo gairdneri*) in fresh and brackish water. *Aquaculture* **77**, 1–10.

van den Thillart, G. and de Bruin, G. (1981). Influence of environmental temperature on mitochondrial membranes. *Biochimica et Biophysica Acta* **640**, 432–447.

van den Thillart, G. and Kesbeke, F. (1978). Anaerobic production of carbon dioxide and ammonia by goldfish, *Carassius auratus*. *Comparative Biochemistry and Physiology* **59A**, 393–400.

van den Thillart, G. and Waarde, A. (1985). Teleosts in hypoxia: aspects of anaerobic metabolism. *Molecular Physiology* **8**, 393–409.

Thorpe, J.E., Morgan, R.J.G., Ottaway, E.M. and Miles, M.S. (1980). Time of divergence of growth groups of potential 1+ and 2+ smolts among sibling Atlantic salmon. *Journal of Fish Biology* **17**, 13–21.

Thorpe, J.E., Talbot, C. and Villarual, C. (1982). Bimodality of growth and smolting in Atlantic salmon. *Aquaculture* **28**, 123–132.

Timeyko, V.N. (1992). Glycogen reserve depletion in the developing eggs of Atlantic salmon under saprolegniosis (In Russian). *Voprosy Ikhtiologii* **32**, 186–190.

Timeyko, V.N. and Novikov, G.G. (1991). Influence of temperature on the utilisation of stored carbohydrates during the embryological development of Atlantic salmon (In Russian). *Voprosy Ikhtiologii* **31**, 851–859.

Timokhina, A.F. (1974). On the food requirement of blue whiting in the Norwegian Sea and at Pockyolain Shoal (In Russian). *Gydrobiologicheskii Zhurnal* **10**, 57–63.

Tinsley, I.J., Krueger, H.M. and Saddler, J.B. (1973). Fatty acid content of coho salmon, *Oncorhynchus kisutch* – a statistical approach to changes produced by diet. *Journal of the Fisheries Research Board of Canada* **30**, 1661–1666.

Tirri, R., Vornanen, M. and Cossins, A.R. (1978). The compensation of ATPase activities in brain and kidney microsomes from cold- and warm-acclimated carp. *Journal of Thermal Biology* **3**, 131–135.

Tochilina (Rakitskaya), L.V. (1990). Morphological characteristics of marine fish blood (In Russian). *In* "Bioenergetics of Aquatic Organisms" (G.E. Shulman and G.A. Finenko, eds), pp.166–178. Naukova Dumka, Kiev.

Tochilina (Rakitskaya), L.V. (1994). Leucocytic formula of marine fish (In Russian). *Gydrobiologicheskii Zhurnal* **30**, 50–58.

Totland, G.K., Kryvi, H., Bone, Q. and Flood, P.R. (1981). Vascularisation of the lateral muscle of some elasmobranchiomorph fishes. *Journal of Fish Biology* **18**, 223–237.

Trenkler, I.V. and Semenkova, T.B. (1990). Hormonal regulation of fish growth in aquaculture (In Russian). *Gidrobiologicheskii Zhurnal* **26**, 49–59.

Trout, G.C. (1954). Otolith growth of the Barents Sea cod. *Rapports et Procès-Verbaux des Réunions, Conseil Permanent International pour l'Exploration de la Mer* **136**, 89–102.

Trusevich, V.V. (1978). Phosphorus metabolism of fish during swimming (In Russian). *In* "Elements of Physiology and Biochemistry in Total and Active Metabolism of Fish" (G.E. Shulman, ed.), pp.145–167. Naukova Dumka, Kiev.

Trusevich, V.V. (1985). Organic matter in the body of some zooplankton organisms in waters of subequatorial divergence (In Russian). *In* "Ecological Systems of Active Dynamic Zones of the Indian Ocean" (T.S. Petipa, ed.), pp.167–172. Naukova Dumka, Kiev.

Trusevich, V.V. and Anninsky, B.E. (1987). Macroerg phosphate compounds in fish muscle during burst swimming (In Russian). *In* "Proceedings of 1st Symposium on Ecological Biochemistry of Fish", Yaroslavl Polytechnic Institute Publishing House, pp.192–194.

Tseitlin, V.B. (1983). Energy demands and growth of deepwater animals (In Russian). *Zhurnal Obshchey Biologii* **44**, 71–77.

Tseitlin, V.B. (1988). Generative production of aquatic animals (In Russian). *Okeanologiya* **28**, 493–497.

Tsintsadze, Z.A. (1991). Adaptational capabilities of various size-age groups of rainbow trout in relation to gradual changes of salinity. *Journal of Ichthyology* **31**, 31–38.

Tsukuda, H. (1975). Temperature dependency of the relative activities of liver lactate dehydrogenase isozymes in goldfish acclimated to different temperatures. *Comparative Biochemistry and Physiology* **52B**, 343–345.

Tsukuda, H. and Ohsawa, W. (1971). Effects of acclimation temperature on the composition and thermostability of tissue proteins in the goldfish, *Carassius auratus* L. *Annotationes Zoologicae Japonenses* **44**, 90–98.

Tsuyuki, H., Roberts, E. and Vanstone, W.E. (1965). Comparative zone electrophorograms of muscle myogens and blood haemoglobins of marine and freshwater vertebrates and their application to biochemical systematics. *Journal of the Fisheries Research Board of Canada* **22**, 203–213.

Tucker, V.A. (1970) Energetic cost of locomotion in animals. *Comparative Biochemistry and Physiology* **34**, 841–846.

Tunnicliffe, V. (1991) The biology of hydrothermal vents: ecology and evolution. *Oceanography and Marine Biology Annual Review* **29**, 319–407.

Turetsky, V.I. (1983). Study of the rates of substance and energy metabolism in fish larvae reared in warm waters (In Russian). *Izvestiya Gosudarstvennogo Nauchno-issledovatelskogo Instituta Ozernogo i Rechnogo Rybnogo* **194**, 61–69.

Tytler, P. (1978). The influence of swimming performance on the metabolic rate of gadoid fish. *In* "Physiology and Behaviour of Marine Organisms" (D.S. McLusky and A.J. Berry, eds), pp.83–92, Pergamon Press, Oxford.

Tyurin, V.A. and Gorbunov, N.V. (1984). The study of state of lipid matrix of plasma membranes of synaptosomes of brain in gorbusha during migration from sea to river (In Russian). *Zhurnal Evolutsionnoy Biokhimii i Physiologii* **20**, 341–345.

Tyurin, V.A., Berman, A.L., Ryzhova, M.P., Chelomin, V.P. and Korchagin, V.P. (1982). Fatty acid composition of phospholipids of the binary layer of photoreceptor membranes and aminophospholipids of rhodopsin microenvironment in cold and warm blooded vertebrates (In Russian). *Zhurnal Evolutsionnoy Biochimii i Physiologii* **18**, 101–105.

Tåning, Å.V. (1952). Experimental study of meristic characters in fishes. *Biological Reviews of the Cambridge Philosophical Society* **27**, 169–193.

Ueda, T. (1967). Fatty acid composition of oils from 33 species of marine fish. *Journal of Shimonoseki University of Fisheries* **16**, 1–10.

Ugolev, A.M. and Kuzmina, V.V. (1993). "Digestive Processes and Adaptations in Fish" (In Russian). St Petersburg, 239 pp.

Umminger, B.L. (1968). Life below zero. *Yale Scientific Magazine* **42**, 6–10.

Umminger, B.L. and Gist, D.H. (1973). Effects of thermal acclimation on physiological responses to handling stress, cortisol and aldosterone injections in the goldfish, *Carassius auratus*. *Comparative Biochemistry and Physiology* **44A**, 967–977.

Ushakov, B.P. (1963). On the classification of animal and plant adaptations and on the role of cytoecology in elaboration of the adaptation concept (In Russian). *In* "The Problems of Animal Cytoecology" (B.P. Ushakov, ed.), pp. 5–20. Academii Nauk, Moscow, Leningrad.

Utida (Uchida), S. and Hirano, T. (1973). Effects of changes in environmental salinity on salt and water movement in the intestine and gills of the eel, *Anguilla japonica*. *In* "Responses of Fish to Environmental Changes" (W. Chavin, ed.), pp. 240–267. Charles C. Thomas, Springfield, Illinois.

Valkirs, A. (1978). Temperature and pH effects on catalytic properties of lactate dehydrogenase from pelagic fish. *Comparative Biochemistry and Physiology* **59A**, 31–36.

Vanstone, W.E., Markert, J.R., Lister, D.B. and Giles, M.A. (1970). Growth and chemical composition of chum and sockeye salmon fry produced in spawning channel and natural environments. *Journal of the Fisheries Research Board of Canada* **27**, 371–382.

Varnavsky, V.S. (1990a). "Smoltification of Salmon" (In Russian). 180 pp. Vladivostok.

Varnavsky, V.S. (1990b). Electroconductivity of blood plasma as a test for the adaptation to sea water in salmon (In Russian). *Doklady Akademii Nauk SSSR* **311**, 1497–1499.

Varnavsky, V.S., Varnavskaya, N.V., Kalinin, S.V. and Kinas, N.M. (1991). RNA/DNA index as a characteristic of growth rate during the early marine period of life in coho salmon (In Russian). *Voprosy Ikhtiologii* **31**, 783–789.

Vellas, F. (1965). Effects of several ecological factors on the activity of uricolytic enzymes in the liver of mirror carp (*Cyprinus carpio* L.). *Annales de Limnologie* **1**, 435–442.

Vdovin, A.N. and Shvydky, G.V. (1993). Physiological aspects of one-finned greenling yellowfish growth in water of Primorye (The Maritime Province). (In Russian). *Voprosy Ikhtiologii* **33**, 156–160.

Veltishcheva, I.F. (1970). Study of metabolism in fish using radioactive carbon (In Russian). *Trudy VNIRO* **69**, 9–18.

Verzhbinskaya, N.A. (1953). Cytochrome system of the brain in phylogenesis of vertebrates (In Russian). *Physiologicheskii Zhurnal, SSSR* **39**, 17–20.

Verzhbinskaya, N.A. (1968). Biochemical evolution of enzyme systems as the base of functional evolution of vertebrate animals (In Russian). *In* "Abiogenesis and Primary Stages of the Evolution of Life" (A.I. Oparin ed.), pp.169–180, Nauka, Moscow.

Vetvitskaya, L.V., Maldov, D.G., Tikhomirov, A.M., Kozlov, A.B. and Nikanorov, S.I. (1992). Enzyme induction and changes in the behaviour of Russian sturgeon fry caused by diverse food (In Russian). *Doklady Akademii Nauk SSSR* **323**, 1186–1192.

Videler, J.J. and Weihs, D. (1982). Energetic advantages of burst-and-coast swimming of fish at high speeds. *Journal of Experimental Biology* **97**, 169–178.

Vieira, V.L.A. and Johnston, I.A. (1992). Influence of temperature on muscle-fibre development in larvae of the herring *Clupea harengus*. *Marine Biology* **112**, 333–341.

Vinberg, G.G. (1956). "The Intensity of Metabolism and Food Requirements of Fish" (In Russian). Belorussian University Press, Minsk, 253 pp.

Vinberg, G.G. (1962). The energy principle in the study of food relations and productivity of environmental systems (In Russian). *Zoologicheskii Zhurnal* **41**, 13–15.

Vinberg, G.G. (1968). The method of estimation of values of production in aquatic animals. Introduction. (In Russian). *In* "Methods of Production Determination in Aquatic Animals" (G.G. Vinber, ed.), pp.9–19. Vysshya Shkola, Minsk.

Vinberg, G.G. (1979). The production of populations of Aquatic invertebrates and general principles of its determination (In Russian). *In* "General Principles in the Study of Aquatic Ecosystems" (G.G. Vinberg, ed.), pp.114–119, Nauka, Leningrad.

Vinberg, G.G. (1983). Van't Hoff temperature coefficient and Arrhenius equation as applied in biology (In Russian). *Zhurnal Obshchey Biologii* **44**, 31–42.

Vinberg, G.G. (1986). Growth efficiency and production in aquatic animals (In Russian). *In* "The Growth Efficiency of Aquatic Organisms" (N.N. Khmeleva, ed.), pp. 20–61, Gomel University Press.

Vinogradov, M.E. (1990). Characteristics of formation of the lower maximum of meso-plankton concentration in the Black Sea (In Russian). *Doklady Akademii Nauk SSSR* **310**, 977–980.

Vinogradov, M.E. and Shushkina, E.A. (1980). The characteristic features of vertical distribution of Black Sea zooplankton (In Russian). *In* "Ecosystems of Black Sea Pelagic Zone" (M.E. Vinogradov, ed.), pp. 179–191. Nauka, Moscow.

Vinogradov, M.E., Arashkevich, E.G. and Ilchenko, S.V. (1992a). The ecology of *Calanus ponticus* population in the deeper layer of its concentration in the Black Sea (In Russian). *Journal of Plankton Research* **14**, 447–458.

Vinogradov, M.E., Sapozhnikov, V.V. and Shushkina, E.A. (1992b) "Ecosystem of the Black Sea" (In Russian). Nauka, Moscow, 112 pp.

Vinogradova, Z.A. (1967). Biochemical aspects of marine plankton research (In Russian). *In* "Problems of Bio-Oceanography" (K.A. Vinogradov, ed.), pp. 52–58.

Vitvitsky, V.N. (1977). Comparative investigation of thermoresistance and electrophoretic lability of muscle proteins in fish habituated to different depths (In Russian). *Ecologya* **1977** (6), 88–92.

Viviani, R. (1968). Lipid metabolism of marine fish with particular reference to utilisation of the product. *Minerva Medica* **59**, 2161–2162.

Viviani, R., Cortesi, P., Crisetig, G. and Borgatti, A.R. (1973). Seasonal variations in lipid constituents of *Clupea sprattus* from the Adriatic Sea. *Rapports et Procès-Verbaux des Réunions. Commission International pour l'Exploration Scientifique de la Mer Méditerranée* **22**, 15–16.

Vladimirov, V.I. (1974). Dependence of the quality of carp embryos and larvae on female age, amino acid content of eggs and the introduction of amino acid additives to water at early developmental stages (In Russian). *In* "Different Quality of Early Ontogenesis of Fish" (V.I. Vladimirov, ed.), pp. 227–254, Naukova Dumka, Kiev.

Volovik, S.P. (1975). Influence of climatic and anthropogenic factors on the structure and production of pelagic fish communities in the Azov Sea (In Russian). *Trudy VNIRO* **103**, 107–119.

Volskis, R.S. (1973). Perspectives for studying the biology of fish and aquatic invertebrate species in their area (In Russian). *Gydrobiologicheskii Zhurnal* **15**, 28–36.

Vorobyev, V.P. (1945). "Anchovy" (In Russian), pp.3–14. Krymizdat, Simferopol.

Vyskrebentsev, B.V. (1975). The comparative ecological principle in studies of fish behaviour (In Russian). *In* "Problems of Zoophysiology, Ethology and Comparative Physiology" (I.A. Shilov, ed.), pp. 28–30, Moscow University Press, Moscow.

Vyskrebentsev, B.V. and Savchenko, N.V. (1970). Experimental investigation of burst speeds of fish locomotion (In Russian). *Rybnoye Khozyaistvo* 1970 (6), 15–17.

Waarde, A. (1983). Aerobic and anaerobic ammonia production by fish. *Comparative Biochemistry and Physiology* **74B**, 675–684.

Walesby, N.J. and Johnston, I.A. (1980). Temperature acclimation in brook trout muscle: adenine nucleotide concentrations, phosphorylation state and adenylate energy charge. *Journal of Comparative Physiology* **139B**, 127–133.

Walker, M.G. and Emerson, L. (1978). Sustained swimming speeds and myotomal muscle function in the trout, *Salmo gairdneri*. *Journal of Fish Biology* **13**, 475–481.

Walker, M.G., Horwood, J. and Emerson, L. (1980). On the morphology and function of red and white skeletal muscle in the anchovies *Engraulis encrasicholus* and *E. mordax*. *Journal of the Marine Biological Association UK* **60**, 31–37.

Walker, R.M. and Johansen, P.H. (1977). Anaerobic metabolism in goldfish (*Carassius auratus*). *Canadian Journal of Zoology* **55**, 1304–1311.

Wallace, J.C. (1973). Observations on the relationship between the food consumption and metabolic rate of *Blennius pholis*. *Comparative Biochemistry and Physiology* **45A**, 293–306.

Waller, U. (1992). Factors influencing routine oxygen consumption in turbot, *Scophthalmus maximus*. *Journal of Applied Ichthyology* **8**, 62–71.

Walsh, P.J. (1981). Purification and characterisation of glutamate dehydrogenases from three species of sea anemones: adaptation to temperature within and among species from different thermal environments. *Marine Biology Letters* **2**, 289–299.

Walsh, P.J., Moon, T.W. and Mommsen, T.P. (1985). Interactive effects of acute changes in temperature and pH on metabolism in hepatocytes from sea raven. *Physiological Zoology* **58**, 727–735.

Walton, M.J. and Cowey, C.B. (1982). Aspects of intermediary metabolism in salmonid fish. *Comparative Biochemistry and Physiology* **73B**, 59–79.

Wang Tsu-Hsiung, Sun Mei-Chuan, Gao Hanjiao, Chu Lan-Fei and Chao Ming-ji (1964). Changes in the biochemical composition of the ovaries of pond-reared *Hypophthalmichthys molitrix* during the reproductive cycle. *Acta Hydrobiologica Sinica* **5**, 103–114.

Wang, Y.L., Buddington, R.K. and Doroshov, S.I. (1987). Influence of temperature on yolk utilisation by the white sturgeon, *Acipenser transmontanus*. *Journal of Fish Biology* **30**, 263–271.

Wardle, C.S. (1978). Non-release of lactic acid from anaerobic swimming muscle of plaice: a stress reaction. *Journal of Experimental Biology* **77**, 141–155.

Wardle, C.S. (1980). Effects of temperature on the maximum swimming speed of fishes. *In* "Environmental Physiology of Fishes" (M.A. Ali, ed.), pp.519–531, NATO Advanced Study Institute Series, No. 35, Plenum Press, New York.

Wardle, C.S. and Videler, J.J. (1980). How do fish break the speed limit? *Nature, Lond.* **284**, 445–447.

Ware, D.M. (1975). Growth, metabolism and optimal swimming speed of a pelagic fish. *Journal of the Fisheries Research Board of Canada* **32**, 33–41.

Ware, D.M. (1980). Bioenergetics of stock and recruitment. *Canadian Journal of Fisheries and Aquatic Science* **37**, 1012–1024.

Warman, A.W. and Bettino, N.R. (1978). Lipogenic activity of catfish liver. Lack of response to dietary changes and insulin administration. *Comparative Biochemistry and Physiology* **59B**, 153–161.

Watanabe, T. (1982). Lipid nutrition in fish. *Comparative Biochemistry and Physiology* **73B**, 3–15.

Watanabe, T. and Takeuchi, T. (1976). Evaluation of pollock liver oil as a supplement to diets for rainbow trout. *Bulletin of the Japanese Society of Scientific Fisheries* **42**, 893–906.

Watanabe, T., Ohhashi, S., Itoh, A., Kitajima, C. and Fujita, S. (1984). Effect of nutritional composition of diets on chemical components of red sea bream broodstock and eggs produced. *Bulletin of the Japanese Society of Scientific Fisheries* **50**, 503–515.

Waversveld, J., Addink, A.D.F. and van den Thillart, G. (1989). The anaerobic energy metabolism of goldfish determined by simultaneous direct and indirect calorimetry during anoxia and hypoxia. *Journal of Comparative Physiology* **159B**, 263–268.

Weatherley, A.H. and Gill, H.S. (1987). "The Biology of Fish Growth". Academic Press, London, New York, 443 pp.

Webb, P.W. (1971). The swimming energetics of trout. 1. Thrust and power output at cruising speeds. *Journal of Experimental Biology* **55**, 489–520.

Webb, P.W. (1975). Hydrodynamics and energetics of fish propulsion. *Bulletin of the Fisheries Research Board of Canada* **190**, 1–159.

Wedemeyer, G.A., Meyer, F.P. and Smith, L. (1976). Diseases of Fishes. *In* "Environmental Stress and Fish Diseases" (S.F. Snieszko and H.R. Axelrod, eds), Vol. 5, pp. 1–192, TFH Publications, Neptune, New Jersey.

Wells, R.M. (1978). Respiratory adaptation and energy metabolism in Antarctic nototheniid fishes. *New Jersey Zoologist* **5**, 813–815.

White, A., Fletcher, T.C. and Pope, J.A. (1986). Seasonal changes in serum lipid composition of the plaice, *Pleuronectes platessa. Journal of Fish Biology* **28**, 595–606.

Wilder, J.B. and Stanley, J.G. (1983). RNA–DNA ratio as an index to growth in salmonid fishes in the laboratory and in a stream contaminated by carbamyl. *Journal of Fish Biology* **22**, 165–172.

Wilson, F.R., Somero, G.N. and Prosser, C.L. (1974). Temperature–metabolism relations of two species of *Sebastes* from different thermal environments. *Comparative Biochemistry and Physiology* **47B**, 485–491.

Wiser, C.J. (1970). Cold resistance and injury in woody plants. *Science, New York* **169**, 1269–1278.

Wissing, T.E. (1974). Energy transformations by young-of-the-year white bass in Lake Mendota, Wisconsin. *Transactions of the American Fisheries Society* **103**, 32–37.

Wittenberger, C. (1971). "The Evolution of Muscular Function in Vertebrates"(In Romanian). Bucharest, 192 pp.

Wittenberger, C. (1973). Metabolic interaction between isolated white and red carp muscle. *Revue Roumaine de Biologie Serie de Zoologie* **18**, 71–76.

Wodtke, E. (1974). Effects of acclimation temperature on the oxidative metabolism of the eel (*Anguilla anguilla*). 1. Liver and red muscle. Changes in the mitochondrial content and in the oxidative capacity of isolated coupled mitochondria. *Journal of Comparative Physiology* **91**, 309–332.

Wodtke, E. (1978). Lipid adaptation in liver mitochondrial membranes of carp acclimated to different environmental temperatures. Phospholipid composition, fatty acid pattern and cholesterol content. *Biochimica et Biophysica Acta* **529**, 280–291.

Wohlschlag, D.E. (1960). Metabolism of an Antarctic fish and the phenomenon of cold adaptation. *Ecology* **41**, 287–292.

Woo, N.Y.S., Bern, H.A. and Nishioka, R.S. (1978). Changes in body composition associated with smoltification and premature transfer to sea water in coho salmon and ring salmon. *Journal of Fish Biology* **13**, 421–428.

Wood, R.J. (1958). Fat cycles of North Sea herring. *Journal du Conseil International pour l'Exploration de la Mer* **23**, 390–398.

Woodhead, A.D. (1974a). Ageing changes in the Siamese fighting fish, *Betta splendens* – I. The testis. *Experimental Gerontology* **9**, 75–81.

Woodhead, A.D. (1974b). Ageing changes in the Siamese fighting fish, *Betta splendens* – II. The ovary. *Experimental Gerontology* **9**, 131–139.

Woodhead, A.D. (1979). Senescence in fishes. *In* "Fish Phenology. Anabolic Adaptivness in Teleosts", *Symposia of the Zoological Society of London* **44**, 179–205.

Woodhead, A.D. and Woodhead, P.M.J. (1965a). Seasonal changes in the physiology of the Barents Sea cod, *Gadus morhua* L., in relation to its environment. I. Endocrine changes particularly affecting migration. *Special Publications of the International Commission on North-West Atlantic Fisheries* **6**, 691–715.

Woodhead, P.M.J. and Woodhead, A.D. (1965b). Seasonal changes in the physiology of the Barents Sea cod, *Gadus morhua* L., in relation to its environment. II. Physiological reactions to low temperatures. *Special Publications of the International Commission on North-West Atlantic Fisheries* **6**, 717–734.

Wootton, R.J. (1977). Effect of food limitation during the breeding season on the size, body components and egg production of female sticklebacks. *Journal of Animal Ecology* **46**, 823–834.

Wootton, R.J. (1979). Energy costs of egg production and environmental determinants of fecundity in teleost fishes. In "Fish Phenology, Anabolic Adaptivity in Teleosts" (P.J. Miller, ed.), pp.133–159, Academic Press, London.

Wootton, R.J. (1990). "Ecology of Teleost Fishes". Chapman and Hall, London, 404 pp.

Wootton, R.J., Allen, J.R.M. and Cole, S.J. (1980). Energetics of the annual reproductive cycle in female sticklebacks. *Journal of Fish Biology* **17**, 387–394.

Yablonskaya, E.A. (1951). Some data about the growth and metabolism in Verkhovka at spawning season (In Russian). *Trudy vsesoyuznogo Gidrobiologicheskogo Obshchestva* **3**, 140–154.

Yada, T. and Hirano, T. (1992). Influence of seawater adaptation on production and growth hormone release from organ-cultured pituitary of rainbow trout. *Zoological Science* **9**, 143–148.

Yakovenko, B.V. and Yavonenko, A.F. (1991). The influence of acclimation temperature on the rate of glycine deamination in white muscle and hepatopancreas of carp (In Russian). *Gidrobiologicheskii Zhurnal* **27**(3), 111.

Yakovleva, K.K. (1969). The dynamics of polyunsaturated fatty acids in some Black Sea fishes. 1. Scorpion fish (In Russian). *Voprosy Ikhtiologii* **9**, 741–747.

Yakovleva, K.K. and Shulman, G.E. (1977). Relationship between protein growth and fat accumulation in Black Sea scorpion fish (In Russian). *Biologiya Morya* (Sevastopol) **1977** (1), 78–81.

Yamada, K., Kobayashi, K. and Yone, T. (1980). Conversion of linolenic acid to 3 highly unsaturated fatty acids in marine fishes and rainbow trout. *Bulletin of the Japanese Society of Scientific Fisheries* **46**, 1231–1238.

Yamaguchi, K., Lavéty, J. and Love, R.M. (1976). The connective tissues of fish. VIII. Comparative studies of hake, cod and catfish collagens. *Journal of Food Technology* **11**, 389–399.

Yamamura, Y. and Muto, S. (1961). Change of vitamin A and carotenoids during the development of salmon eggs. *Bulletin of the Tokai Regional Fisheries Research Laboratory* **19**, 171–179.

Yarzhombek, A.A. (1964). Some results of the biochemical investigations of salmon (In Russian). In "Salmon Culture in the Far East" (P.A. Moiseev, ed.), pp.142–144, Moscow.

Yarzhombek, A.A. (1975). The nature of fish swimming velocities (In Russian). *Rybnoye Khozyaistvo* 1964 (8), 26–27.

Yarzhombek, A.A. (1996). "Biological Resources of Fish Growth" (In Russian). VNIRO, Moscow, 168 pp.

Yarzhombek, A.A. and Klyashtorin, L.B. (1974). On the relationship between standard and resting metabolism of fish (In Russian). *Voprosy Ikhtiologii* **14**, 508–513.

Yarzhombek, A.A., Shcherbina, T.V., Shmakov, N.F. and Gusseinov, A.G. (1983). Specific dynamic effect of food on metabolism of fish (In Russian). *Voprosy Ikhtiologii* **23**, 639–645.

Yoshida, Y. (1970). Studies on the efficiency of food conversion to fish body growth. *Bulletin of the Japanese Society of Scientific Fisheries* **36**, 156–164.

Yoshikawa, J.S.M., McCormick, S.D., Young, G. and Bern, H.A. (1993). Effects of salinity on chloride cells and Na^+, K^+-ATPase activity in the teleost *Gillichthys mirabilis*. *Comparative Biochemistry and Physiology* **105A**, 311–317.

Yuneva, T.V. (1990). Seasonal dynamics of the fatty acid content of lipids in Black Sea anchovy and sprat (In Russian). In "Bioenergetics of Aquatic Organisms" (G.E. Shulman and G.A. Finenko, eds), pp. 196–207, Naukova Dumka, Kiev.

Yuneva, T.V. and Svetlichny, L.S. (1996). Hypoxia as a necessary factor in the accumulation of large amounts of lipid in planktonic copepods. In "6th International Conference on Copepoda," Oldenburg/Bremerhaven, Germany, July 29–August 3, p.110 (Abstract).

Yuneva, T.V., Shulman, G.E. and Chebotareva, M.A. (1986). Seasonal dynamics of fatty acid composition of lipids in Black Sea sprat (In Russian). *Voprosy Ikhtiologii* **26**, 113–118.

Yuneva, T.V., Shulman, G.E., Chebanov, N.A. and Shchepkina, A.M. (1987). The docosohexaenoic acid content in muscle of male pink salmon during the spawning season (In Russian). *Zhurnal Evolutsionnoy Biokhimii i Physiologii* **23**, 707–710.

Yuneva, T.V., Shulman, G.E., Chebanov, N.A., Shchepkina, A.M., Vilenskaya, N.I. and Markevich, N.B. (1990). The relationship between the docosohexaenoic acid content in the bodies of spawners and survival of eggs and prolarvae of pink salmon (In Russian). *Biologicheskii Nauki* **1990** (10), 85–89.

Yuneva, T.V., Shulman, G.E. and Shchepkina, A.M. (1991). The dynamics of lipid characteristics of horse-mackerel during swimming (In Russian). *Zhurnal Evolutsionnoy Biochimii i Physiologii* **27**, 730–735.

Yuneva, T.V., Shulman, G.E., Shchepkina, A.M. and Melnikov, V.V. (1992). The lipid composition of euphausids from the equatorial Atlantic (In Russian). *Gydrobiologicheskii Zhurnal* **28**, 61–67.

Yuneva, T.V., Shchepkina, A.M. and Shulman, G.E. (1994). The lipid content of tissues of squid from the tropical Atlantic (In Russian). *Gydrobiologicheskii Zhurnal* **30**, 78–86.

Yurovitsky, Yu.G. and Sidorov, V.S. (1993). Ecologo-biochemical monitoring and ecologo-biochemical testing of the threatened environmental areas (In Russian). *Izvestia Akademii Nauk SSSR Seriya Biologiya* 1993 (1), 74–82.

Zabelinsky, S.A. and Shukolyukova, E.P. (1989). Comparative study of the phospholipid composition of the brain in vertebrates with different normal internal temperatures of the body (In Russian). *Zhurnal Evolutsionnoy Biokhimii i Physiologii* **25**, 143.

Zabelinsky, S.A., Chebotareva, M.A., Brovtsina, N.B. and Kravchenko, A.I. (1995). On the 'adaptive specialisation' of fatty acid content and conformation condition in the membrane lipids of fish gills (In Russian). *Zhurnal Evolutsionnoy Biokhimii i Physiologii* **31**, 29–36.

Zagorskich, O.M. and Kirsipuu, A.A. (1990). Comparative characteristics of muscle and liver composition in bream after wintering and summer feeding (In Russian). *In* "Biochemistry of Ecto- and Endothermal Organisms of Normal and Pathological Condition", pp.27–33. Petrozavodsk.

Zaika, V.E. (1972). "Specific Production of Aquatic Invertebrates" (In Russian). Naukova Dumka. Kiev, 148 pp.

Zaika, V.E. (1983). "Comparative Productivity of Aquatic Organisms" (In Russian). Naukova Dumka, Kiev, 206 pp.

Zaika, V.E. (1985). "Balance Theory of Animal Growth" (In Russian). Naukova Dumka, Kiev, 193 pp.

Zaitsev, Yu. P. (1992). Ecological condition of the Black Sea shelf near Ukraine's coast (In Russian). *Gydrobiologicheskii Zhurnal* **28**, 3–18.

Zaks, M.G. and Sokolova, M.M. (1961). On the mechanisms of adaptation to the water salinity in sockeye salmon (In Russian). *Voprosy Ikhtiologii* **1**, 331–338.

Zanuy, S. and Carrillo, M. (1985). Annual cycles of growth, feeding rate, gross conversion efficiency and haematocrit levels of sea bass adapted to two different osmotic media. *Aquaculture* **44**, 11–25.

Zaporozhets, O.M (1991). Estimation of physiological condition of salmon fry from the swimming performance in hydrodynamic tube (In Russian). *Voprosy Ikhtiologii* **31**, 165–168.

Zenkevich, L.A. (1963). "Biology of the Seas of the USSR" (In Russian). Academy of Sciences of USSR, Moscow, 740 pp. [English translation (1963) Allen and Unwin, London, 954pp.]

Zharov, V.L. (1965). Body temperature of tuna (Thunnidae) and other perciform fishes of the tropical Atlantic (In Russian). *Voprosy Ikhtiologii* **5**, 157–160.

Zhidenko, A.A., Grubinko, V.V., Smolsky, A.S. and Yavonenko, A.F. (1994). Influence of abiotic factors of the environment on the formation of ketone bodies in fish (In Russian). *Gydrobiologicheskii Zhurnal* **30**, 87–92.

Zhirmunsky, A.V. (1966). The problems of cytoecology (In Russian). *In* "Text-book of Cytology" (Yu.B. Vakhtin, S.A. Krolenko and T.A. Chernobryadskaya, eds), pp.623–687, Moscow, Leningrad.

Zhukinsky, V.N. (1986). "The Influence of Abiotic Factors on the Quality and Viability of Fish during early Ontogenesis" (In Russian). Agropromizdat, Moscow, 244 pp.

Zhukinsky, V.N. and Gosh, R.I. (1973). Viability of embryos in relation to the intensity of oxidative phosphorylation and activity of ATPase in ovulatory ovocytes of roach and bream (In Russian). *Dopovidi Akademiyi Nauk Ukrayins'koyi RSR Ser.* **B 35**, 1044–1047.

Zhukinsky, V.N. and Gosh, R.I. (1988). Biological quality of the ovulated eggs of commercial fish: criteria and a method for express estimation (In Russian). *Voprosy Ikhtiologii* **28**, 773–781.

Zotin, A.I. (1974). "Thermodynamic Approach to the Problems of Development, Growth and Ageing" (In Russian). Nauka, Moscow, 183 pp.

Zotin, A.I. and Krivolutsky, D.A. (1982). The rate and direction of evolutionary progress of organisms (In Russian). *Zhurnal Obshchey Biologii* **43**, 3–13.

Zotin, A.I., Vladimirov, I.G. and Kirpichnikov, A.A. (1990). Energy metabolism and the trend of evolution in the class Mammalia (In Russian). *Zhurnal Obshchey Biologii* **51**, 760–767.

Zotina, R.S. and Zotin, A.I. (1967). Quantitative correlations between weight, length, age, egg size and fecundity of animals (In Russian). *Zhurnal Obshchey Biologii* **28**, 82–92.

Zusser, S.G. (1971). "Diurnal Vertical Migrations of Fish" (In Russian). Pishchevaya Promyshlennost, Moscow, 224 pp.

Zwaan, A. de and Zandee, D.I. (1972). Body distribution and seasonal changes in the glycogen content of the common sea mussel *Mytilus edulis*. *Comparative Biochemistry and Physiology* **43B**, 53–58.

Zwingelstein, G., Malak, A.N. and Brichon, G. (1978). Effect of environmental temperature on biosynthesis of liver phosphatidyl choline in the trout. *Journal of Thermal Biology* **3**, 229–233.

APPENDIX

Nomenclature of Fish Species Mentioned in the Text, Tables and Figures

Common name used	Scientific name
Alaska pollock	*Theragra chalcogramma* (Pallas)
American plaice	*Hippoglossoides platessoides* (Fabricius)
Anchovy	*Engraulis encrasicholus* (L.)
Annual fish	*Cynolebias adloffi* Ahl
Annular bream	*Diplodus annularis* (L.)
Annular gilthead	*Diplodus annularis* (L.)
Arctic char	*Salvelinus alpinus* (L.)
Armoured grenadier	*Nematonurus armatus* (Hector)
Atlantic cod	*Gadus morhua* (L.)
Atlantic herring	*Clupea harengus* (L.)
Atlantic mackerel	*Scomber scombrus* (L.)
Atlantic salmon	*Salmo salar* (L.)
Australian tuna	*Allothunnus fallai* (Serventy)
Azov Sea anchovy	*Engraulis encrasicholus* L.
Azov sturgeon	*Acipenser güldenstädti* Brandt et Ratzenberg
Baltic cod	*Gadus callarias* L. = *Gadus morrhua* L.
Baltic herring	*Clupea harengus* L.
Baltic sprat	*Sprattus sprattus* (L.)
Bighead	*Aristichthys nobilis* (Richardson) ex Gray
Black bullhead	*Ameiurus melas* (Rafinesque)
Black Sea anchovy	*Engraulis encrasicholus* L.
Bleak	*Alburnus alburnus* (L.)
Blenny	*Enedrias nebulosus* (Temminck et Schlegel)
Blue shark	*Prionace glauca* (L.)
Borchgrevinks notothenid	*Pagothenia borchgrevinki* (Boulenger)
Brook stickleback	*Culaea inconstans* (Kirtland)
Brook trout	*Salvelinus fontinalis* (Mitchill)
Black bullhead	*Ameiurus melas* (Rafinesque)
Brown rockfish	*Sebastes auriculatus* Girard
Brown trout	*Salmo trutta* L.
Bullhead	*Myoxocephalus scorpius* (L.)
Burbot	*Lota lota* (L.)
Californian anchovy	*Engraulis mordax* Girard
Californian sardine	*Sardinops caerulea* (Girard)
Capelin	*Mallotus villosus* (Müller)
Carp	*Cyprinus carpio* L.

Caspian kilka	*Clupeonella engrauliformis* (Borsdin)
Caspian sturgeon	*Acipenser güldenstädti* Brandt et Ratzenberg
Castor oil fish	*Ruvettus pretiosus* Cocco
Catfish	*Mystus armatus* (Day)
Catfish	*Corydoras aeneus* (Cuvier et Valenciennes)
Channel catfish	*Ictalurus punctatus* (Rafinesque)
Cod	*Gadus morhua* L.
Coelacanth	*Latimeria chalumnae* Smith
Coho salmon	*Oncorhynchus kisutch* (Walbaum)
Deep-sea rat-tail	*Nematonurus armatus* (Hector)
Dolphin fish	*Coryphaena hippurus* L.
Eel	*Anguilla anguilla* (L.)
Featherback	*Notopterus notopterus* (Pallas)
Flounder	*Pleuronectes flesus* L.
Frogfish	*Histrio histrio* (L.)
Giant gourami	*Colisa fasciata* (Bloch et Schneider)
Gilthead sea bream	*Sparus aurata* L.
Goby	*Neogobius melanostomus* (Pallas)
Gray sole	*Glyptocephalus cynoglossus* (L.)
Green notothenia	*Notothenia gibberifrons* Lönnberg
Grey shark	*Gymnothorax johnsoni* Smith
Goldfish	*Carassius auratus* (L.)
Grayling	*Thymallus thymallus* (L.)
Grenadier	*Coryphaenoides* spp.
Grey mullet	*Mugil cephalus* L.
Gudgeon	*Gobio gobio* (L.)
Guppy	*Poecilia reticulata* Peters
Haddock	*Melanogrammus aeglefinus* (L.)
Herring, Atlantic	*Clupea harengus* L.
Horse-mackerel, Atlantic	*Trachurus trachurus* (L.)
Horse-mackerel, Black Sea	*Trachurus mediterraneus ponticus* Aleev
Humpback salmon	*Oncorhynchus gorbuscha* (Walbaum)
Icefish	*Chaenocephalus aceratus* (Lönnberg)
Ivasi	*Sardinops melanostictus* (Temminck et Schlegel)
Japanese eel	*Anguilla japonica* Temminck et Schlegel
Japanese sardine	*Sardinops melanostictus* (Temminck et Schlegel)
Jelly cat	*Anarhichas denticulatus* Krøyer
Kilka	*Clupeonella cultiventris* (Nordmann)
Killifish	*Fundulus heteroclitus* (L.)
Kokanee salmon	*Oncorhynchus nerka* (Walbaum)
Ladyfish	*Elops hawaiensis* Regan
Lampfish	*Lampanictus australis* Tåning
Lemon sole	*Microstomus kitt* (Walbaum)
Longfin cod	*Antimora rostrata* Günther
Long-jawed mudsucker	*Gillichthys mirabilis* Cooper
Mackerel, Atlantic/Mediterranean	*Scomber scombrus* L.
Mackerel tuna	*Euthynnus alletteratus* (Rafinesque)
Marbled notothenia	*Notothenia rossii marmorata* Fischer
Masked greenling	*Hexagrammos octogrammus* (Pallas)
Medaka	*Oryzias latipes* (Temminck et Schlegel)
Mediterranean anchovy	*Engraulis encrasicholus* L.

Mediterranean pickerel	*Spicara maena* (L.)
Milkfish	*Chanos chanos* (Forskål)
Mongolian grayling	*Thymallus brevirostris* Kessler
Moray eel	*Muraena helena* L.
Mudfish	*Ophicephalus striatus* Bloch
Mud skipper	*Boleophthalmus boddaerti* (Pallas)
Mullet	*Mugil cephalus* L.
Navaga (Wachna cod)	*Eleginus navaga* Pallas
Northern pike	*Esox lucius* L.
Ocean sunfish	*Mola mola* (L.)
Ombre	*Sciaena umbra* L.
One-finned greenling	*Hexagrammos* sp.
Pacific anchovy	*Engraulis japonicus* Temminck et Schlegel
Pacific herring	*Clupea pallasii* Valenciennes
Pacific sardine	*Sardinops caerulea* (Girard)
Pacific saury	*Cololabis saira* (Brevoort)
Painted comber	*Serranus scriba* (L.)
Patagonian toothfish	*Dissostichus eleginoides* Smitt
Perch	*Perca fluviatilis* L.
Pickerel, Black Sea	*Spicara smaris* (L.)
Pike	*Esox lucius* L.
Pike-perch	*Sander lucioperca* (L.)
Pilchard	*Sardina pilchardus* (Walbaum)
Pink salmon	*Oncorhynchus gorbuscha* (Walbaum)
Plaice	*Pleuronectes platessa* L.
Plains killifish	*Fundulus zebrinus kansae* Garman
Polar cod	*Boreogadus saida* (Lepechin)
Rainbow trout	*Salmo gairdnerii* Richardson = *Oncorhynchus mykiss* (Walbaum)
Rasbora, common	*Rasbora daniconius* (Hamilton)
Red mullet	*Mullus barbatus* L.
River goby	*Neogobius fluviatilis* (Pallas)
Roach	*Rutilus rutilus* (L.)
Round goby	*Neogobius melanostomus* (Pallas)
St Peter's fish	*Oreochromis mossambica* (Peters)
Saithe	*Pollachius virens* (L.)
Sand eel	*Ammodytes* spp.
Sardine	*Sardina pilchardus* (Walbaum)
Saury-pike	*Sander saurus* (Walbaum)
Scorpionfish, Black Sea	*Scorpaena porcus* L.
Sea bass	*Dicentrarchus labrax* (L.)
Sea scorpion	*Scorpaena porcus* L.
Shad	*Alosa sapidissima* (Wilson)
Sheepshead minnow	*Cyprinodon variegatus* Lacépède
Shore rockling	*Gaidropsarus mediterraneus* (L.)
Siamese fighting fish	*Betta splendens* Regan
Silver carp	*Hypophthalmichthys molitrix* (Valenciennes)
Silver salmon	*Oncorhynchus kisutch* (Walbaum)
Silverside, Black Sea	*Atherina mochon pontica* Eichwald
Smelt	*Osmerus eperlanus* L.
Smooth hammerhead shark	*Sphyrna zygaena* (L.)

Snailfish	*Nectoliparis* sp.
Snapper	*Lutianus sebae* (Cuvier)
Snook	*Centropomus undecimalis* (Bloch)
Sockeye salmon	*Oncorhynchus nerka* (Walbaum)
Spiny dogfish	*Squalus acanthias* L.
Sprat	*Sprattus sprattus* (L.)
Steelhead salmon/trout	*Oncorhynchus mykiss* (Walbaum)
Stickleback	*Pungitius pungitius* (L.)
Sting ray	*Dasyatis pastinaca* (L.)
Striped bass	*Morone saxatilis* (Walbaum)
Striped greenling	*Hexagrammos agrammus* (Temminck et Schlegel)
Syrman goby	*Neogobius syrman* (Nordmann)
Three-spined stickleback	*Gasterosteus aculeatus* L.
Thresher shark	*Alopias vulpinus* (Bonaterre)
Tilapia	*Tilapia = Sarotherodon = Oreochromis mossambicus* (Peters)
Toad goby	*Mesogobius batrachocephalus* (Pallas)
Turbot	*Scophthalmus maximus* (L.)
Turbot, Black Sea	*Scophthalmus maximus maeoticus* (Pallas)
Vermilion rockfish	*Sebastes miniatus* (Jordan et Gilbert)
Viviparous blenny	*Zoarces viviparus* (L.)
Wall-eye pollock	*Theragra chalcogramma* (Pallas)
West Pacific sardine	*Sardinops melanostictus* (Temminck et Schelegel)
Whitefish	*Coregonus* spp.
White sea flounder	*Pleuronectes flesus* L.
Whiting, Black Sea	*Merlangius merlangus euxinus* (Nordmann)
Whiting, North Sea	*Merlangius merlangus* L.
White perch	*Morone americana* (Gmelin)
Wolf-fish	*Anarhichas lupus* L.
Yellowfin sole	*Limanda aspera* (Pallas)
Yellowfin tuna	*Thunnus albacares* (Bonnaterre)
Yellowfish	*Pleurogrammus azonus* (Jordan et Metz)

Subject Index

Notes: 1. Page numbers in *italics* refer to illustrations and **emboldened** pages refer to tables. There are occasionally textual references on the same pages. 2. Fish are entered under their common name, except where there is none, where their Latin species name is given. (A list of common names with Latin equivalents is given on pages 327–30)

Cumulative Index of Titles

Notes: **Titles of papers** have been converted into subjects and a specific article
may therefore appear more than once

Cumulative Index of Authors